The Handbook of Environmental Chemistry

Editor-in-Chief: O. Hutzinger

Volume 5 Water Pollution
Part P

Handbook of Environmental Chemistry

Recently Published and Forthcoming Volumes

Environmental Specimen Banking
Volume Editors: S. A. Wise and P. P. R. Becker
Vol.3/S, 2006

Polymers: Chances and Risks
Volume Editors: P. Eyerer, M. Weller and C. Hübner
Vol. 3/V, 2006

The Rhine
Volume Editor: T. P. Knepper
Vol. 5/L, 03.2006

Persistent Organic Pollutants in the Great Lakes
Volume Editor: R. Hites
Vol. 5/N, 2006

Antifouling Paint Biocides
Volume Editor: I. Konstantinou
Vol. 5/O, 2006

Estuaries
Volume Editor: P. J. Wangersky
Vol. 5/H, 2006

The Caspian Sea Environment
Volume Editors: A. Kostianoy and A. Kosarev
Vol. 5/P, 2005

Marine Organic Matter: Biomarkers, Isotopes and DNA
Volume Editor: J. K. Volkman
Vol. 2/N, 2005

Environmental Photochemistry Part II
Volume Editors: P. Boule, D. Bahnemann and P. Robertson
Vol. 2/M, 2005

Air Quality in Airplane Cabins and Similar Enclosed Spaces
Volume Editor: M. B. Hocking
Vol. 4/H, 2005

Environmental Effects of Marine Finfish Aquaculture
Volume Editor: B. T. Hargrave
Vol. 5/M, 2005

The Mediterranean Sea
Volume Editor: A. Saliot
Vol. 5/K, 2005

Environmental Impact Assessment of Recycled Wastes on Surface and Ground Waters
Engineering Modeling and Sustainability
Volume Editor: T. A. Kassim
Vol. 5/F (3 Vols.), 2005

Oxidants and Antioxidant Defense Systems
Volume Editor: T. Grune
Vol. 2/O, 2005

The Caspian Sea Environment

Volume Editors:
Andrey G. Kostianoy · Aleksey N. Kosarev

With contributions by
A. I. Ginzburg · V. B. Goryunova · A. G. Gul · M. G. Karpinsky
D. N. Katunin · A. N. Korshenko · A. N. Kosarev · A. G. Kostianoy
Y. R. Nalbandov · N. P. Nezlin · N. A. Sheremet · T. A. Shiganova
V. S. Tuzhilkin · I. S. Zonn

Springer

Environmental chemistry is a rather young and interdisciplinary field of science. Its aim is a complete description of the environment and of transformations occurring on a local or global scale. Environmental chemistry also gives an account of the impact of man's activities on the natural environment by describing observed changes.
The Handbook of Environmental Chemistry provides the compilation of today's knowledge. Contributions are written by leading experts with practical experience in their fields. The Handbook will grow with the increase in our scientific understanding and should provide a valuable source not only for scientists, but also for environmental managers and decision-makers.
The Handbook of Environmental Chemistry is published in a series of five volumes:

Volume 1: The Natural Environment and the Biogeochemical Cycles
Volume 2: Reactions and Processes
Volume 3: Anthropogenic Compounds
Volume 4: Air Pollution
Volume 5: Water Pollution

The series Volume 1 The Natural Environment and the Biogeochemical Cycles describes the natural environment and gives an account of the global cycles for elements and classes of natural compounds.
The series Volume 2 Reactions and Processes is an account of physical transport, and chemical and biological transformations of chemicals in the environment.
The series Volume 3 Anthropogenic Compounds describes synthetic compounds, and compound classes as well as elements and naturally occurring chemical entities which are mobilized by man's activities.
The series Volume 4 Air Pollution and Volume 5 Water Pollution deal with the description of civilization's effects on the atmosphere and hydrosphere.
Within the individual series articles do not appear in a predetermined sequence. Instead, we invite contributors as our knowledge matures enough to warrant a handbook article.
Suggestions for new topics from the scientific community to members of the Advisory Board or to the Publisher are very welcome.

ISBN 978-3-642-06649-8 e-ISBN 978-3-540-31505-6

ISSN 1433-6863

DOI 10.1007/b138238

Springer is a part of Springer Science+Business Media

springeronline.com

Softcover reprint of the hardcover 1st edition 2005

Cover design: E. Kirchner, Springer-Verlag

Printed on acid-free paper 02/3141 YL – 5 4 3 2 1 0

Prof. Dr. M. A. K. Khalil
Department of Physics
Portland State University
Science Building II, Room 410
P.O. Box 751 Portland, Oregon 97207-0751, USA
aslam@global.phy.pdx.edu

Prof. Dr. D. Mackay
Department of Chemical Engineering
and Applied Chemistry
University of Toronto
Toronto, Ontario, Canada M5S 1A4

Prof. Dr. A. H. Neilson
Swedish Environmental Research Institute
P.O. Box 21060
10031 Stockholm, Sweden
ahsdair@ivl.se

Prof. Dr. J. Paasivirta
Department of Chemistry
University of Jyväskylä
Survontie 9
P.O. Box 35
40351 Jyväskylä, Finland

Prof. Dr. Dr. H. Parlar
Institut für Lebensmitteltechnologie
und Analytische Chemie
Technische Universität München
85350 Freising-Weihenstephan, Germany

Prof. Dr. S. H. Safe
Department of Veterinary
Physiology and Pharmacology
College of Veterinary Medicine
Texas A & M University
College Station, TX 77843-4466, USA
ssafe@cvm.tamu.edu

Prof. P. J. Wangersky
University of Victoria
Centre for Earth and Ocean Research
P.O. Box 1700
Victoria, BC, V8W 3P6, Canada
wangers@telus. net

The Handbook of Environmental Chemistry Also Available Electronically

For all customers who have a standing order to The Handbook of Environmental Chemistry, we offer the electronic version via SpringerLink free of charge. Please contact your librarian who can receive a password or free access to the full articles by registering at:

springerlink.com

If you do not have a subscription, you can still view the tables of contents of the volumes and the abstract of each article by going to the SpringerLink Homepage, clicking on "Browse by Online Libraries", then "Chemical Sciences", and finally choose The Handbook of Environmental Chemistry.

You will find information about the

- Editorial Board
- Aims and Scope
- Instructions for Authors
- Sample Contribution

at springeronline.com using the search function.

Preface

Environmental Chemistry is a relatively young science. Interest in this subject, however, is growing very rapidly and, although no agreement has been reached as yet about the exact content and limits of this interdisciplinary discipline, there appears to be increasing interest in seeing environmental topics which are based on chemistry embodied in this subject. One of the first objectives of Environmental Chemistry must be the study of the environment and of natural chemical processes which occur in the environment. A major purpose of this series on Environmental Chemistry, therefore, is to present a reasonably uniform view of various aspects of the chemistry of the environment and chemical reactions occurring in the environment.

The industrial activities of man have given a new dimension to Environmental Chemistry. We have now synthesized and described over five million chemical compounds and chemical industry produces about hundred and fifty million tons of synthetic chemicals annually. We ship billions of tons of oil per year and through mining operations and other geophysical modifications, large quantities of inorganic and organic materials are released from their natural deposits. Cities and metropolitan areas of up to 15 million inhabitants produce large quantities of waste in relatively small and confined areas. Much of the chemical products and waste products of modern society are released into the environment either during production, storage, transport, use or ultimate disposal. These released materials participate in natural cycles and reactions and frequently lead to interference and disturbance of natural systems.

Environmental Chemistry is concerned with reactions in the environment. It is about distribution and equilibria between environmental compartments. It is about reactions, pathways, thermodynamics and kinetics. An important purpose of this Handbook, is to aid understanding of the basic distribution and chemical reaction processes which occur in the environment.

Laws regulating toxic substances in various countries are designed to assess and control risk of chemicals to man and his environment. Science can contribute in two areas to this assessment; firstly in the area of toxicology and secondly in the area of chemical exposure. The available concentration ("environmental exposure concentration") depends on the fate of chemical compounds in the environment and thus their distribution and reaction behaviour in the environment. One very important contribution of Environmental Chemistry to the above mentioned toxic substances laws is to develop laboratory test methods, or mathematical correlations and models that predict the environmental fate of new chemical compounds. The third purpose of this Handbook is to help in the basic understanding and development of such test methods and models.

The last explicit purpose of the Handbook is to present, in concise form, the most important properties relating to environmental chemistry and hazard assessment for the most important series of chemical compounds.

At the moment three volumes of the Handbook are planned. Volume 1 deals with the natural environment and the biogeochemical cycles therein, including some background information such as energetics and ecology. Volume 2 is concerned with reactions and processes in the environment and deals with physical factors such as transport and adsorption, and chemical, photochemical and biochemical reactions in the environment, as well as some aspects of pharmacokinetics and metabolism within organisms. Volume 3 deals with anthropogenic compounds, their chemical backgrounds, production methods and information about their use, their environmental behaviour, analytical methodology and some important aspects of their toxic effects. The material for volume 1, 2 and 3 was each more than could easily be fitted into a single volume, and for this reason, as well as for the purpose of rapid publication of available manuscripts, all three volumes were divided in the parts A and B. Part A of all three volumes is now being published and the second part of each of these volumes should appear about six months thereafter. Publisher and editor hope to keep materials of the volumes one to three up to date and to extend coverage in the subject areas by publishing further parts in the future. Plans also exist for volumes dealing with different subject matter such as analysis, chemical technology and toxicology, and readers are encouraged to offer suggestions and advice as to future editions of "The Handbook of Environmental Chemistry".

Most chapters in the Handbook are written to a fairly advanced level and should be of interest to the graduate student and practising scientist. I also hope that the subject matter treated will be of interest to people outside chemistry and to scientists in industry as well as government and regulatory bodies. It would be very satisfying for me to see the books used as a basis for developing graduate courses in Environmental Chemistry.

Due to the breadth of the subject matter, it was not easy to edit this Handbook. Specialists had to be found in quite different areas of science who were willing to contribute a chapter within the prescribed schedule. It is with great satisfaction that I thank all 52 authors from 8 countries for their understanding and for devoting their time to this effort. Special thanks are due to Dr. F. Boschke of Springer for his advice and discussions throughout all stages of preparation of the Handbook. Mrs. A. Heinrich of Springer has significantly contributed to the technical development of the book through her conscientious and efficient work. Finally I like to thank my family, students and colleagues for being so patient with me during several critical phases of preparation for the Handbook, and to some colleagues and the secretaries for technical help.

I consider it a privilege to see my chosen subject grow. My interest in Environmental Chemistry dates back to my early college days in Vienna. I received significant impulses during my postdoctoral period at the University of California and my interest slowly developed during my time with the National Research Council of Canada, before I could devote my full time of Environmental Chemistry, here in Amsterdam. I hope this Handbook may help deepen the interest of other scientists in this subject.

Amsterdam, May 1980 *O. Hutzinger*

Twenty-one years have now passed since the appearance of the first volumes of the Handbook. Although the basic concept has remained the same changes and adjustments were necessary.

Some years ago publishers and editors agreed to expand the Handbook by two new open-end volume series: Air Pollution and Water Pollution. These broad topics could not be fitted easily into the headings of the first three volumes. All five volume series are integrated through the choice of topics and by a system of cross referencing.

The outline of the Handbook is thus as follows:

1. The Natural Environment and the Biochemical Cycles,
2. Reaction and Processes,
3. Anthropogenic Compounds,
4. Air Pollution,
5. Water Pollution.

Rapid developments in Environmental Chemistry and the increasing breadth of the subject matter covered made it necessary to establish volume-editors. Each subject is now supervised by specialists in their respective fields.

A recent development is the accessibility of all new volumes of the Handbook from 1990 onwards, available via the Springer Homepage springeronline.com or springerlink.com.

During the last 5 to 10 years there was a growing tendency to include subject matters of societal relevance into a broad view of Environmental Chemistry. Topics include LCA (Life Cycle Analysis), Environmental Management, Sustainable Development and others. Whilst these topics are of great importance for the development and acceptance of Environmental Chemistry Publishers and Editors have decided to keep the Handbook essentially a source of information on "hard sciences".

With books in press and in preparation we have now well over 40 volumes available. Authors, volume-editors and editor-in-chief are rewarded by the broad acceptance of the "Handbook" in the scientific community.

Bayreuth, July 2001 *Otto Hutzinger*

Contents

Hdb Env Chem Vol. 5, Part P (2005): 1–3
DOI 10.1007/698_5_001

Published online: 4 November 2005

Introduction

Aleksey N. Kosarev[1] (✉) · Andrey G. Kostianoy[2]

[1]Geography Department, Lomonosov Moscow State University, Vorobievy Gory, 119992 Moscow, Russia
kosarev@ocean.geogr.msu.su

[2]P.P. Shirshov Institute of Oceanology, Russian Academy of Sciences, 36 Nakhimovsky Pr., 117997 Moscow, Russia
kostianoy@online.ru

Abstract The Caspian Sea, the largest enclosed basin on the planet, is distinguished by special natural conditions, contains rich natural resources (biological and mineral), and plays an important geopolitical role in the region. Therefore, its study is central to the interests not only of near-Caspian states, but far beyond them as well. Numerous publications are devoted to issues related to the Caspian Sea, including the principal issues of interannual sea-level oscillations, the fate of sturgeon, and the "oil and sea" problem.

The natural regime of the Caspian Sea is formed under the dominant influence of external factors such as riverine runoff and atmospheric activity. These factors define the particular features of the hydrological structure and circulation of the waters of the sea, which demonstrate a rapid and significant long-term variability far greater than the variations of the corresponding parameters in open seas. This necessitates comprehensive monitoring of the natural environmental conditions of the Caspian Sea and changes in its main parameters. The results of such monitoring should be regularly published to enable tracing of the characteristic stages in the evolution of the Caspian Sea manifested in its regime. Meanwhile, starting in the 1990s, this kind of generalization has faced great difficulties. These difficulties are related to the fact that most Caspian Sea studies from the last century were conducted by Soviet scientists (expeditionary operations, processing and systematization of the materials obtained, publication of monographs and articles, etc.). With the fall of the Soviet Union, the intensity of Caspian research has been sharply reduced—expeditions have become rare and many coastal hydrometeorological stations have been closed. More than a decade has passed since the publication of the last monograph with a multidisciplinary description of the natural conditions of the Caspian Sea (A.N. Kosarev, E.A. Yablonskaya *The Caspian Sea*, SPB Academic Publishing, The Hague, 1994). Meanwhile, recently, significant events have occurred in the Caspian ecosystem that require close attention and monitoring. For example, changes in the riverine runoff regime and sea-level trend have resulted in an essential transformation of the hydrological structure and distribution of chemical parameters. The increase in the anthropogenic impact has led to increased pollution of a series of near-shore regions, which is directly reflected in the functioning of biological communities. The spontaneous invasion of the ctenophore mnemiopsis has aggravated the problem.

To be sure, the above-mentioned difficulties in the collection of data hamper the preparation of scientific publications; nevertheless, they cannot be used to justify avoiding such publications. The efforts on generalization of the materials on present-day conditions of the sea regime and their comparison with the results obtained in earlier studies deserve special attention and support. The development of hydrocarbon exploration on the shelf and in the open regions of the Caspian Sea makes it necessary to take an increasingly ecological approach in studies of the structure and dynamics of the Caspian Sea waters (currents, storm surges, sea-level changes, wind, waves, etc.). A thorough assessment of these factors is absolutely necessary while performing prospecting and engineering operations in the sea since these factors determine the conditions of propagation and transformation of polluting substances and influence the dwelling conditions of organisms. Since the amount of current oceanographic information on the Caspian Sea has sharply decreased, it is especially important to use satellite data and to develop computational technologies and numerical modeling.

The solution to the principal problems in studies of the Caspian Sea faces uncertainty with respect to its legal statute. The five near-Caspian states disagree on the organization and coordination of marine expeditions and on the mutual exchange of available materials. Therefore, international regulation of the legal statute of the Caspian Sea now represents one of its most urgent problems.

This book presents a systematization and description of current knowledge on the physical oceanography, marine chemistry and pollution, and marine biology of the Caspian Sea. Special attention is paid to socioeconomic, legal, and political problems in the Caspian Sea region. The book is based on numerous observational data collected by the contributing authors during sea expeditions, on archival data of the Moscow State University, the State Oceanographic Institute, the Russian Research Institute for Fisheries and Oceanography, the Shirshov Institute of Oceanology, and the Caspian Research Institute for Fisheries, as well as on a wide-ranging body of scientific literature published mainly in Russian. These data are complemented by the results of a series of projects conducted under the auspices of two long-term federal programs of the Ministry of Education and Science of the Russian Federation: The World Ocean and Investigations in the Priority Fields of Science and Technology, where extensive research has been carried out over the past 6 years.

Chapter 2 is devoted to a brief description of physicogeographical features of the Caspian Sea. Chapter 3 presents the general particularities of the thermohaline structure and circulation of the waters of the sea. Chapter 4 describes the seasonal and interannual variability of the sea surface temperature in various parts of the sea based on satellite data. Chapters 5 and 6 discuss the natural chemistry of the Caspian Sea and general information on the chemical contamination of the seawater and bottom sediments. Chapter 7 is devoted to existing research on the seasonal and interannual variability of remotely sensed chlorophyll patterns. Chapters 8–10 review the marine biology of the Caspian Sea focusing on its biodiversity, introduced species, and resources. Chapter 11 addresses the history and present condition of Kara-Bogaz-Gol Bay. Chapter 12 deals with the general ecological problems in the Caspian Sea. Economic and international legal dimensions are discussed in Chap. 13. Chapter 14 concludes with a review of modern environmental conditions of the Caspian Sea.

This book is intended for specialists working in various fields of physical oceanography, marine chemistry, pollution studies, and biology and for specialists studying a litany of problems: from regional climate to mesoscale processes and from remote sensing of the seas to numerical and laboratory modeling. It may also be useful to undergraduate and graduate students of oceanography. The authors hope that this monograph will complement knowledge of the nature of the Caspian Sea, especially the present-day state of this extremely interesting basin. More information on particular issues may be obtained from the reference lists at the end of each chapter.

On behalf of the authors, we would like to thank the editors at Springer-Verlag for their timely interest in the Caspian Sea and their support for the present publication. This study was supported by two federal programs of the Ministry of Education and Science of the Russian Federation: The World Ocean and Investigations in the Priority Fields of Science and Technology, by a series of grants from the Russian Foundation for Basic Research, and by the NATO SfP Project N981063 Multidisciplinary Analysis of the Caspian Sea Ecosystem (MACE).

Hdb Env Chem Vol. 5, Part P (2005): 5–31
DOI 10.1007/698_5_002

Published online: 12 October 2005

Physico-Geographical Conditions of the Caspian Sea

Alexey N. Kosarev

Geographic Department, Lomonosov Moscow State University,
Vorobievy Gory, 119992 Moscow, Russia
akosarev@mail.ru

Abstract The main physico-geographical features of the Caspian Sea are under consideration. They include brief description of the coasts, bottom relief and sediments, peculiarities of the deltas areas of Volga, Ural and other rivers of the Caspian Sea. General information on the Caspian Sea region hydrometeorological characteristics (wind, waves, storm surges, sea ice) is given. Special attention is paid to main factors of the sea-water balance (river runoff, precipitation, evaporation, outflow to the Kara-Bogaz-Gol Bay). Long-term variability of these parameters is shown for the past century resulting in the analysis of the Caspian Sea level variability. The main reasons of significant changes of the sea level are discussed.

Keywords Bottom relief · Climate · Coasts · Sea level · Sediments · River deltas · Water balance

1 Introduction

The Caspian Sea, the greatest enclosed basin of the world, is located inside the Eurasian continent. It occupies a vast depression in the earth's crust. Its level is below the level of the World Ocean; at present, the increment is 27 m. Under these conditions, the area of the sea exceeds 390 000 km^2 and the water volume reaches 78 000 km^3 at a mean depth of 208 m; the maximum sea depth is 1025 m. The sea extends over 1030 km from the north to the south being from 200 to 400 km wide. The large size and the great meridional extension of the sea define the natural diversity of its different regions.

With respect to the physico-geographical conditions and the character of the bottom topography, the sea can be subdivided into three parts: the North, Middle, and South Caspian (Fig. 1). The shallow-water northern part of the sea (with sea depths less than 15–20 m) completely refers to the shelf. The Mangyshlak Swell separates it from the Derbent Basin of the Middle Caspian with a maximum depth of 788 m. The underwater Apsheron Swell with sea depths up to 160–180 m separates the Derbent Basin from the South Caspian depression, where the maximum depth of the Caspian Sea was recorded. The conventional boundary between the North and Middle Caspian runs along the line Chechen' Island–Cape Tyub Karagan, while that between the Middle and South Caspian runs along the line Zhiloi Island–Cape Kuuli (see Fig. 1). The water volumes in these three parts of the sea comprise 0.5, 33.9, and 65.6% of the total volume of the sea.

The Caspian Sea is characterized by a small number of islands; their total area is about 2000 km^2. The majority of the islands are located in the North Caspian. Here, in the western part of the region, Chechen' Island (120 km^2, the largest island of the sea) and Tuylenii Island (68 km^2) are located; the largest island of its eastern part is Kulaly Island (73 km^2). Numerous low islands are scattered over the near-mouth area off the Volga River. In the Middle Caspian, off the Apsheron Peninsula, one finds the islands of the archipelago with the same name. In the South Caspian, south of Baku Bight, minor islands of the Baku Archipelago (Bulla, Los', Svinoi, and others) are located. Ogurchinskii Island is located off the eastern coast of the sea.

About 130 rivers enter the Caspian Sea; a few of them are large and numerous rivers are small. The catchment basin of the sea covers an area of 3.5 million km^2. The value of the ratio between the areas of the sea and the catchment basin (1 : 10) explains the significant influence of the processes proceeding in the entire basin on its natural conditions. In the feeding of the Caspian Sea, the most important is the Volga River basin, whose area equals 1.4 million km^2, i.e., about 40% of the total catchment area.

The riverine network around the Caspian Sea is extremely irregular. All the principal rivers either enter the North Caspian or are confined to the western coast of the sea. The mean annual runoff of these rivers (Volga, Ural, Terek, Sulak, and Kura) reaches more than 90% of the total runoff of all the rivers into the sea. The rest of the runoff is related to the Iranian rivers and minor streams of the western coast of the Caspian Sea. The eastern coast of the sea is almost free from permanent riverine runoff into the sea. Over the past decades, the intensive use of the water resources of the rivers of the Caspian

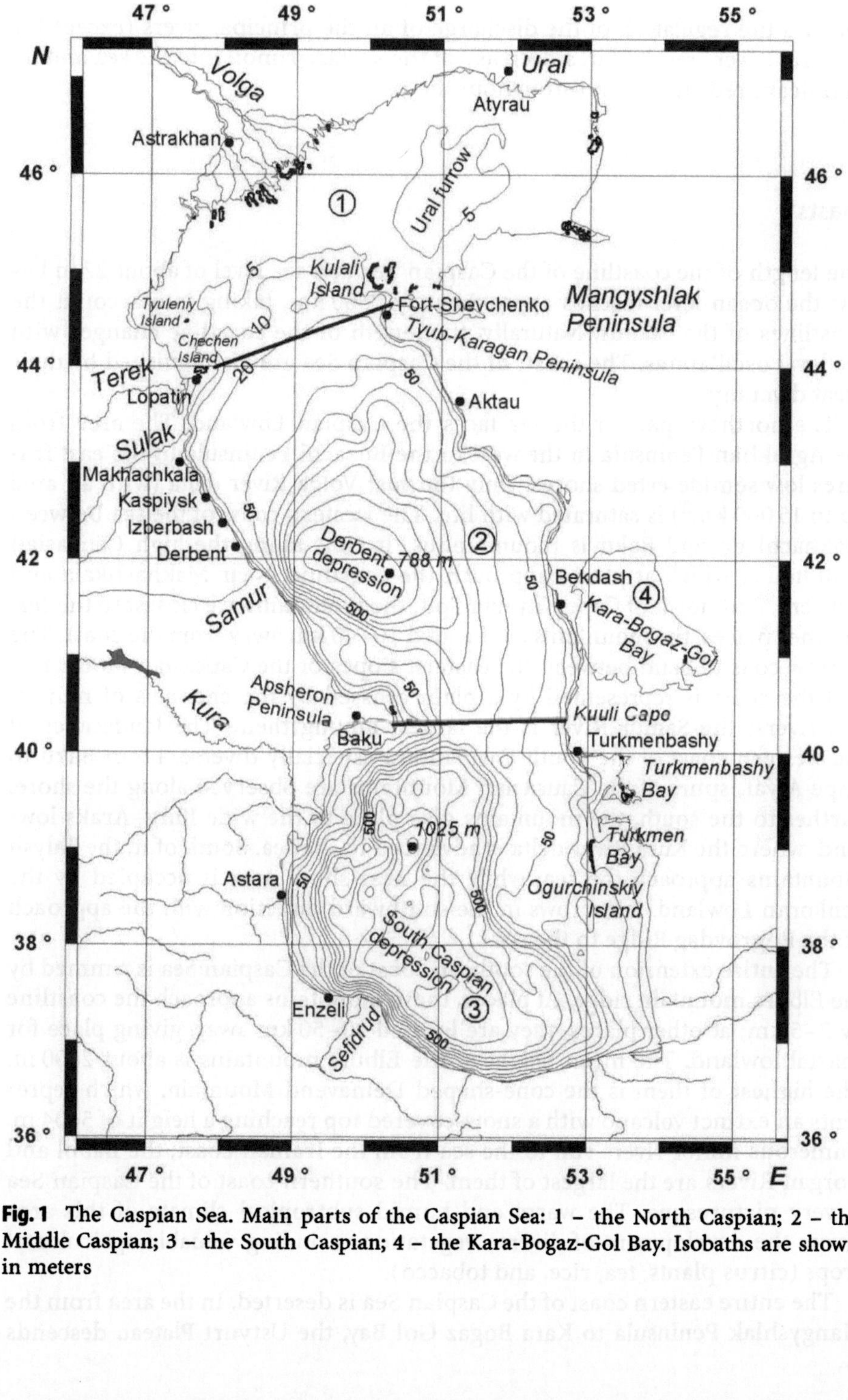

Fig. 1 The Caspian Sea. Main parts of the Caspian Sea: 1 – the North Caspian; 2 – the Middle Caspian; 3 – the South Caspian; 4 – the Kara-Bogaz-Gol Bay. Isobaths are shown in meters

Sea and the regulation of the discharge of all the principal rivers (except for the Ural River) resulted in a decrease of the surface runoff into the sea and its significant redistribution throughout the year.

2 Coasts

The length of the coastline of the Caspian Sea at a sea level of about 27 m below the ocean level reaches approximately 7500 km, taking into account the coastlines of the islands. Naturally, the length of the coastline changes with sea level oscillations. The coasts of the Caspian Sea are distinguished by their great diversity.

The northern part of the sea faces the Caspian Lowland. The area from the Agrakhan Peninsula in the west to the Buzachi Peninsula in the east features low semideserted shores. Only the vast Volga River delta (with an area up to 15 000 km^2) is saturated with life. The western coast of the sea between Makhachkala and Baku is mountainous. It runs along the high Caucasian Mountains, which at places approach the coastline. Near Makhachkala and Derbent, and north of Cape Kilyazin Spit, the mountains are closest to the sea. At other places, the mountains are located 10–40 km away from the coast. The narrow coastal band between the eastern slopes of the Caucasian Mountains and the coast is represented by a plain crossed by the channels of numerous rivers; the Samur River is the largest among them. The landscapes of the western coast of the South Caspian are especially diverse. From Baku to Cape Alyat, spurs of the Caucasian Mountains are observed along the shore. Farther to the south, the mountains give place to the wide Kura–Araks lowland, where the Kura River delta is advanced to the sea. South of it, the Talysh Mountains approach the sea, while the near-shore area is occupied by the Lenkoran Lowland. It narrows in the southward direction with the approach of the Bogrovdag Ridge to the sea.

The entire extension of the southern coast of the Caspian Sea is rimmed by the Elburs mountain ridge. At places, these mountains approach the coastline by 2–5 km; at other places, they are located 30–50 km away, giving place for coastal lowland. The mean height of the Elburs mountains is about 2000 m. The highest of them is the cone-shaped Demavend Mountain, which represents an extinct volcano with a snow-covered top reaching a height of 5604 m. Numerous minor rivers run to the sea from the Iranian coast; the Babol and Gorgan Rivers are the largest of them. The southern coast of the Caspian Sea is very picturesque. The warm and humid subtropical climate of this area favors the development of diverse vegetation including valuable agricultural crops (citrus plants, tea, rice, and tobacco).

The entire eastern coast of the Caspian Sea is deserted. In the area from the Mangyshlak Peninsula to Kara Bogaz Gol Bay, the Ustyurt Plateau descends toward the sea forming steep escarpments (chinks). South of Kara Bogaz Gol Bay, one finds the low mountains of the Kubadag ridge and Turkmenbashy Bay bound in the south by the Cheleken Peninsula. The eastern coast of the South Caspian is low; here, the sands of the western part of the Kara Kum desert approach the sea.

A few major bays are distinguished in the topography of the Caspian Sea coasts. In particular, they are Agrakhan and Kizlyar bays on the western coast of the Middle Caspian and Kyzylagach Bay in the South Caspian. In the northeast of the sea, the vast Mangyshlak Bay is located between the Buzaachi and Tyub Karagan Peninsulas. The dissected eastern coast of the Caspian Sea contains large Kazakh, Kara Bogaz Gol, and Turkmen bays. The coastline of the Caspian Sea is subjected to significant changes related to the periodic interannual sea level oscillations. In so doing, the coastline of the shallow-water northern part of the sea suffers especially strong changes [1].

3
Bottom Relief and Sediments

In the tectonic respect, the depression of the Caspian Sea is closely related to the geological structures of the surrounding land. The North Caspian area is located within the limits of the Russian and the younger Scythian–Turanian platforms. The western part of the Middle Caspian and the entire South Caspian lie within the Alpine folded area. The differences in the geological structure of the Caspian depression regions are manifested in the features of the bottom topography of the sea.

The floor of the North Caspian is represented by a shallow-water plain with numerous islands and extinct river channels. In the bottom topography, one can trace branched ancient channels of the Volga, Ural, and Terek Rivers. In the eastern part of the North Caspian, the Ural'skaya Borozdina trough with a depth increment up to 5 m is distinguished.

The North and Middle Caspian are separated by the Mangyshlak Swell. In the bottom topography, it is manifested by a shallow-water area extended from Chechen' Island to Kulaly Island and farther to the Tyub Karagan Peninsula. The southern slope of the swell conventionally bound by the 20-m depth contour is gentle.

In the bottom topography of the Middle Caspian, one can easily distinguish the principal morphological elements such as the shelf, the slope, and the floor. The depression of the Middle Caspian is asymmetric; its western part (except for the northernmost area) features a narrow shelf and a steep slope, while the eastern part is characterized by a wide shelf and a gentle slope. In the west, the principal role in the formation of the shelf belongs to the accumulation processes, while in the east, abrasion dominated.

The width of the western shelf of the Middle Caspian ranges from 130 km in its northern part adjacent to the Mangyshlak Swell to 15 km off the Samur River. Farther southward, the shelf becomes wider again (up to 60 km). The edge of the shelf is located at depths from 70 to 110 m. The width of the eastern shelf varies from 50 to 130 km.

The upper part of the continental slope is separated from the shelf by a bend. In the middle part of the sea, the western slope features a width of 20–60 km. The eastern slope is represented by a slightly inclined leveled plain 40–150 km wide.

Starting from sea depths of 600–700 m, the underwater slope turns into the abyssal plain of the Middle Caspian named the Derbent Basin. It extends along the western coast of the sea from the northwest to the southeast over 250 km, being from 40 to 80 km in width. In this basin, the depth mark equal to 788 m greatest in the middle part of the sea is noted. The floor of the basin is represented by a smooth surface slightly inclined from the northwest to the southeast.

The location of the South Caspian in a geosyncline area causes the significant complicacy and irregularity of its topography. The western shelf features a width from 15 to 60 km, and its outer edge is located at depths from 60 to 150 m. In the northern part of the western shelf, there are numerous islands and banks, whose origin is mostly related to mud volcanism. The eastern shelf of the South Caspian is significantly wider than the western shelf and reaches up to 190 km in width. The outer edge of the shelf is located at depths 100–130 m. Off the southern coast of the Caspian Sea, the shelf is very narrow (5–10 km) and steep.

In the southern part of the sea, the structure of the continental slope is distinguished by its great complicacy. On the western slope, a series of rises up to 500 m in height is traced. Their feet lie at depths of 700–800 m. The tops of many rises are crowned by mud volcanoes. The eastern slope of the South Caspian Basin, similar to its western slope, is rather steep and features a stepwise profile. The foot of the slope is located at depths of 750–800 m. The southern slope of the basin is distinguished by greater inclination angles and significant fragmentation.

The deep-water South Caspian Basin features a more complicated structure than the Middle Caspian (Derbent) Basin. Its central part is occupied by an abyssal plain, where submarine rises alternate with depressions. In the central region of the South Caspian, one finds the Abikha Basin with the depth mark greatest in the entire sea (1025 m). In the southernmost part of the South Caspian Basin, the Elburs foredeep is located; in the bottom topography, it is manifested by a depression.

The distribution of the bottom sediments in the Caspian Sea is controlled by the bottom topography, water dynamics, and hydrochemical conditions. More than 90% of the liquid and 75% of the solid discharges into the sea are delivered to the North and Middle Caspian. The principal enrichment of the waters with the terrigenous matter occurs in the near-mouth zones; its further propagation over the sea is governed by the currents. The terrigenous matter is also supplied to the sea due to the coastal abrasion and driven with the winds. Because of the intensive water circulation in the shelf zone, the transport of the major part of the terrigenous particulate matter proceeds precisely here, while in the central deep-water basins, conditions favorable for its precipitation are formed.

The principal role in the formation of the bottom sediments of the Caspian Sea belongs to the detrital and carbonate matter of biogenic and chemogenic origins. The hydrochemical regime of the waters of the sea is characterized by an oversaturation by carbonates, high alkalinity, and enhanced pH values; they provide the conditions favorable for the precipitation of carbonates, especially in shallow-water areas.

The bottom sediments of the Caspian Sea are represented by calcareous and terrigenous deposits. In the North Caspian, coarse-grained sediments such as silts and terrigenous sands dominate. The abundant delivery of nutrients by the Volga River creates conditions favorable for the development of bottom mollusk biocoenoses, which provides the carbonate supply to the sediments. Meanwhile, due to the active water dynamics, fine-grained matter is transported to the Middle Caspian forming a tongue of sandy matter along the western slope of the Middle Caspian depression. On the western shelf and slope of the Middle Caspian, down to sea depths of 30–60 m, terrigenous silts dominate. The Terek, Sulak, and Samur Rivers deliver great amounts of fine-grained matter, which is partly accumulated in the narrow shelf zone. In the eastern part of the Middle Caspian, where the shelf is wider, under the arid climatic conditions, an intensive accumulation of carbonate sediments dominated by the biogenic component is observed. The rise of the abyssal waters enriched with nutrients, which occurs in the summertime under the conditions of the upwelling on the eastern shelf, favors the development of plankton and benthos, which supply great amounts of biogenic carbonate to the bottom sediments. On the eastern slope, at depths of 200–400 m, one observes an enrichment of the sediments with silica, which is also related to the upwelling of the waters rich in silicon. In addition, in the eastern part of the Middle Caspian, under the conditions of intensive evaporation, the chemogenic precipitation of carbonates in the form of oolite grains from the oversaturated seawater is of great significance. The floor of the Middle Caspian Basin is covered by weakly calcareous clayey silts gray or grayish green in color.

In the South Caspian, on its western shelf, a gradual replacement of sands by weakly calcareous silts is observed in the direction from the coast to the open sea. The floor of the deep-water part of the South Caspian Basin is covered with weakly calcareous clayey silts. On selected rises, one can observe outcrops of baserocks, which are related to landslides and mud volcanism. The calcareous clayey silts also cover the foot of the eastern underwater slope of the South Caspian, while calcareous silts with high carbonate contents are encountered on its slope and shelf.

Despite the high sedimentation rates, on the floor of the sea, one can also find zones of restricted or "zero" sediment accumulation. They are spread along shelf edges and at selected sites of the continental slope, on the tops of rises, and on the Apsheron Swell. At the shelf edge in the axial zone of the Apsheron Swell, the sediments are absent due to the intensive water dynamics; on the continental slopes, sediment sliding plays the principal role [2].

4 River Deltas

The near-mouth areas of the rivers entering the Caspian Sea play an important role in its physico-geographical setting. In these boundary areas formed at the river–sea interaction, the changes in the environmental conditions proceed at especially high rates. Meanwhile, precisely these regions are of the greatest interest when mastering the biological resources of the sea. The principal information on the deltas of the largest rivers entering the Caspian Sea is presented in Table 1.

The Volga River delta is the second large among the deltas of the rivers of Russia and is one of the most strongly branched in the world (Fig. 2).

The present-day delta of the Volga River represents a plain slightly inclined toward the sea with absolute height marks of about – 20 m at the top of the delta and – 26.5 m near its seaward edge; so, the delta is located above the present-day mean sea level. The area of the above-water part of the delta ex-

Table 1 Present-day characteristics of the near-mouth areas of the principal rivers entering the Caspian Sea [3]

River	State attribution of the delta	Delta area, km^2	Delta length over the main channel, km	Length of the marine edge of the delta, km	Number of mouths of natural streams	Character of the near-mouth area
Volga	Russia, Kazakhstan	11 000	160	175	> 500	Shallow-water
Ural	Kazakhstan	500	44	–	5	Shallow-water
Terek	Russia	8900	170	~ 200	1	Deep-water (from August 1977)
Sulak	Russia	43.7	6.3	27	1	Deep-water
Kura	Azerbaijan	136	18.0	62	2	Deep-water

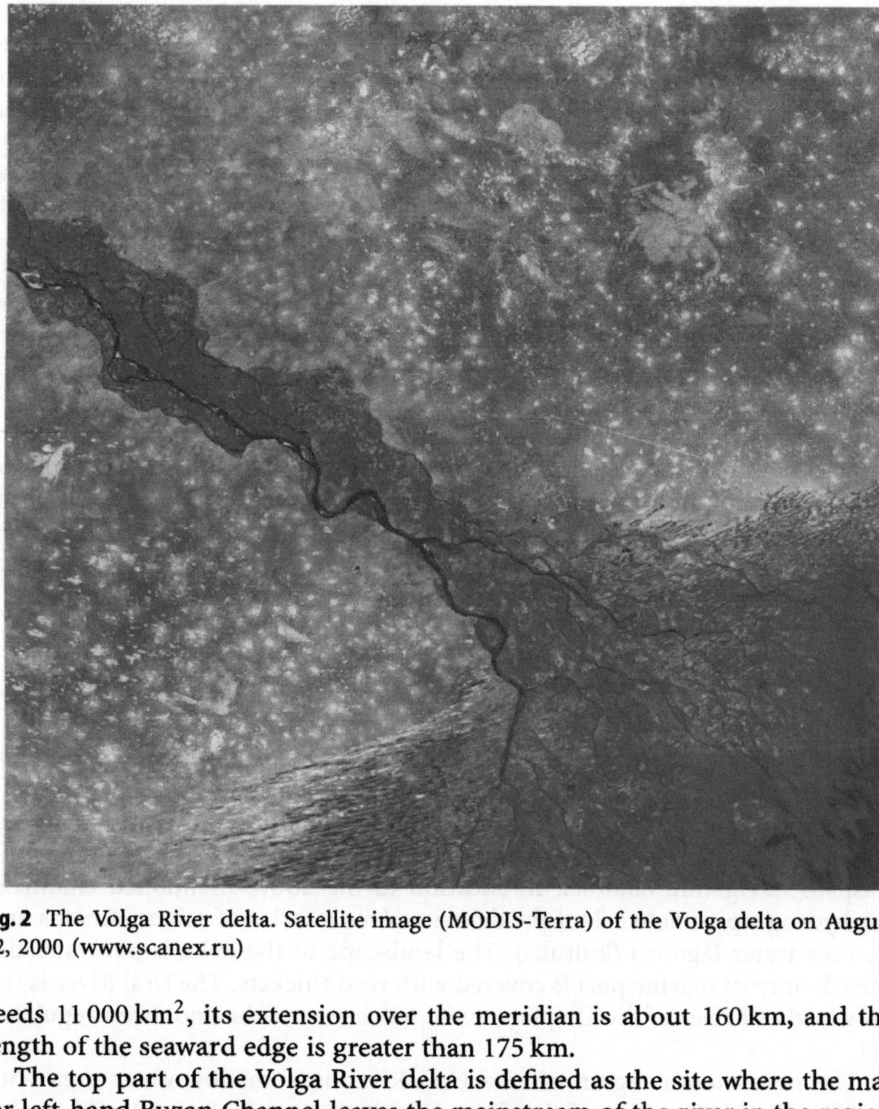

Fig. 2 The Volga River delta. Satellite image (MODIS-Terra) of the Volga delta on August 22, 2000 (www.scanex.ru)

ceeds 11 000 km^2, its extension over the meridian is about 160 km, and the length of the seaward edge is greater than 175 km.

The top part of the Volga River delta is defined as the site where the major left-hand Buzan Channel leaves the mainstream of the river in the region of Verkhnee Lebyazh'e village. The major floodplain Akhtuba Channel flows into the Buzan Channel. Near Astrakhan', large Bolda, Kamyzyak, and Staraya Volga channels separate from the main branch of the delta. The further continuation of the main branch of the delta is named the Bakhtemir Channel. In its marine part, this channel is continued by the Volga–Caspian navigation channel about 100 km in length. With respect to its topography and vegetation, the delta of the Volga River is subdivided into three zones – the upper, middle, and lower (seaside) zones. The network of channels and waterways is most branched in the lower zone. The Volga River delta is distinguished by the extreme diversity of natural and agricultural landscapes. In the middle zone, a unique topographic feature is represented by the so-called Ber hills of eolian origin. Between them, numerous lakes (il'mens) are located. The youngest lower zone is characterized by the permanent interaction of riverine and marine processes.

The shallow-water near-mouth area off the Volga River extends far toward the sea from the seaward edge of the delta. Its average meridional length exceeds 100 km and its area is about 25 000 km^2. The seaward edge of the delta is bound by a belt of underwater vegetation well recognized in satellite images. It extends over more than 200 km being 10–15 km in width. With respect to the salinity distribution, the coastal area off the Volga River mouth can be subdivided into two zones – the zone of transport of the riverine waters, where salinity never exceeds 1–2‰, and the zone of mixing between the riverine and marine waters, where the salinity rapidly grows from 1–2 to 9–10‰. South of this frontal zone, seawaters with a salinity of 11.5–12‰ and greater are spread. The location of the frontal zone, its width, and the values of the salinity gradient depend on the volume of the Volga River runoff, the character of its distribution over the sea, and the water exchange with the Middle Caspian; in different years, they may significantly differ.

Large-scale changes play the leading role in the evolution of the Volga River delta. The shape and size of the delta vary depending on the sea level rises and falls.

In the delta of the Ural River, the principal node of its channel network is represented by the site of splitting the river into two major channels – the Yaik Channel consisting of a system of individual streams and the Zolotoi Channel, which is the main waterway of the delta; its seaward extension is the Ural–Caspian navigation channel. In addition to the above-mentioned channels, the hydrographic network of the delta involves a series of minor streams and shallow-water lagoons (kultuks). The landscape of the delta is poor and deserted; only its marine part is covered with reed thickets. The Ural River is the only lowland river of the Caspian basin, whose runoff has not been regulated yet.

The near-mouth region off the Ural River is a shallow-water area with small inclination angles of the floor. Within it, the Tyulen'i Islands are located. The seaward boundary of this area runs approximately along the 3–5 m depth contours. Meanwhile, due to the flatness of the sea floor, at the offset and onset level changes reaching 1 m, the boundary of this region may move within a wide range. In selected years, the mouth bar may be dried and occupied by plants. The continuation of the Ural–Caspian channel leads to the Ural'skaya Borozdina trough, which is the deepest area of the North Caspian. The Ural'skaya Borozdina trough serves as an extension of the underwater channel of the Ural River and was eroded by the river under the conditions of a lower sea level standing.

In April–June, a zone of desalinated is clearly expressed in the coastal waters off the Ural River mouth. Meanwhile, starting from June, no noticeable influence of the Ural River runoff on the salinity of the eastern part of the North Caspian has been observed. In the low water period, marine waters may approach the river mouth and, under favorable winds, even penetrate inside the river channel. The influence of the nutrient runoff of the river is traces only over a small area of the near-shore waters.

The delta of the Terek River begins 170 km away from the sea and represents a vast plain with a cellular topography. The hydrographic network of the delta is complicated by numerous branches, artificial channels, lakes, and pools. In the delta, two waterworks facilities were constructed (Kargaly and Kopaisk control structures), which redistribute the Terek River runoff between the mainstream of the delta (Kargaly waterway) and the irrigation channels in its northern part. By the middle 1970s, the Kargaly waterway had flown into the shallow-water Agrakhan Bay; after August 1977, it enters directly to the Middle Caspian via an artificial waterway across the Agrakhan Peninsula. By 1991, the length of the Kargaly waterway had exceeded 100 km. In the delta of the Terek River, waterlogged swamps, steppes, and tugai forests dominate the landscapes along the channels, while closer to sea, sand dunes are encountered.

The near-mouth area off the Terek River delta is shallow-water with no clearly expressed shelf edge. It includes Kizlyar and Agrakhan bays and the southwestern coastal region of the North Caspian. Over the greater part of this area, sea depths are smaller than 5 m.

The delta of the Sulak River represents an alluvial plain composed mostly of riverine deposits. It is located close to the earlier extinct delta of the river. The Sulak Spit separates Sulak Bight from the sea. At present, the Sulak River flows into the sea in the form of a single channel up to 150 m wide. The characteristic landscapes in the delta are represented by meadows, reed thickets, salt bottoms, sand beaches, and dunes.

The runoff of the Sulak River is regulated by the dams of hydroelectric power stations and its natural regime is distorted. Since then, the sediment runoff significantly has reduced and the flooding volume has decreased, while the wintertime runoff of the river has grown. At the end of the 1950s, the runoff of the Sulak River was pointed eastward via an artificial channel and now the river flows into the open relatively deep-water part of the Middle Caspian. At the place of the river entering to the sea, a new delta is formed and the branches of the old Sulak River delta, which has previously separated the bight from the sea, died out. The changes in the direction of the river channel and location of its mouth significantly influenced the physico-geographical conditions of the delta and the hydrology of the near-shore area of the sea.

The delta of the Kura River is the fourth largest delta of the Caspian Sea; it represents a part of the Kura–Araks Lowland with steppe and semidesert landscapes and salinated soils. The top of the Kura River delta, where the river splits into two branches – the southeastern (the Navigable Kura) and the northeastern (the Old Kura) – is located a few kilometers away from the mouth. Almost the entire runoff is concentrated in the Navigable Kura, while the Old Kura dies out. In the 1950s, the Kura River runoff was regulated by the Mingechaur Reservoir, which significantly changed the hydrological regime of the mouth area. In particular, during the flood period, the runoff is reduced and the volume of the sediments delivered to the mouth area decreased by more than 50%. Within the present-day delta, dry channels of extinct branches and dried lagoons and lakes are recognized.

The sea area off the Kura River delta is rather deep-water; it is bound by the shelf edge and has a characteristic shape of a fan advanced far toward the sea. The principal transformation of the riverine waters proceeds within a narrow offshore band a few kilometers wide. During the flood period, the influence of the Kura River runoff on the salinity distribution is traced over a distance up to 15 km from the mouth.

On the southern coast of the Caspian Sea, the Sefidrud River delta is distinguished. It represents a plain that begins in the region of Resht 60 km away from the Caspian Sea. The widely spread deltaic deposits cover an area of about 1800 km^2. The significant sediment runoff provided by the river favored the land accretion in this region and the displacement of the site of the river entering to the sea, which accelerated the delta evolution. At present, the Sefidrud River flows into the Caspian Sea near Kiyashakhr. The single-channel delta of the Sefidrud River occupies an area of about 40 km^2. The width of the main channel of the delta is 100–120 m at a depth of 1.5–2 m. The region of the delta is represented by a rushy waterlogged lowland.

The sea areas off the mouths of all the Caspian rivers represent open extensions of their deltas. The areas off the Volga and Ural Rivers are shallow-water, while off the rest of the rivers they are relatively deep-water. In the top parts of the Ural, Sulak, and Kura deltas, selected parts of the rivers are subjected to the action of storm surges. Regular (not catastrophic) surges do not reach the top parts of the Volga and Terek deltas.

The natural resources of the Caspian deltas are widely used in fishery and agriculture. Important navigation routes run over the deltas of the Volga and Ural Rivers.

In order to protect land from floods, waterways of many deltas are bordered. In the upper parts of the Volga and Terek deltas, large regulating facilities such as the Volga flow dividing valve and the Kargaly control structure are constructed. In these deltas, there are numerous irrigation and supply channels. A series of navigation channels (Volga–Caspian, Ural–Caspian) and many fish bypass channels cross the offshore areas of the Volga and Ural.

The deltas of the principal Caspian rivers, in addition to their economic value, have the first-rate ecological importance as the areas of spawning of valuable fish species, rich vegetation (for example, reed and lotus), wintering and nesting of birds, etc. In the deltas of the Volga and Kura Rivers, biospheric reserves are located.

The response of the Caspian deltas to the sea level oscillations is defined by a series of factors; among them, in addition to the magnitude and intensity of the sea level changes, other factors are no less important. First, they are the morphometry (width and depth) of the near-mouth sea area and the intensity of the liquid and solid runoff of the river and its channels along with the topography of the seaward part of the delta, the inclinations of the water surface in it, sea waves, and local hydrotechnical measures. The further fate of the deltas of the Caspian rivers depends first on the character of the multiannual oscillations of the Caspian Sea level and on the natural and anthropogenic changes in the riverine runoff and its regulation in the deltas [3, 4].

5 Climate

Because of its great meridional extension, the Caspian Sea finds itself within several climatic zones. Its northern part lies in the zone of a temperate continental climate, the western coasts feature a moderately warm climate, while the southwestern and southern regions of the sea refer to the subtropical zone. The eastern coasts of the sea are characterized by a desert climate.

In the wintertime, the weather in the North and Middle Caspian is controlled by the continental polar air related to the influence of the Siberian anticyclone and by the Arctic air masses propagating from the Kara and Barents seas. In the South Caspian, penetration of southern cyclones is frequently observed. In the west, the weather is instable and rainy, while in the east, the weather is dry. In the summertime, the sea is affected by the spurs of the Azores maximum mostly in the west and north-west, while the southeastern part of the sea are subjected to the influence of the vast Iranian–Afghan minimum. In this season, stable and dry weather is maintained over the Caspian Sea.

In the winter (January–February), the air temperature ranges from − 10 °C (in the most severe winters, from − 30 °C) in the northeastern part of the sea to 8–12 °C in the south. In the summer (July–August), the mean monthly air temperature over the entire sea equals 24–26 °C; off the eastern coasts, the maximum air temperature exceeds 40 °C.

The atmospheric precipitation is different over different sea regions: from 100 mm per year off the arid eastern coasts to 1700 mm per year in the southwestern part of the sea (Lenkoran region). In the open sea, the mean precipitation rate is equal to 200 mm per year.

6 Wind and Waves

The character of the winds plays an important role in the formation of the field of the currents and in the propagation and transformation of admixtures in the upper layer of the sea; therefore, let us assess it in more detail. The wind regime of the Caspian Sea is defined by the action of three principal factors: regional atmospheric activity, topography of the coasts (orography), and local circulation induced by the thermal increments between the land and the sea. The following types of stable wind fields may be distinguished with respect to the direction of the air mass transfer: the northerly (northwesterly, northerly, and northeasterly), southeasterly, and vortical. During the major part of the year, the most stable winds of northerly and southeasterly directions dominate over the sea area. Throughout the year, the field of the northerly winds is observed in 40% of the cases; in the summertime they dominate (up to 50%) and almost half of the winds are northwesterly. The southeasterly winds feature a recurrence of about 36%; in the winter and spring, they are more frequent (about 40%). The vortical type is observed in 4–7% of the cases, while the average annual share of calm days (wind speed less than 5 m s^{-1}) is 20%, they are mostly observed in the summer season.

When northwesterly and southeasterly winds dominate (about 10–12% of the cases), their speed values mostly comprise 5–9 m s^{-1} (moderate winds); in the summer season, this value somewhat increases. Strong winds of these directions (with a speed greater than 10 m s^{-1}) do not exceed 4–6%. The mean recurrence rate of strong storms (more than 25 m s^{-1}) is very low; it equals once per a few years, as far as the entire sea area is considered. The mean annual wind speed over the area of the Caspian Sea is 5.7 m s^{-1}. The greatest mean speeds equal to 6–7 m s^{-1} are observed in the Middle Caspian; in the region of the Apsheron Peninsula, these values are 8–9 m s^{-1}. Over the entire eastern coast, the mean annual wind speed is approximately the same equal to 5–6 m s^{-1}, with its maximum value observed off the Mangyshlak Peninsula. In the South Caspian, where strong winds are rarely observed, the mean annual wind speed is 3–4 m s^{-1}, and the recurrence rate of weak winds here reaches 90%.

In the southern part of the sea, the number of days with storms (wind speed greater than 15 m s^{-1}) is not greater than 20–30 per year; in the North Caspian and on the eastern coasts of the Middle Caspian, 30–40 storms are observed every year. The most intensive storm activity is confined to the region of the Apsheron Peninsula (50–60 days per year); there, the orographic effect is especially strong. While flowing over the Caucasian Mountains, which here approach close to the sea, the dominating northwesterly winds acquire clearly expressed northerly direction and reach a speed of 20–25 m s^{-1}. These are the well known "Baku Nords." South of the Apsheron Peninsula, wind speeds gradually decrease down to 2–3 m s^{-1} in the southernmost regions of the sea. This phenomenon is especially well manifested in the summertime.

Over the North Caspian, easterly winds dominate. Their average annual recurrences in the western and eastern parts of the area are 50 and 36%, respectively. Westerly and northwesterly winds are also rather common (12%) and the recurrence rate of calm weather equals 14%. The mean annual wind speed in the North Caspian is 5.8 m s^{-1}; the lowest values (3.5–3.8 m s^{-1}) are noted in the summertime. Strong winds (greater than 15 m s^{-1}) are mostly related to the easterly and southeasterly directions; their annual recurrence does not exceed 1–2%.

According to the wind fields dominating over the Caspian Sea, in the open regions of the sea, waves mostly propagate from the north or the northwest (32%) or, alternatively, from the southeast and the south (36%). More rare (about 12%), waves of easterly directions are observed. In approximately 20% of the cases, at small or zero wind speeds, waves are weak and unsteady. In the open part of the sea, large swells are observed, coming most often from the north or the northwest.

With the wind speed growth, the recurrence rates of the corresponding wave fields decrease; therefore, waves corresponding to wind speeds less than 10 m s^{-1} are the most frequent, while those corresponding to speeds greater than 25 m s^{-1} are the most rare. At wind speeds less than 15 m s^{-1}, in the summer, northerly waves dominate, while in the winter, the waves feature southerly directions. In so doing, the height of the waves with a 5% probability does not exceed 3 m. Storm waves more frequently develop in the winter and spring under northerly winds with speeds up to 20 m s^{-1}. Strong long-term storms are most characteristic of the open areas of the Middle Caspian, where they feature mostly northwesterly and southeasterly directions. The wave field patterns are also affected by the orographic factor: the storm activity is displaced toward the western coast of the Middle Caspian. Generally, the waves off the eastern coasts of the sea are twice as weak as those off its western coasts.

At the northerly storms, the greatest waves are observed in the region of the Apsheron Archipelago; they are most frequent in the wintertime. At the epicenter of the maximum waves in the region of the Neftyanye Kamni Bank, wave height may reach 8 m and even 9–10 m during extreme events. Most often, off the Apsheron Peninsula, waves with a height less than 2 m were observed. Off the Turkmenian coast, moderate and strong northwesterly winds (5–15 m s^{-1}) induce waves up to 1 m high, while under storm winds, their height may reach 2–3 m.

Under southeasterly winds, the largest waves develop in the north of the Middle Caspian, in the Makhachkala–Derbent region, and off the Mangyshlak Peninsula, where the height of the waves with a 5% probability may reach 6–7 m. A similar wave height is observed in the open part of the sea at southeasterly storms. Easterly winds even of greatest speeds induce waves with a height not greater than 2–3 m.

In the North Caspian, wind wave development is restricted by the small sea depths in the region. Most often (about 70%), northwesterly, easterly, and southeasterly waves are observed. The greatest possible wave heights increase with the depth growth from the north to the south. At a wind with a speed of 15–20 m s^{-1}, the height of the waves with a 5% probability increases from 0, 5 m off the Volga River delta to 4 m over the shelf edge at the boundary with the Middle Caspian. The calmest season in the North Caspian is the summer, when windless weather often dominates over vast areas [5].

7 Storm Surges

In the enclosed Caspian Sea, the coupled action of the atmospheric pressure and wind upon the water surface cause sharp local oscillations of sea level known as storm surges. Their magnitude (height) is defined as the deviation of the level at a given point from its mean monthly value. In selected regions of the Caspian Sea, surges may reach significant values. The knowledge of their features and characteristics is of great practical value, in particular, in terms of ecology, since the surge onsets and subsequent offsets of the level may result in the delivery of pollutants from the adjacent land used for industrial or agricultural purposes into the marine environment.

The characteristics of the onsets and offsets (height, duration, rate of the level changes) depend on the features of the wind field and local physico-geographical conditions. The surges are best manifested in the coastal zones of the sea, in its bays and bights. The highest onsets are characteristic of the shallow-water North Caspian, where, in extreme cases, they can reach a height of 3–4 m. In the middle and southern parts of the sea, surges are smaller though still significant. According to the data of long-term observations, here, the greatest heights of onsets are 50–70 cm, while offsets can reach values of 30–100 cm. Thus, the range of the level variations makes 100–150 cm. The recurrence frequency of storm surges in different regions of the sea ranges from one to five times per month, their duration varies from a few hours to a day and greater (1.5–2.5 days on the average), and the rate of the level change may be very great. In the wintertime, the ice cover may reduce the onset and offset magnitudes.

The development of storm surges in the North Caspian is favored by the small sea depths and low inclination angles of the floor in the near-shore area. The southeasterly and easterly winds, which are frequent in the North Caspian, cause an onset off the western and northwestern coasts and in the near-mouth area of the Volga River and an offset off the eastern coast of the sea. The northwesterly and westerly winds produce an opposite effect in the short-term level variations. In the western part of the North Caspian, the greatest onsets were noted off Kaspiiskii settlement (more than 4 m); in its eastern part, the greatest heights were observed near Zhilaya Kosa settlement (2.5 m). The maximum offsets were noted near the Volga–Caspian floating lighthouse (2.3 m) and off Zyuidvestovaya Shalyga Island (1.5 m), correspondingly.

Due to the flatness of the shores of the North Caspian and of the near-shore floor areas, strong storm surge onsets lead to a flooding of vast land areas, while during offsets, great shallow-water areas are dried. The width of the flooded zones reach 30–50 km and that of the dried band is 10–15 km. In the eastern part of the North Caspian, onsets more than 40 cm in height are registered from 5 to 20 times per year. At winds up to 15 m s^{-1}, sea level rise reaches 50–80 cm, while winds up to 25 m s^{-1} cause a rise of 90–150 cm.

In the deep-water basin of the sea, northwesterly and northerly winds cause onsets in the South Caspian and offsets in the Middle Caspian. Southeasterly winds produce an opposite situation resulting in level falls in the southern part of the sea and level rises in its middle part. Meanwhile, the general pattern of the onset–offset phenomena is complicated by the excitation of compensation currents and influence of local conditions. Off the northern coast of the Apsheron Peninsula, strong and durable northwesterly winds cause an onset effect, while southerly winds result in offsets. Their maximum ranges are 70 and 60 cm, respectively.

In the near-shore regions of the South Caspian from Astara in the west to Turkmenbashi in the east, the heights of the onsets caused by the winds from the northerly sector equal 50–80 cm. At southeasterly winds, the offset magnitude in the region of Turkmenbashi and Aladzha exceeds 1 m. In the eastern part of the Middle Caspian, the greatest offsets are caused by northwesterly winds, while the highest onsets occur at southeasterly winds. Off the Mangyshlak Peninsula and in the region of Bekdash settlement, the range of these oscillations exceeds 1 m.

The existence of close relations between the atmospheric pressure fields, winds, and storm surges provides a good basis for developing various methods for their forecasts. There exist statistical methods for forecasting storm surges from the actual field of the atmospheric pressure and hydrodynamical modeling, which allows one to calculate the height of the maximum level onset and the width of the zone of possible flooding [6].

8
Sea Ice

The Caspian Sea is a partially frozen aquatic area. In the middle part of the sea, ice covers a small area and, in mild winters, it is absent at all. The North Caspian is frozen every year; in so doing, the major part of its area is covered with fixed ice. A band of floating ice 20–30 km in width borders the fast ice zone. The southern boundary of the mean ice propagation runs approximately over the shelf edge in the North Caspian forming an arc from Chechen' Island to Kulaly Island and farther to Cape Tyub Karagan.

In moderate winters, the ice formation begins in the middle November in the shallow-water northeastern regions of the sea; by the end of this month, ice is developed off the entire northern coast of the sea. At the beginning of January, the entire area of the North Caspian finds itself covered with ice. In the Middle Caspian, ice appears in December starting from shallow-water bays and bights of the eastern coast and in January it appears in the region of Makhachkala. In warm winters, ice formation in near-shore zones begins 10–20 days later than in moderate winters; in the open sea, this delay may reach a month and greater. In severe winters, ice formation in the North Caspian proceeds 20–30 days earlier than usually. Off the eastern coast of the Middle Caspian, ice formation is possible as early as at the end of October–November, while off its western coast it starts at the end of November–beginning of December.

In the North Caspian, the maximum thickness of the fast ice is observed in the northeast in January (40–50 cm), while in the western part of the area and off the Volga River delta it reaches its maximum by the end of the winter, in February (20–30 cm). In very cold winters, the thickness of the fast ice may reach 80–90 and 60–70 cm, respectively. The height of the hummocks makes 1–1.5 m, while in selected cases it may reach 2–3 m.

During severe winters, events of wind-driven transport of great masses of floating ice along the western coast of the sea far to the south up to the Apsheron Peninsula are observed. Under such extreme conditions, ice can block the approaches to the port of Makhachkala and threaten marine hydraulic constructions.

Starting from the second half of February, intensive destruction of the ice cover begins. First, the near-shore areas of the Middle Caspian are released from ice, followed by the northwestern part of the sea, the open regions of the North Caspian, and, at last, the extreme northeast. The final release of the sea from ice occurs at the end of March to the beginning of April. In mild winters, the northern part of the sea is free from ice as early as in the middle March, while during severe winters, the processes of the ice cover destruction are decelerated and the time of the complete release from ice is shifted toward the middle April.

The greatest duration of the ice cover period (120–140 days on the average) is observed in the northeastern regions of the sea and in the eastern part of the Volga River near-mouth area. Over the areas with sea depths of 2–5 m, ice is retained over 80–90 days; the ice cover periods off Tuylenii and Kulaly islands and off Cape Tyub Karagan are 70 and 50–60 days, respectively. During mild winters, the number of the days with ice cover is significantly smaller: from 100–150 days in the northeast of the sea to only 15–20 days in the southern regions of the North Caspian. In severe winters, the duration of the ice cover existence in the northeastern region of the sea, off Cape Tyub Karagan, and in the rest of the regions of the North Caspian reaches 140–170, 100, and 100–150 days, respectively.

The studies of the physicochemical properties of the ice from the North Caspian showed that the temperature of the surface layer of the ice is commonly by 1–3 °C lower than the air temperature; meanwhile, with cooling, this increment grows. The average ice salinity equals about 1‰. The ice contains 2–5-fold elevated amounts of phosphates and nitrates as compared to the near-ice water. The ice melting provides a significant enrichment of the waters of the North Caspian with these nutrients [1, 6, 7].

9 Water Balance

For the enclosed basins of a lacustrine type such as the Caspian Sea, the notion of the water balance (or, more precisely, water budget) defined the relation between the water input and output to the basin. When the sum of the input terms of the water balance equation exceeds the sum of the output terms, the volume of the water in the basin increases and its level rises. When the input part of the balance equation is smaller than its output part, the volume of the water decreases and the level falls. The increment between the output and the input components of the water balance is referred to as the resulting value.

In the Caspian Sea, the input component of the balance consists of the surface riverine runoff, atmospheric precipitation, and groundwater supply, while the output component includes the evaporation from the sea surface and water discharge to Kara Bogaz Gol Bay (where the water is evaporated). In the water balance of the Caspian Sea, the most important are the riverine runoff and evaporation; relations between their intensities mostly control the interannual changes in the water volume and the sea level of the basin.

The water balance of the Caspian Sea for different periods was repeatedly estimated by many scientists; however, certain discrepancies in the result obtained are observed. This is related to the character of the materials used and the technique of the calculations. The greatest difficulties are connected with the determination of the evaporation intensity and the value of the groundwater supply, since there are no direct measurements of these components. Therefore, all the estimates of the water balance presented above should be regarded as approximate values obtained under certain assumptions.

The riverine runoff represents the principal input item of the water balance providing up to 80% of the total water delivery to the sea. Meanwhile, the riverine runoff is subjected to strong variability. In the past century, it ranged from 335 km^3/year in 1900–1929 to 240 km^3/year in 1970–1977 at an average value of 300 km^3/year. About 80% of the total riverine runoff is provided by the Volga River, whose mean annual runoff makes about 240 km^3/year. In so doing, the greatest runoff equal to 368 km^3/year was recorded in 1926, while the minimum and extremely small values equal 150 and 163 km^3/year were noted in 1921 and 1973, respectively. Thus, during the past century, the differences in the runoff of the Volga River exceeded 200 km^3/year. Up to 25% of the Volga River runoff is supplied to the sea in May–June during the flood periods. About 15% of the runoff to the Caspian Sea is provided by the Ural River and the rivers of the western coast – Terek, Sulak, Samur, and Kura. The runoff of minor rivers including those of the Iranian coast makes about 5%.

The above-mentioned noticeable variations in the riverine runoff to the Caspian Sea cause significant changes in the freshwater balance of the sea (up to 50 km^3/year) and corresponding sea level oscillations. In the riverine runoff dynamics of the Caspian Sea, one should take into account its consumption for various economic purposes; they can hardly be estimated. According to the data available, the magnitude of the anthropogenic removal of the surface runoff to the Caspian Sea may reach 40 km^3/year; 25 km^3/year of them refer to the Volga River. At the absence of these losses, the Caspian Sea level in 1955–1990 might be located 1.6 m higher with respect to its actual position.

The seasonal level variations are also controlled by the amounts of the river waters supplied into the sea, mostly by the Volga River. In this context, the minimum values of the sea level height are observed in the winter, while the maximum values are confined to the summer. The range of the intra-annual sea level oscillations comprises about 30 cm.

The volume of the atmospheric precipitation is significantly smaller as compared to the riverine runoff and evaporation; therefore, its influence on the Caspian Sea level oscillations is not very strong. The relative contribution of the atmospheric precipitation to the input component of the water balance changed from 18% at the beginning of the past century to 25% in the 1970s–1990s, when about 85 km^3/year of the atmospheric precipitation was supplied to the sea surface. This value significantly exceeds the mean annual value characteristic of the 1900s–2000s. The latter value is equal to 75 km^3/year, which is equivalent to a water layer 20 cm thick. From the beginning of the past century, a tendency to the growth in the atmospheric precipitation intensity over the surface of the Caspian Sea was observed.

The role of the groundwater discharge in the water balance of the Caspian Sea is insignificant. According to indirect data, its average value over the perimeter of the sea is accepted to be about 4 km^3/year.

Evaporation from the sea surface represents the major output component of the water balance. As compared to the riverine runoff, the interannual variations of the evaporation intensity are essentially smaller. In the 1930s, the evaporation processes over the Caspian Sea were most intensive. In 1900–1929, about 390 km^3/year were evaporated from the sea surface (96-cm water layer); in 1930–1941 was 397 km^3/year (100-cm layer). In the recent years (1978–1995), evaporation was about 349 km^3/year (92-cm layer). Thus, the changes in the evaporation intensity over the 20th century exceeded approximately 40 km^3/year. Throughout the year, the highest evaporation is observed from June to December, when about 70% of the annual volume is evaporated. Over the sea area, the evaporation processes are most intensive in the North Caspian, where it comprises an equivalent of a 100-cm water layer.

The output component of the water balance of the Caspian Sea also includes the seawater discharge to Kara Bogaz Gol Bay (see Chap. 11). The intensity of this discharge depends on the height increment between the levels in the sea and in the bay and on the bottom morphology of the strait that connects them. At the same area of the strait cross section, the greater the level difference the greater the discharge of the Caspian seawater into the bay. Before the beginning of the level fall in the 1930s, the annual seawater supply to the bay had been 20–25 km^3, while the level increment had equaled 0, 5 m. The long-term level fall in the Caspian Sea resulted in a decrease in the water discharge to the bay and, at the end of the 1970s, only 5–10 km^3 of seawater per year were delivered. In 1980, in order to reduce the water deficiency in the Caspian Sea balance, the strait was closed with a dam. Meanwhile, by this time, the Caspian level had already started to rise; therefore, in 1984, the dam was partly destroyed to let the water into the bay and, beginning from 1992, a free water delivery of the Caspian seawater to the bay was resumed.

An analysis of the water balance of the Caspian Sea over the past century showed that, before the late 1970s, the balance had been negative mostly at the expense of the riverine runoff. The water balance deficiency caused the general long-term tendency to a sea level fall. Only in 1978, a radical turn in the water regime of the Caspian Sea occurred; from then and up to the present, the water balance has been characterized by a positive resulting value. During recent decades, the excess of the amount of the water delivered to the sea by the riverine runoff and atmospheric precipitation with respect to the water losses caused by the evaporation and discharge to Kara Bogaz Gol Bay comprised approximately 45–50 km^3/year. This provided an increase in the water volume in the sea and its rapid level rise [7, 8].

10
Sea Level Problem

The studies if the long-term sea level oscillations in the Caspian Sea are especially important, since they affect virtually all the principal processes proceeding in the basin and influence the economic activity on its shores.

Among the main factors controlling the long-term oscillations of the Caspian Sea level, one can recognize the geological (changes in the volume of the sea depression due to the tectonic movements) and climatic (variations in the water balance of the sea) factors, although their contributions to the level dynamics are different. Tectonic movements played the decisive role at the initial stages of the formation of the Caspian Basin. Meanwhile, as early as in the Holocene (no less than 10 000 years B.P.), the climatic conditions over the sea catchment area and its basin came to the foreground as the principal reason for the large-scale changes in the Caspian Sea level. According to the historical, cartographic, and paleogeographic studies, the intensive marine regressions and transgressions accompanied by significant changes in the area of the sea could not be caused by the weak tectonic movements characteristic of those times; they were rather related to water-climatic reasons. The differences in the directions of the tectonic movements within the Caspian Sea basin and the values of their increments (a few millimeters per year, which is by 1–2 orders of magnitude smaller than the actual sea level oscillations) do not allow one to accept the "tectonic" theory as that responsible for the level changes in the Caspian Sea during the recent geological epoch.

The range of the level changes over the past 2000 years reaches 7 m. The lowest sea level standing was observed in the 6th–7th centuries; later on, it changed within a narrower range - from - 30 to - 25 m (here, all the heights are presented with respect to the Baltic altitude system). An analysis of the instrument level observations that started in 1830 shows that, from the beginning of the 20th century up to 1929, the level of the Caspian Sea had been located close to a mark of - 26.2 m. Subsequently, it began to rapidly drop and, by 1956, it fell by almost 2 m (Fig. 3). This fall was caused by a strong drought in the Volga River basin, which resulted in a decrease of its runoff. In the 1950s, the humidity in the sea basin increased; meanwhile, in these years, major reservoirs on the Volga River were constructed, which required large water volumes for their filling. In addition, water expenses for economic needs also increased. Therefore, in the 1950s–1960s, the level of the Caspian Sea stabilized rather than rose. In the 1970s, a new level fall was observed caused by the decrease in the Volga River runoff and an increase in the evaporation from the sea surface. In 1977, the level fell down to a mark of - 29 m, which was the lowest over the past 400–500 years. During the 20th century, the total range of the level fall was 3 m. In so doing, a half of this fall is related to the anthropogenic withdrawals from the riverine runoff. The level drop resulted in the decrease in the sea area by approximately 40 000 km^2, mostly due to the drying of the shallow-water North Caspian.

Starting from 1978, a rapid sea level rise began; in 1995, the level reached a mark of - 26.7 m (see Fig. 3). This rise was also caused by the changes in the water balance, whose increment corresponds well to the range of the level rise during this period. The positive resulting value of the water balance is mostly defined by the high runoff of the Volga River, whose relation to the sea level position is reliably established [9]. By 2004, the level of the Caspian Sea fell again by 30 cm down to a mark of - 27 m.

The studies of the level oscillations in the Caspian Sea and of the possibilities for their forecast represent a special complicated problem; it is discussed in numerous publications. The basic fundamental concepts of many authors are close or completely coincide. For example, there are virtually no doubts about the water–climatic control of the long-term level variations in the

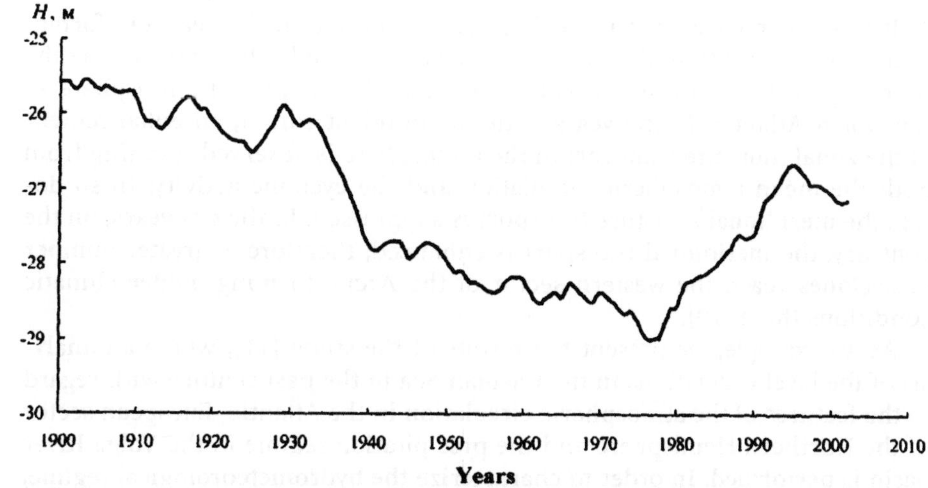

Fig. 3 Long-term changes in the Caspian Sea level in the 20th century

Caspian Sea. Below, we briefly present selected matters discussing the reasons for the recent sea level changes in the basin under consideration.

The oscillations of the Caspian Sea level is a result of interrelated hydrometeorological processes proceeding not only in its catchment area but also far beyond its limits. The principal cause-and-effect relation providing these oscillations is as follows: changes in the atmospheric circulation in the Atlantic–European sector → changes in the hydrometeorological conditions in the Volga River basin → changes in the riverine runoff to the Caspian Sea → sea level variations in the Caspian Sea. In this respect, the relations between the circulation factors in the atmosphere and the meteorological regime in the Volga River basin are the least studied. The formation of the level of the Caspian Sea is also strongly affected by the precipitation regime.

In the studies performed in to the middle of the past century, the relations between the level of the Caspian Sea and the climatic conditions in the Volga River basin were established on the basis of an analysis of the circulation activity over Europe and West Asia and on the manifestations of the principal circulation patterns - westerly, easterly, and meridional. During recent decades, the analysis of the factors responsible for the level oscillations in the Caspian Sea involved significantly greater areas considering the large-scale climate-forming factors not only in Northern Hemisphere but in the Southern Hemisphere as well, such as the North Atlantic Oscillation (NAO), El Ni@no–Southern Oscillation (ENSO), and others. As a result, the principal conclusion was justified that the sea level regime in the Caspian Sea is mostly governed by the ratio between the cyclonic and anticyclonic activities over the sea basin and the related regime of the atmospheric precipitation.

It was reliably stated that, during the cold period of the year, the formation of the humidity in the Volga River basin is completely controlled by the processes of the moisture exchange between the ocean and atmosphere in the North Atlantic. In the years with an abundant runoff, an enhancement of the zonal moisture transport in the atmosphere is observed resulting from both the mean atmospheric circulation and the cyclonic activity. In so doing, the meridional moisture transport is suppressed. In the dry years, on the contrary, the meridional transport is enhanced; therefore, a greater number of cyclones reach the western sector of the Arctic forming milder climatic conditions there [10].

As an example, we present the results of the study [11], where an analysis of the level oscillations in the Caspian Sea in the past century with regard to the features of the atmospheric circulation in the Atlantic–European sector of the Northern Hemisphere and the precipitation regime in the Volga River basin is performed. In order to characterize the hydrometeorological regime, the authors used the number of days with anticyclonic conditions in two regions of the European part of Russia in 1900–1993. These estimates showed a distinct descending trend of the number of days with anticyclones during the above period accompanied by a growth in the atmospheric precipitation in the Volga and Ural River basins. Meanwhile, against the background of this trend, in the period of a sea level fall, the number of days with anticyclones was significantly greater than in the periods of a sea level rise. For the main seasons of the year, a close relation was recognized between the dominating type of the atmospheric circulation, the amount of atmospheric precipitation, and the air temperature anomalies over the Volga River catchment area. These conclusions agree with the results of [12], where it was shown that, in the relatively dry period of 1966–1975, the number of cyclones of various origins in the Volga River basin was by 27–38% of the norm smaller than in the water-abundant period of 1976–1985. The recurrence frequencies of the cyclones (anticyclones) in the periods considered play a significant role in the formation of the hydrometeorological conditions in the watershed areas of the Volga and Ural Rivers and of the level regime in the Caspian Sea. The correlation between the number of days with anticyclones and the total precipitation in the Volga River basin during 1901–1997 was found to be negative; in so doing, it was best manifested in the wintertime.

Reliable materials confirming the dominating role of the climatic factors in the level oscillations of the Caspian Sea have recently been acquired in the Hydrometeorological Center of the Russian Federation. E.S. Nesterov has analyzed the changes in the fundamental parameters of the ocean and the atmosphere in the Atlantic–European region over 40 years (1957–1996). He assessed the ocean surface temperature, its meridional gradient, atmospheric pressure at the sea surface, and geopotential at the 500 GPa level for three regions of the North Atlantic. The most common feature of these characteristics is the growth in their meridional gradients that started in the early 1970s. This resulted in an enhancement of the western transport in the atmosphere. After 1975, precisely the preservation of the high values of the meridional gradients of the above characteristics has been the reason for the domination of zonal circulation patterns in the atmosphere. This caused a growth in the number of cyclones driven from the North Atlantic to the Volga River basin, an increase in the precipitation there, and in the corresponding level rise in the Caspian Sea [13].

For the sake of a more detailed estimation of the influence of the atmospheric circulation on the Caspian Sea level, this study also included an analysis of the variations of the NAO index characterizing by the pressure difference between the two well-known atmospheric centers - the Azores maximum and the Iceland minimum. The analysis showed that, beginning from the middle 1950s to 1979, a negative phase of NAO had dominated, while the interval from 1980 to 1995 was characterized by a positive phase. During the periods of the negative phase of NAO, over the majority of the regions of Russia, a smaller number of cyclones driven from the Atlantic were observed as compared to the periods with positive NAO values. The fact that the negative phase of NAO coincided with a fall in the level of the Caspian Sea and the positive phase, on the contrary, was confined to a level rise supports the conclusion that precisely the displacements of the routes of the wintertime cyclones related to the changes in the phase of NAO represent one of the principal reasons for the radical change in the trend of the level variations in the Caspian Sea that occurred in the late 1970s [13].

The forecasts of the level changes are of great scientific and practical values; meanwhile, this task is extremely difficult. It is related to the necessity of elaboration of a climatic forecast for a vast region covering the entire catchment of the Caspian Sea. Therefore, at present, only probability estimates of the level changes in the Caspian Sea for the forthcoming decades are possible based on the paleogeographic analysis in combination with the long-term data on the water balance. According to the estimates available, under the present-day physico-geographical conditions, the basic position of the mean level of the Caspian Sea is confined to a mark of -28 m ($\pm 1.5-2$ m). The multiannual oscillations of the Caspian level is a regular phenomenon, which reflects the "respiration" of the basin. The possibility of the level changes by 1–1.5 m over a few decades should be taken into account when developing and implementing economic measures in the coastal zone of the sea.

The significant level variations of the Caspian Sea result in the changes in the structure of its coastal zone, which is now under active economic use. Generally, level falls lead to land accretion and drying of the near-shore marine areas, while the level rises, on the contrary, are accompanied by landward displacement of the coastline, abrasion, and flooding of sea shores. Meanwhile, these tendencies are differently manifested in different areas of the coasts depending on the type of the shore and on the influence of many other factors. The materials available allow us to briefly characterize the changes that had occurred in the coastal zone of the Caspian Sea during the latest sea level rise from 1978 to 1995. In so doing, we will consider only the western coast of the sea referring to Russia and Azerbaijan.

Within the Russian coastal zone, with respect to geomorphology, three major regions may be distinguished: the Volga River delta, the Terek–Kuma Lowland (within Kalmykia), and the narrow near-shore band with marine terraces along the coast of Dagestan. During the sea level fall from 1929 up to the end of the 1950s, on the average, the seaward edge of the delta of the Volga River had advanced by about 10–20 km. The end of the period of the regression was characterized by a rather stable position of the delta outlines. The recent rise of the Caspian Sea level did not produce the expected changes in the marine part of the delta. For the time being, its influence has been significantly suppressed by the presence of the vast shallow-water area off the river mouth. During the two decades of the level rise, it resulted only in an increase in the sea depths in the near-mouth area off the Volga River delta by about 1 m and in a reconstruction of the system of the offshore accumulative islands (some of them disappeared). However, if the sea level will reach a mark of - 26.5 m, the influence of the sea may noticeably grow. In particular, this may be manifested in a freer penetration of level onsets into the delta and in an enhancement of flooding during the corresponding periods [4].

In the above-mentioned portion of the Terek–Kuma plain, the influence of the recent Caspian level increases in the southward direction with the distance from the delta of the Volga River. In the southern part of the Kalmykian coast, this influence is manifested in the landward shift of the narrow coastal band and in the flooding of its outer edge by the sea. The receding of the marine edge of the shore proceeds at a rate of approximately 200 m per year. Farther to the south over this coast, the sea level rise caused erosion of coastal spits and formation of low abrasive escarpments.

In the coastal plain of Dagestan, the level rise resulted in an enhancement of the coastal abrasion including the coastal terraces in the regions of Makhachkala, Kaspiisk, and Derbent. During the period of the level rise, the area of the beaches significantly decreased. Due to various anthropogenic reasons, the erosion of the marine edge off the delta of the Sulak River also strengthened. On the whole, during the 20-year-long period of the recent transgression of the Caspian Sea, processes of coastline recession began to dominate over the Russian coasts; meanwhile, their mechanisms might be different. The accretion of the coasts was partly retained in the extreme north of the coast, especially within the Volga River delta [14].

Estimates of the flooding of the coast of Azerbaijan (from the Samur River delta to Astara) were made for the level positions at marks of - 26.5 and - 25.0 m. It was shown that the level rise should provide the greatest flooding in the southern coastal territories of Azerbaijan between the Kura and Astara Rivers. According to these estimates, the total flooded area in the coastal zone of Azerbaijan made greater than 480 km^2. At present, many municipal and economic objects are under the menace of being flooded in the case of a further sea level rise [15].

Acknowledgements This study was supported by the Russian Foundation for Basic Research Grant No 03-05-64279.

References

1. Baidin SS, Kosarev AN (eds) (1986) The Caspian Sea. Hydrology and hydrochemistry. Nauka, Moscow (in Russian)
2. Krylov NA (ed) (1987) The Caspian Sea. Geology and oil and gas resources. Nauka, Moscow (in Russian)
3. Mikhailov VN, Kravtsova VI, Magritskii DV, Mikhailova MV, Isupova MV (2004) Vestnik Kaspiya 2:60 (in Russian)
4. Mikhailov VN (1997) Mouths of the rivers of Russia and adjacent countries: present, past, and future. Geos, Moscow (in Russian)
5. Koshinskii SD (1975) Regime characteristics of strong winds over the seas of the USSR. Gidrometeoizdat, Leningrad (in Russian)
6. Zonn IS (2004) The Caspian Encyclopaedia. Mezhdunarodnye otnosheniya, Moscow (in Russian)
7. Terziev FS, Kosarev AN, Aliev AA (1992) (eds) Hydrometeorology and hydrochemistry of the seas, Vol. 6. The Caspian Sea, Issue 1. Hydrometeorological conditions. Gidrometeoizdat, St. Petersburg (in Russian)
8. Mikhailov VN, Dobrovol'skii AD, Dobrolyubov SA (2005) Hydrology. Vysshaya shkola, Moscow (in Russian)
9. Kosarev AN, Kuraev AV, Nikonova RE (1996) Vestnik Mosk. un-ta, Ser 5, Geografiya 5:47 (in Russian)
10. Malinin VM (1994) The problem of the Caspian Sea level forecast. St. Petersburg (in Russian)
11. Meshcherskaya AV, Golod MP, Belyankina IG (2002) In: Climatic changes and their aftereffects. Nauka, St. Petersburg, p 180 (in Russian)
12. Babkin VI, Postnikov AN, Smyslov SV, Mineeva IV (1992) Tr. GGI, 360:48 (in Russian)
13. Abuzyarov EK, Nesterov ES (1999) In: 70 years of the Hydrometeorological Center of the Russian Federation. Gidrometeoizdat, St. Petersburg, p 216 (in Russian)
14. Lukyanova SA, Solov'eva GD (2001) In: Coastal zone of seas, lakes, and reservoirs. vol 1, Novosibirsk, p 166 (in Russian)
15. Aliev AS (2001) Rise of the Caspian Sea level and flooding of the coastal zone of the Azerbaijan Republic. Elm, Baku (in Russian)

Hdb Env Chem Vol. 5, Part P (2005): 33–57
DOI 10.1007/698_5_003

Published online: 6 October 2005

Thermohaline Structure and General Circulation of the Caspian Sea Waters

Valentin S. Tuzhilkin[1] (✉) · Alexey N. Kosarev[2]

[1]State Oceanographic Institute, 6 Kropotkinsky Per., 119838 Moscow, Russia
valver@orc.ru

[2]Geographic Department, Lomonosow Moscow State University, Vorobievy Gory, 119899 Moscow, Russia

Abstract The results of statistical and physical analysis of historical ship (more then 56 000 stations) and coastal temperature data and salinity measurements are presented. The thermohaline structure of the Caspian Sea waters is characterized by a significant spatial inhomogeneity, especially in the near-mouth areas of the Caspian Sea rivers and in the area of the summertime upwelling off the eastern coast of the Middle Caspian. The seasonal variability of water temperature and salinity is confined to the upper 100 m and 20 m layers, respectively. The thermohaline structure of the entire water column of the Caspian Sea is subject to significant interannual variability, including multiannual trends that dominate other kinds of variability (seasonal and synoptic) in the intermediate and abyssal layers of the sea. Owing to enhanced river discharge in the 1980s and 1990s, a hydrostatically stable vertical salinity stratification formed in the Caspian Sea deep-water areas, which replaced the previously existing uniform salinity over the entire water column and caused changes in the summertime vertical thermal structure. The multiannual trends in the variability of the thermohaline structure of the Caspian Sea waters may exert significant influence upon other components (abiotic and live) of its ecosystem.

Keywords Currents · Salinity · Temperature · Structure · Variability

Abbreviations

HMS Hydrometeorological station
ENSO El Niño–Southern Oscillation cycle

NCAR/NCEP The National Center for Atmospheric Researches/The National Center for Environmental Predictions

1
Introduction

This chapter covers the major large-scale features of the Caspian Sea physical oceanography: the thermohaline structure of the waters, their general circulation, and seasonal and interannual variability. To a large extent, these determine the condition and functioning of many other components of the Caspian ecosystem, particularly the chemical properties of the waters and marine flora and fauna.

This chapter is mostly based on the results of statistical and physical analyses of ship and coastal measurement data. Their spatial and temporal distribution over the years is presented in Fig. 1. The total number of simultaneous ship measurements of temperature and salinity vertical profiles exceeds 56 000. Geographically, they are mostly concentrated over standard cross sections and near coastal hydrometeorological stations (HMSs). In Fig. 1a the standard cross sections are shown by bold dots and HMSs by asterisks. The roman numerals indicate the numbers of the standard cross sections that were most frequently visited in the 1950s–1990s. The majority of the shipborne thermohaline observations fall precisely within this interval (see Fig. 1b).

Among these types of previous studies, the monographs [1–3] and atlases [4, 5], which helped to form the general picture of the physical and oceanographic conditions of the Caspian Sea are notable. This study is distinguished for the greater amount of the observational data used and for the more up-to-date techniques of their processing and analysis, close to those used in the widely known Climatological Atlas of the World Ocean [6]. This provides the possibility of significant refinement and improvement of existing concepts of the physical oceanography of the Caspian Sea waters.

Owing to the isolation of the Caspian Sea from the World Ocean, the formation of its thermohaline and circulation regime proceeds only under the action of atmospheric processes over the sea basin and its vast drainage area. Together with river runoff, the fluxes of heat and freshwater across the sea surface caused by this action control the large-scale features of the thermohaline structure and its temporal variability. The impact of the wind in the form of the fluxes of momentum and relative vorticity generates the three-dimensional general circulation that redistributes the heat and freshwater supplied from the atmosphere over the sea area and down the water column.

In the most general way, the isolation of the Caspian Sea is manifested in the thermohaline diagram shown in Fig. 2, in which every point represents

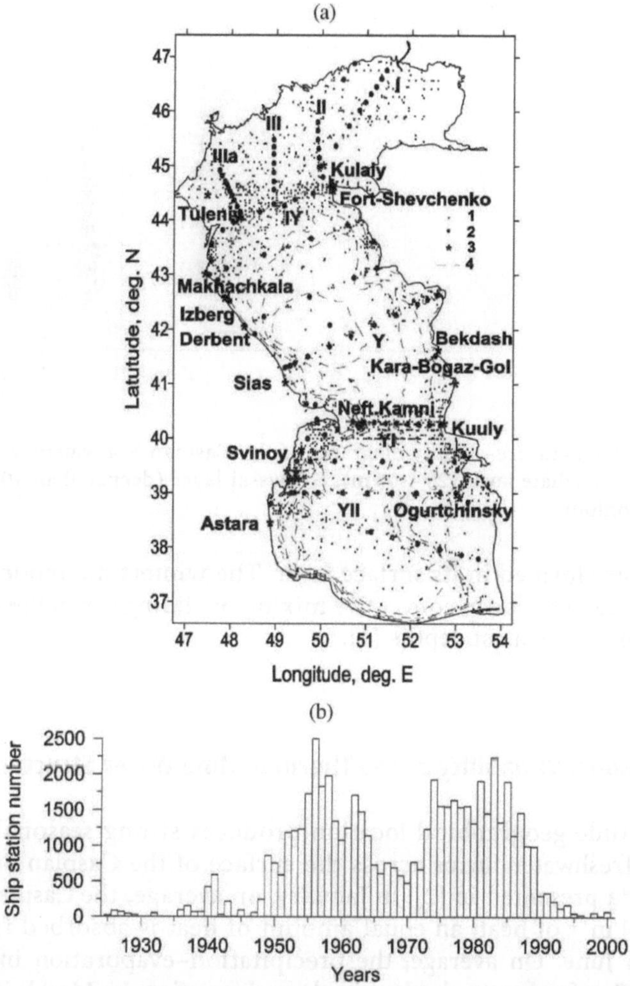

Fig. 1 **a** Location of oceanographic stations in the Caspian Sea. *1* The stations of all ship thermohaline measurements, *2* the ship stations of standard sections, *3* coastal hydrometeorological stations (HMS), *4* depth contours of 20, 50, 100 and 500 m. **b** Distribution of the all stations of ship observations over the years

a fixed water volume (three-dimensional cell) with particular climatic thermohaline properties. For example, the sparse points in the left-hand part of Fig. 2a represent the small-in-volume and strongly freshened waters of the Northern Caspian in the near-mouth area of the Volga River. This figure suggests that in the Caspian Sea, the ranges of the thermohaline properties of the intermediate (Fig. 2b) and abyssal (Fig. 2c) waters lie completely within the range of the properties typical of its surface waters (Fig. 2a). Figure2 thus illustrates the well-known fact that the waters of the entire water column of the

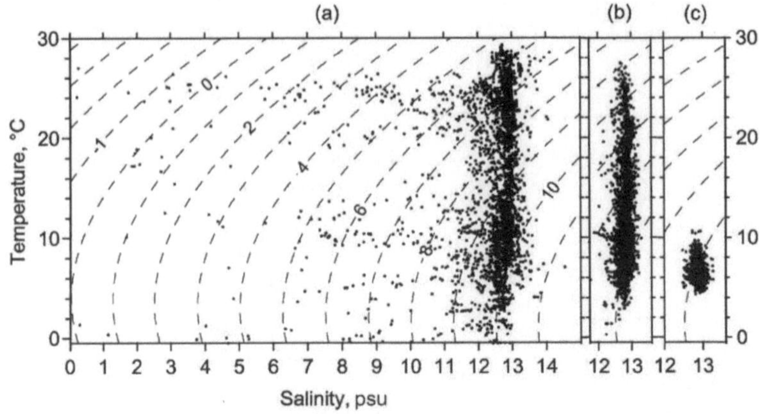

Fig. 2 Climatic temperature–salinity diagrams of the Caspian Sea waters. **a** Surface layer (0–20 m). **b** Intermediate layer (20–100 m). **c** Abyssal layer (deeper than 100 m). *dashed lines* Sigma-t contours

Caspian Sea are formed in its surface layer. The wintertime mode of the surface waters, due to the deep convective mixing or sliding down the continental slope, reaches the greatest depths [1].

2 Climatic Seasonal Variability of the Thermohaline Water Structure

The mid-latitude geographical location produces strong seasonal variations in heat and freshwater fluxes across the surface of the Caspian Sea. According to the data presented in [2], in January, on average, the Caspian Sea loses about 400 MJ m^{-2} of heat; an equal amount of heat is absorbed from the atmosphere in June. On average, the precipitation–evaporation increment in the Caspian Sea freshwater budget is always negative. In March it comprises – 0.06 m of the mean freshwater layer thickness, while in August it is – 1.04 m. More than 40% of the river discharge to the sea is supplied in May–June.

These facts control the clear seasonal character of the thermal regime of the surface water layer of the Caspian Sea down to 100 m (Fig. 3a, b). Deeper, the temperature is quite steady with respect to time and uniform over the vertical. Significant seasonal variations in the salinity in the surface layer (up to 3–5 practicalsalinityunit (‰)) occur in the near-mouth areas of the Caspian Sea. Over the rest of the sea area, in terms of its multiannual regime, salinity variations are not greater than 0.5 and the salinity is quasihomogeneous over the entire water column from the surface to the bottom (Fig. 3c). The large range of its peak values seen in Fig. 3c is caused by the significant multiannual variability of the Caspian Sea salinity, similar to that of its mean surface level (see Kosarev, in this volume).

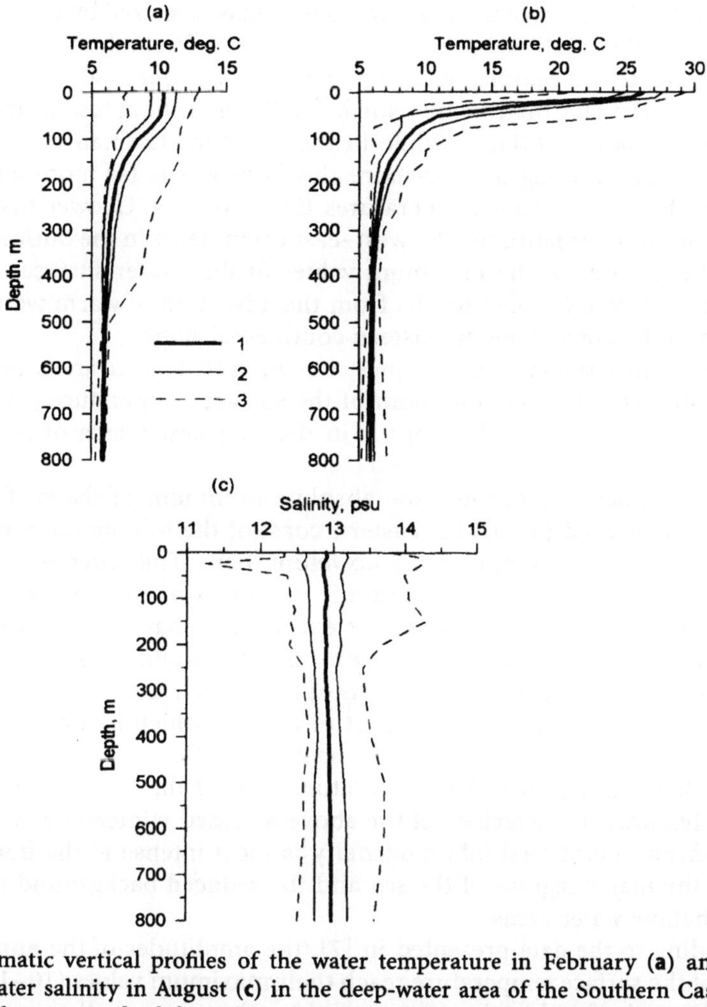

Fig. 3 Climatic vertical profiles of the water temperature in February (**a**) and August (**b**) and water salinity in August (**c**) in the deep-water area of the Southern Caspian Sea. *1* Mean values, *2* standard deviations, *3* extreme values. ‰ Practical salinity units

The climatic temperature and salinity fields of the Caspian Sea are obtained at standard depths every month, following a technique similar to that in [6]. The best horizontal resolution in the areas with enhanced observation density (see Fig. 1a) were 12' and 16' for latitude and longitude respectively. Over the rest of the area, the grid cells were two to three times as large.

The climatic temperature and salinity fields of the surface water layer of the Caspian Sea are presented in Figs. 4 and 5, respectively, for February, April, August, and November, that is, for the months when regular observations over the standard cross sections were performed. They are notable for having the best statistical reliability and characterizing all the seasons of the year.

The wintertime surface temperature field is characterized by its significant increase from the north to the south, especially manifest at the boundaries between the Northern, Middle, and Southern Caspian areas (see Fig. 4a). This is caused by the fact that, in the Northern Caspian, heat losses during the winter reach 600–800 MJ m^{-2}, while in the Southern Caspian they are only 200 MJ m^{-2}. The cooling of near-shore shoals proceeds far more efficiently, so that in the winter, water temperatures there are 2–3 °C lower than in the open sea at the same latitude. The west–east asymmetry in the outlines of the tongue-shaped area of the maximum values of the winter surface temperatures in the Middle Caspian results from the advection of warm waters from the south to the north along its eastern continental slope.

In the spring, shallow-water zones are the best-heated areas of the sea (see Fig. 4b). The absolute minimum of the surface temperature (the area of the most weak springtime heating) is in the deep-water area of the Middle Caspian.

In the summer (see Fig. 4c), the absolute minimum of the surface temperature is displaced toward the eastern coast of the Middle Caspian due to the development of a summertime upwelling here. The latter is one of the most interesting physical–oceanographic processes of the Caspian Sea and will be paid particular attention at the end of this chapter. Over the rest of the Caspian Sea, the surface temperature is quite uniform due to the relatively uniform heat supply from the atmosphere in the summer. This does not apply to the extreme northeastern part of the sea, which is characterized by a negative surface heat balance as early as August [2].

In the fall (see Fig. 4d), the temperature field of the Caspian Sea surface acquires features characteristic of the above-assessed wintertime conditions: the significant meridional inhomogeneity is most intense at the boundaries between the major regions of the sea and the reduced background temperature in shallow-water areas.

According to the data presented in [7], the amplitudes of the annual harmonics of the surface temperature reach their maximum values (10–12 °C) on the northern shelf and in the western and southwestern shallow-water areas of the Caspian Sea, while the minimum values (6–7 °C) are confined to the area of the summer upwelling in the Middle Caspian. The amplitude of the annual harmonic of the water temperature at a depth of 50 m is as small as 4 °C and at a depth of 100 m it is not statistically different from zero.

The most large-scale features of the salinity field are retained throughout the year (Fig. 5). This is related to the steady geographical location of the sources of fresh (river mouth) and saline (shallow-water zones off the arid eastern coasts) waters. Only the position and intensity of salinity frontal zones, which bound the areas of direct influence of the above-mentioned sources, are subjected to certain seasonal changes. In particular, these frontal zones are located in the North Caspian (in the Volga River near-mouth area), in the southwestern part of the sea (off the Kura River mouth), and along the southeastern shallow-water zone, where the salinity effect of evaporation is especially significant.

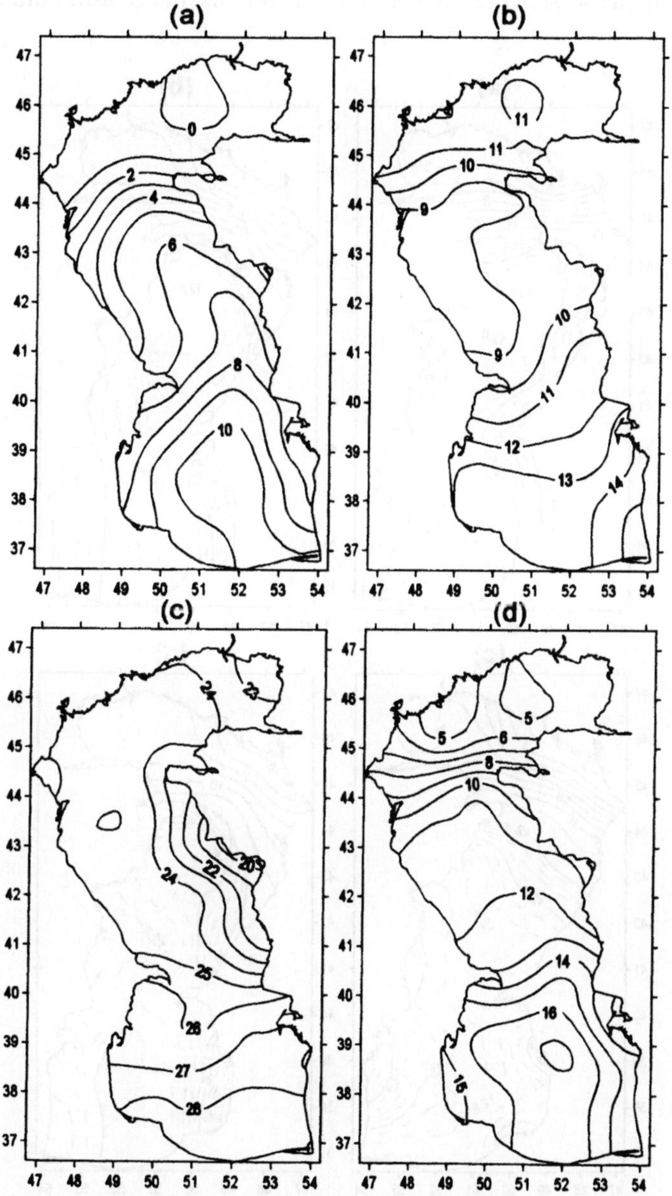

Fig. 4 Climatic fields of the water temperature (degrees celsius) in the surface layer of the Caspian Sea in February (**a**), April (**b**), August (**c**), and November (**d**)

The annual distribution of the evaporation intensity affects the seasonal features of the surface salinity field over the entire deep-water part of the Caspian Sea. In particular, in the winter (see Fig. 5a), when evaporation is minimum, the surface salinity in the Middle and South Caspian is also small, especially off the western coast where the river discharge and enhanced pre-

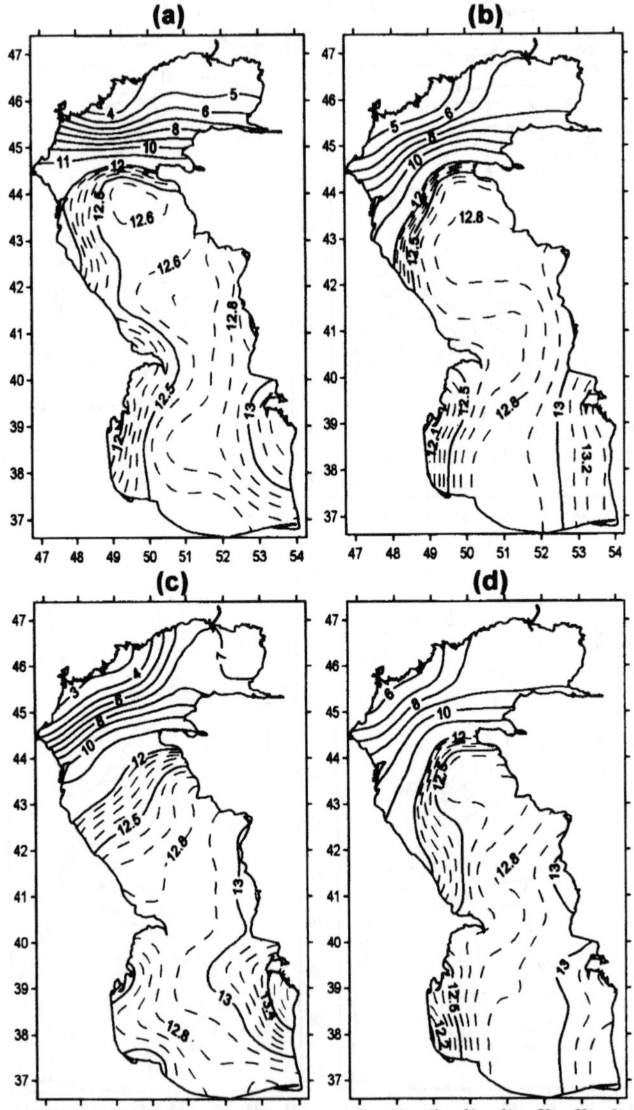

Fig. 5 Climatic fields of the water salinity (practical salinity units) in the surface layer of the Caspian Sea in February (**a**), April (**b**), August (**c**), and November (**d**)

cipitation are concentrated. In the summer (see Fig. 5c), the maximum evaporation prevents propagation of reduced salinity in the waters from the North Caspian, though such propagation is noted both in the spring (see Fig. 5b) and in the fall (see Fig. 5d).

According to the data presented in [7], in the Volga River near-mouth area, the amplitude of the annual harmonic of the surface salinity reaches 1–2 ; off the Kura River mouth it reaches 0.5 , while over the rest of the area it doesn't exceed 0.3 . At a depth of 20 m, its values decrease by a factor of 3–4 and below that do not statistically differ from zero. The field of the minimum phase of the surface salinity in the Caspian Sea is characterized by a frontal zone approximately coinciding with the August position of the 12 salinity contour in Fig. 5c. North of this area, the salinity minimum is observed in June–August, being related to the propagation of freshwater wave from the Volga River mouth. South of this zone, the minimum is confined to January–March in accordance with the minimum evaporation [7].

The changes in the horizontal thermal water structure of the Caspian Sea over the vertical are shown in Fig. 6 by the example of the temperature fields in the upper 200-m layer in August. The strongest irregularities in the field are observed at a depth of 30 m in the core of the seasonal thermocline (see Fig. 6b). It sharply differs from the field at the surface (see Fig. 6a), except for the upwelling area off the eastern coast of the Middle Caspian. This is caused by the fact that, outside the upwelling area, the structure of the sea surface temperature field in August is controlled by the heat flux across the surface of the Caspian Sea; meanwhile, at a depth of 30 m, it is related to the dynamically formed seasonal thermocline topography. The latter means a higher position of the thermocline in the upwelling favorable east parts of the Middle and South Caspian and its deeper position in the downwelling favorable western parts of these Caspian sub-basins. This pattern of the temperature field is retained down to a level of 100 m, though the intensity of inhomogeneities decreases by a factor of 5–10 (see Fig. 6c, d). At a depth of 200 m the temperature maximum in the western part of the Middle Caspian is replaced by a minimum (see Fig. 6e). Meanwhile, the principal feature of this temperature field is represented by the increment between the background temperatures in the Middle (about 6 °C) and southern (more than 7 °C) deep-water areas of the Caspian Sea. The temperature field at a depth of 200 m in August differs very little from the annual mean field (see Fig. 6f). Thus, in the Caspian Sea, below a depth of 100 m, there is virtually no seasonal variability of the temperature structure. This statement applies equally to the salinity field, whose horizontal structure changes negligibly with depth.

In Fig. 6a, we also present (in brackets) the climatic values of the sea surface temperature according to the data from systematic observations at stationary coastal HMSs, whose locations are shown in Fig. 1a. There is a rather good correspondence of the climatic estimates made from the ship and coastal observations data.

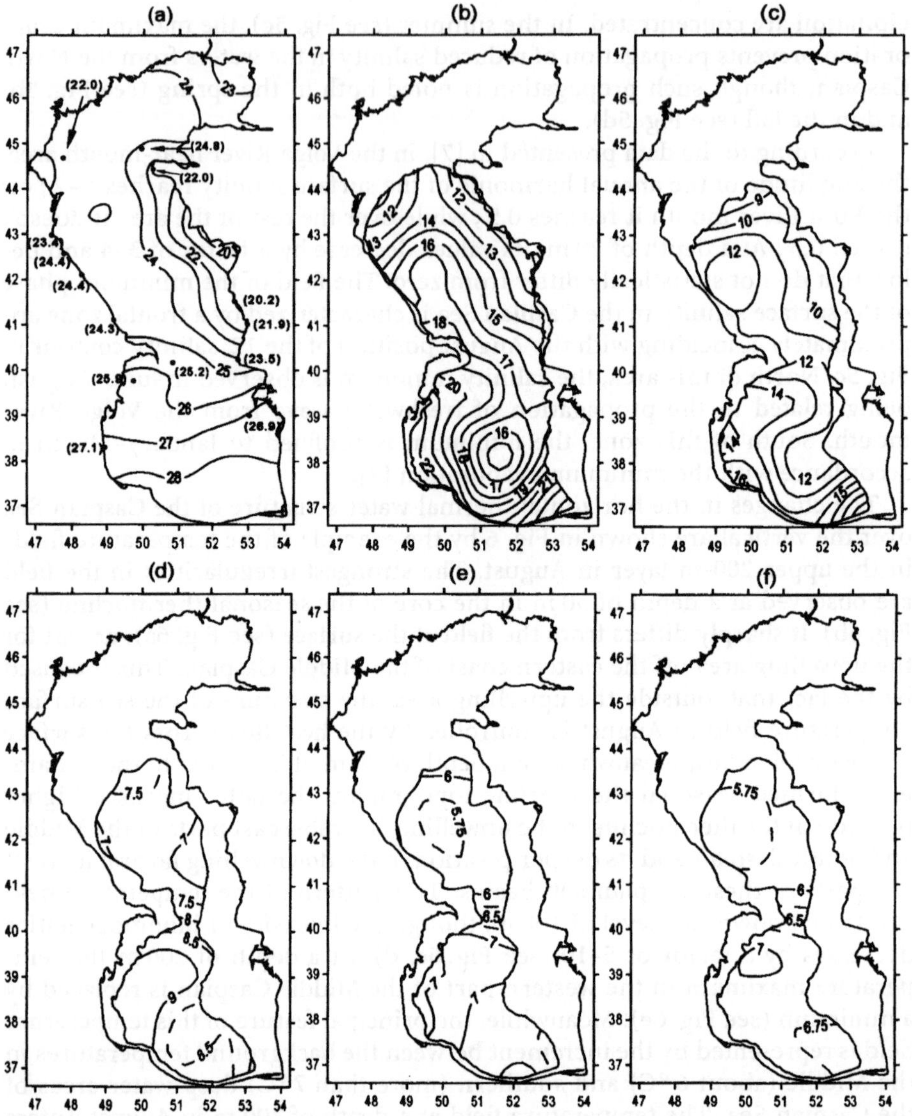

Fig. 6 Climatic fields of the water temperature (degrees celsius) in August in the Caspian Sea at a depth of 0 (**a**), 30 (**b**), 50 (**c**), 100 (**d**), and 200 m (**e**) and annual mean temperature field at a depth of 200 m (**f**). The *numerals in brackets* represent the climatic values of the sea surface temperature in August according to the data of observations at the coastal HMS shown by the dots

A comparison between the temperature and salinity fields in the waters of the Caspian Sea shows significant differences in their structures, which corresponds well to the above-mentioned different mechanisms of the formation of these fields. The absence of correlation between the temperature and salinity over the entire water column of the Caspian Sea is also well illustrated by Fig. 2.

The density field of the waters of the Caspian Sea in near-mouth areas is almost completely defined by the salinity field, while outside these areas, 70–80% is controlled by the temperature field. Therefore, we will not assess it separately in order to avoid repeating the above-listed inferences.

3 General Water Circulation

The instrumental observations available on the currents of the Caspian Sea are mostly concentrated in its shelf zone [2, 8]. All the series were very short-term and can characterize only the synoptic and higher frequency ranges of the water dynamics.

Knowledge of the general water circulation in the Caspian Sea has therefore been based on diagnostic simulations by numerical hydrodynamic models using the temperature and salinity fields, similar to those described above.

The results of the first simulations of this kind performed in the middle 1970s were not very reliable because of the imperfections of the numerical hydrodynamic models and the initial and boundary conditions used. Only as late as the middle 1990s [9, 10], were these difficulties overcome. Simulations using the nonlinear hydrodynamic adaptation model produced three-dimensional seasonal fields of the general circulation of the Caspian Sea waters that agreed with the known facts obtained by observations.

In Fig. 7, we present the mean annual field of the general water circulation obtained by averaging the seasonal fields reproduced in [9, 10]. In the deep-water areas of the Caspian Sea, the general water circulation features a sub-basin eddy character. In the Middle Caspian, throughout the year, there exists a dipole system consisting of a cyclonic gyre in its northwestern part and an anticyclonic gyre in its southeastern part. A similar dipole exists in the Southern Caspian as well; meanwhile, in this region, the anticyclone is located in the northwest of the basin, while the cyclone is confined to its southeastern part.

The seasonal variabilities of the above-mentioned circulation systems are manifested by coupled variations in their positions, sizes, and intensities [10]. In particular, in the wintertime the cyclonic gyre in the Middle Caspian and the anticyclonic gyre in the South Caspian are most intense, while in the summertime, in contrast, the anticyclone in the Middle Caspian and the cyclone in the South Caspian have the greatest intensities. The current velocities at the centers of the sub-basin gyres are 0.05–0.10 m s^{-1} and reach 0.20 m s^{-1} at the interfaces of gyres with opposite vorticity. In the vertical direction, all

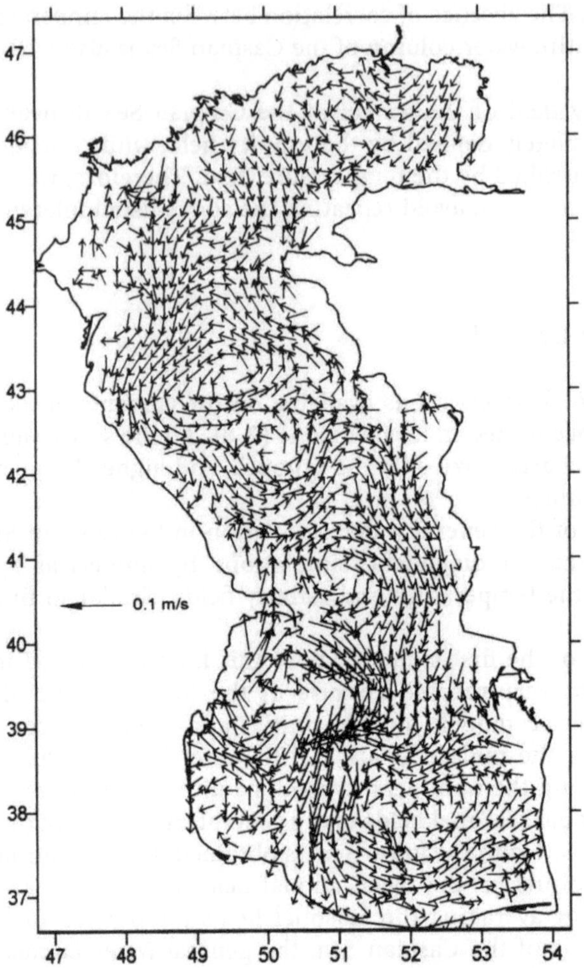

Fig. 7 Annual mean climatic field of the current vectors in the surface layer of the Caspian Sea simulated by the numerical model of hydrodynamic adaptation

the gyres are traced down to a depth of 100 m. Deeper, the character of the circulation becomes less distinct and the current velocities decrease down to 0.05 m s^{-1}.

Under the actual conditions in the surface sea layer, wind-induced currents play a rather significant part, especially in the shallow-water North Caspian. Therefore, in this part of the Caspian Sea, the currents have the least stable direction and velocity. On average, the northeastward and southwestward water transports along the principal North Caspian frontal zone dominate (see Fig. 5). Under extreme storm conditions in the North Caspian, one can expect current velocities near to 1 m s^{-1} and the onset level rises by 2–3 m [2].

In the synoptic range of the current variability in the Caspian Sea, oscillations with periods from 2–3 days to 1–3 weeks prevail [8]. They are related to the synoptic variability of the direct wind impact and to coastal trapped waves [11, 12]. In the higher frequency range, current variability is dominated by inertial gravity waves and seiches [1, 2, 8, 11].

4 Interannual Variability of the Thermohaline Water Structure

There has been increasing interest in the long-term variability of the waters in various regions of the World Ocean in recent decades, with a particular focus on inland seas, since these are the most sensitive to changes in external conditions. In the 1970s–1990s, significant changes in physical and chemical processes and corresponding water properties were discovered in inland seas such as the Mediterranean, Black, Baltic, and Japan Seas. Brief reviews of multiannual changes in these seas are presented in [13, 14]. In the Caspian Sea, a sharp change from a fall to a rise (by more than 2 m) in the mean water surface level (see Kosarev, in this volume) was the best manifestation of the multiannual variability of its hydrological regime [2, 15].

It has been stated that, in parallel with the mean water surface level oscillations of the Caspian Sea, significant changes in the thermohaline water structure proceeded in its surface layer caused by large long-term anomalies in the freshwater budget, winter severity, and atmospheric circulation over the sea area [13, 16].

Changes in the external thermodynamical impact on the Caspian Sea were especially significant in the decades 1968–1977 and 1978–1987. For example, in 1968–1977, the annual river discharge was $240 \pm 13\ km^3$, while in 1978–1987 it reached $306 \pm 11\ km^3$. The respective values of the mean winter severity index (represented by the sum of negative daily mean values of sea-level air temperature from HMS data) in Makhachkala were -154 ± 48 °C and -90 ± 16 °C, and those of the mean index of the meridional air transfer in the summer season (represented by the difference in atmospheric pressure at sea-level between the eastern and western coasts of the Caspian Sea) were 0.01 ± 0.11 and -0.39 ± 0.15 hPa.

This resulted in a decrease in the summertime salinity of the surface water in 1978–87 with respect to the preceding decade by 0.2–0.3 in the South Caspian and 1.0–1.5 in the North Caspian. In parallel, over the entire sea area, the surface water temperature in February increased by 0.5–2.0 °C. Meanwhile, in the eastern parts of the Middle and South Caspian (i.e., in upwelling zones), the surface water temperature in August decreased by 0.8–1.2 °C [13, 16]. Along with this, the above-mentioned large-scale horizontal features of the thermohaline fields of the Caspian Sea as well as the positions and intensities of major frontal zones hardly changed [16].

In this section, we consider the multiannual changes in the vertical thermohaline structure of the Caspian Sea waters according to the data from systematic ship measurements of the vertical temperature and salinity profiles over the standard section Y (Divichi–Kenderli) crossing the most deep-water area of the Middle Caspian (see Fig. 1a). In [13], it is shown that the multiannual changes in the South Caspian are similar.

The multiannual changes in the annual river discharge, the index of winter severity (mean value from North Caspian HMS data), the water salinity at a depth of 0, 100, and 600 m, and the time–vertical section of the water temperature in the 0–600 m layer of the deep-water part of cross section Y over 1956–2000 are presented in Fig. 8. An analysis of Fig. 8 shows the presence of distinct multiannual trends in the changes in the vertical thermohaline structure of the entire water column of the Caspian Sea, alongside quasiperiodic 2–5-year variations. In the late 1970s, the trends began to feature opposing signs, so that the salinity of the Caspian waters decreased in the entire water column from the surface to the bottom. At the same time, a hydrostatically stable (though weak) vertical salinity stratification of the Caspian waters was formed (see Fig. 8c). By the middle 1990s, the thickness of the summer thermocline decreased by a factor of 3–4, while the vertical temperature gradients in it increased by the same factor (see Fig. 8d). The temperature of the abyssal waters in the Middle Caspian also significantly (by more than 0.5 °C) increased.

Undoubtedly, the fundamental change in the vertical salinity structure of the waters should be regarded as the principal event in the multiannual variability of the Caspian Sea hydrophysical condition. This change is presented in the submeridional cross section observed with a 20-year interval (Fig. 9).

The vertical salinity structure of the Caspian Sea that was observed in the middle of the 1970s might be referred to as the subtropical type. During this period, at the end of summer each year, conditions were favorable to the development of deep autumn and winter convection . Owing to the negative freshwater balance, lenses of more saline waters formed in the surface layer. They kept their near-surface position due to the existence of the seasonal thermocline. One such lense can be seen in Fig. 9a in the upper water layer over the Middle Caspian basin. Owing to the autumn erosion of the seasonal thermocline, it should inevitably sink down to the deeper layers of the basin.

By the middle 1990s, the vertical salinity structure of the Caspian Sea was transformed; salinity grew with depth, so that the structure could be classified as the subarctic type. A layer of the subsurface salinity minimum had formed beneath the thin seasonal thermocline (see Fig. 9b). This phenomenon was caused by some increasing of the sea surface salinity in summer, due to high evaporation in that season. Under these conditions, at the autumn destruction of the seasonal thermocline, the surface waters had no chance to sink to depths greater than 200 m. The enhanced accumulation of the wintertime waters in the intermediate layers that started in the middle

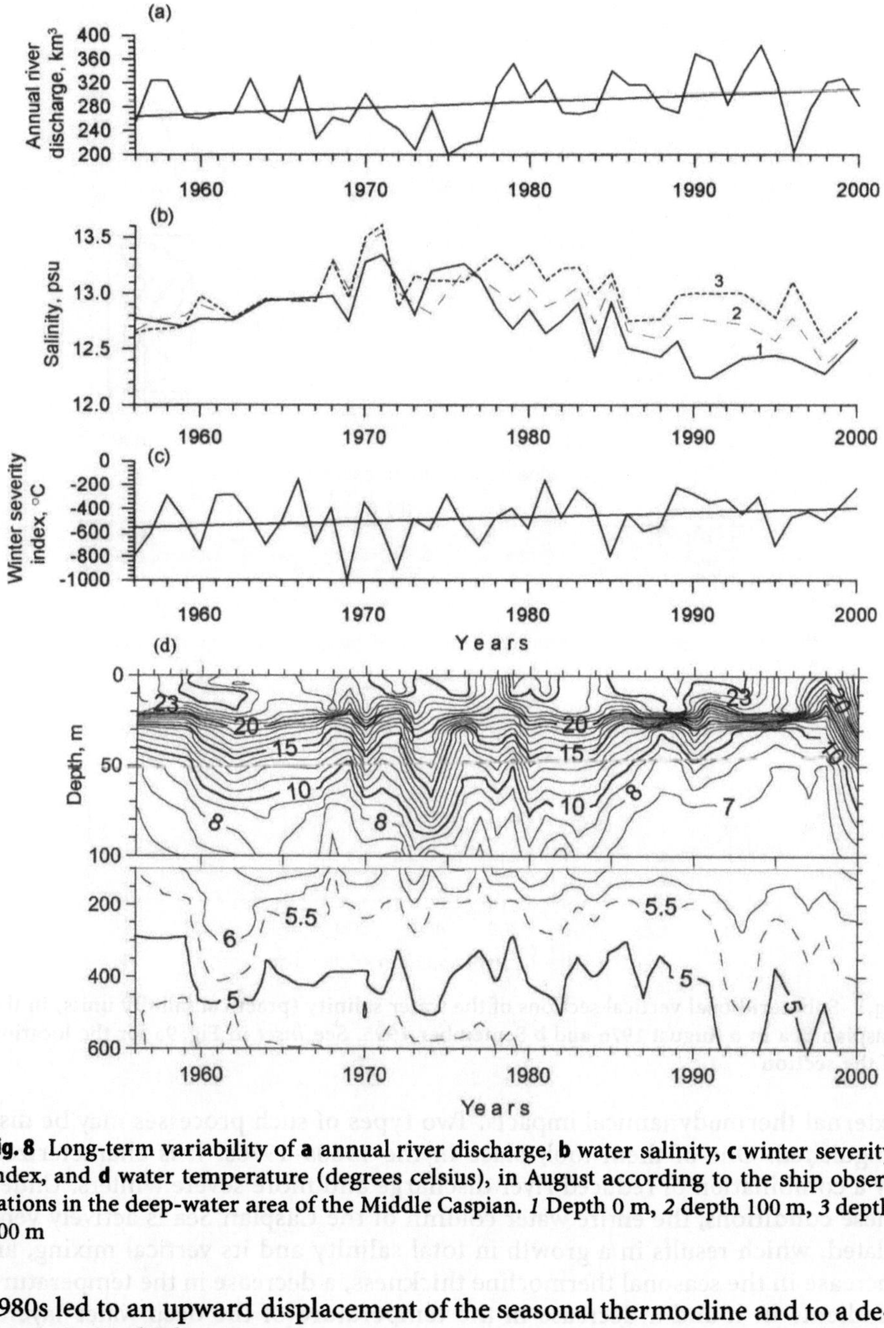

Fig. 8 Long-term variability of **a** annual river discharge, **b** water salinity, **c** winter severity index, and **d** water temperature (degrees celsius), in August according to the ship observations in the deep-water area of the Middle Caspian. *1* Depth 0 m, *2* depth 100 m, *3* depth 600 m

1980s led to an upward displacement of the seasonal thermocline and to a decrease in its thickness.

Thus, all the multiannual trends observed in the thermohaline water structure of the Caspian Sea are linked into a chain of processes controlled by

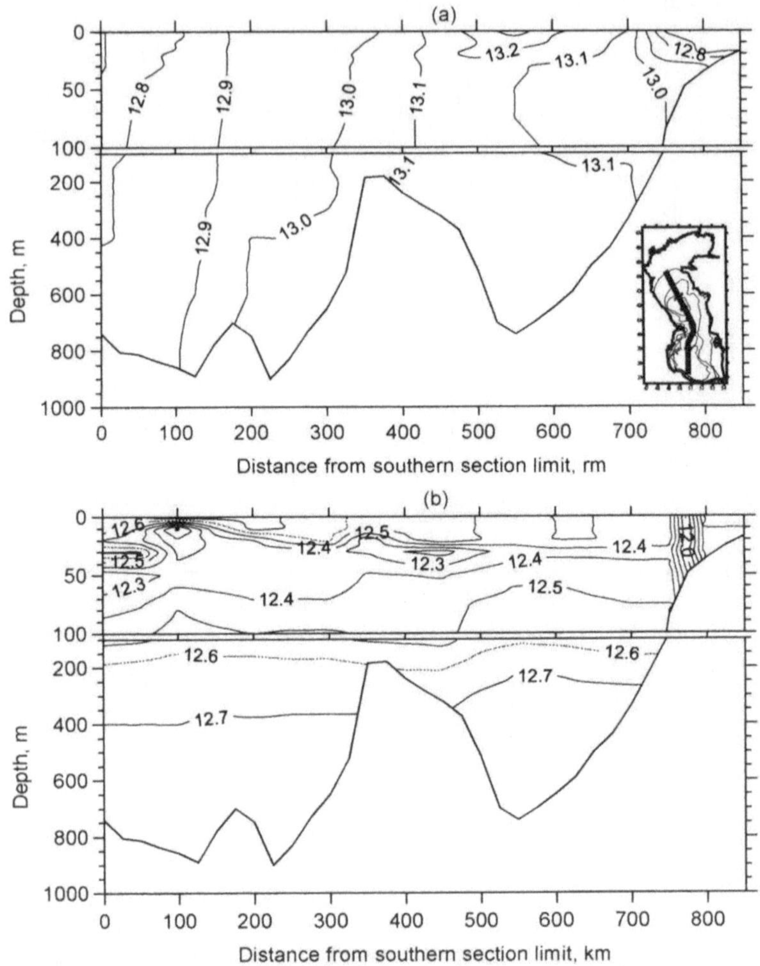

Fig. 9 Submeridional vertical sections of the water salinity (practical salinity units) in the Caspian Sea in **a** August 1976 and **b** September 1995. See *inset* in Fig. 9a for the location of the section

external thermodynamical impacts. Two types of such processes may be distinguished. One of them took place in the 1960s–1970s. It is characterized by a combination of reduced river discharge and more severe winters. Under these conditions, the entire water column of the Caspian Sea is actively ventilated, which results in a growth in total salinity and its vertical mixing, an increase in the seasonal thermocline thickness, a decrease in the temperature gradients in it and a decrease in the temperature of the deep-water layers. The second type of process took place in the 1980s–1990s. It is characterized by an opposite combination of the external factors: enhanced river discharge and prevalence of mild winters. In this case, a hydrostatically stable vertical salinity stratification is formed, which leads to a drop in the vertical diffusive exchange of heat and salt. The systematic ventilation of the near-bottom layers is deranged and the lower boundary of the ventilated zone is located at a depth of 200–300 m. As a result, the temperatures in the deep and near-bottom water layers increase somewhat.

The multiannual changes, in their turn, should influence the hydrochemical water structure and the functioning of the biotic component of the Caspian Sea ecosystem. In this respect, the second of these two hydrological processes is far less favorable, since the growth in the stable, vertical water density stratification leads to a decrease in the oxygen content in the abyssal layers of the Caspian sea (see Tuzhilkin et al., this volume) and to suppression of the supply of the surface euphotic layer with nutrients provided by the turbulent exchange. The weakening of the autumn–winter convection together with the decrease in the thickness of the upper quasihomogeneous layer and the sharpening of the summer thermocline also result in a reduced dilution of the polluted surface waters. These phenomena may also cause very negative aftereffects in the Caspian biota.

5
Summertime Upwelling off the Eastern Coast of the Middle Caspian and its Interannual Variability

Studies of the summer upwelling off the eastern coast of the Middle Caspian are described in numerous publications, and their results overviewed in several monographs [1–3, 8, 11]. Increased interest in this area is explained by the fact that the Middle Caspian upwelling is the most prominent thermal and dynamical phenomenon in the Caspian Sea in the summer season. Similar to most of the analogous regions of the World Ocean [12], the upwelling off the eastern coast of the Middle Caspian features synoptic spatial and temporal scales. The temporal scale is of a few days; the size of the upwelling is about 5–20 km in the direction normal to the coast and many tens of kilometers in the longshore direction [11]. With southerly winds, an upwelling is formed off the western coast of the Middle Caspian as well, but in the summertime such events are rather rare because of the obvious domination of northerly winds over the Caspian Sea. The high summertime recurrence of synoptic upwellings off the eastern coast of the Middle Caspian during the period from May to September makes this phenomenon climatically significant and it is well illustrated in the multiannual mean monthly fields of the water temperature (see Fig. 4c).

In this section, we focus on the climatic thermal manifestations of the upwelling and their interannual variability based on the observation data at the Bekdash HMS (see Fig. 1a). This is the only HMS located within the zone of action of the upwelling, where systematic observations of the hydrometeorological characteristics were performed four times a day during the 1950s–1990s [2].

In the region of the Bekdash HMS, the mean monthly Ekman vertical velocities caused by the wind field vorticity reach their maximum negative (upwelling favorable) values -0.2×10^{-5} m s^{-1} in June–July (Fig. 10, curve 1). In these months, the spring–summer temperature increase stops here (curve 3 in Fig. 10) despite the rather large heat delivery from the atmosphere (curve 2 in Fig. 10). Over a few hours, the actual upward velocities may reach 10^{-4} m s^{-1} (10 m day^{-1}) [11].

In August, the mean monthly climatic surface temperature in the region of the Bekdash HMS is 4 °C lower than off the opposite coast of the Middle Caspian (see Fig. 6a). During the remaining months (from October to April), a downwelling is observed here. In the wintertime, its intensity is higher by an order of magnitude than that of the summertime upwelling. This means that the eastern shelf and slope of the Caspian Sea can be considered a potential source of ventilation for its abyssal waters, complementing that of the northern slope [1].

The summertime upwelling off the eastern coast of the Middle Caspian operates in the upper 50-m layer (Fig. 11). The climatic uplift of the isotherms here never exceeds 20 m, which corresponds to a mean annual Ekman vertical velocity of the water rise of about 6 m per month, as follows from Fig. 10. However, the shallow and sharp seasonal thermocline in the Caspian Sea results in significant thermal manifestations of the upwelling off its eastern coast.

The summertime Caspian upwelling cannot be regarded as a local phenomenon. In Fig. 12, one can clearly see how, from July to August, the irregularity of the water temperature field in the seasonal thermocline increases over the entire deep-water area of the Caspian Sea with the development of the upwelling in the its eastern part and the compensation downwelling in

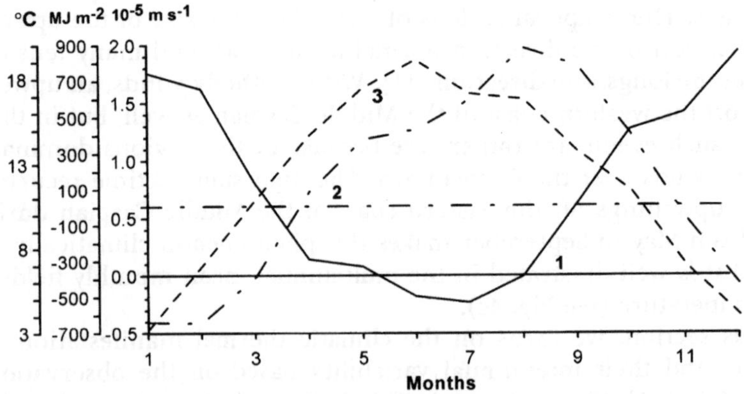

Fig. 10 Climatic annual cycles of the *1* Ekman vertical velocity, *2* heat flux across the sea surface, *3* sea surface temperature at the Bekdash HMS

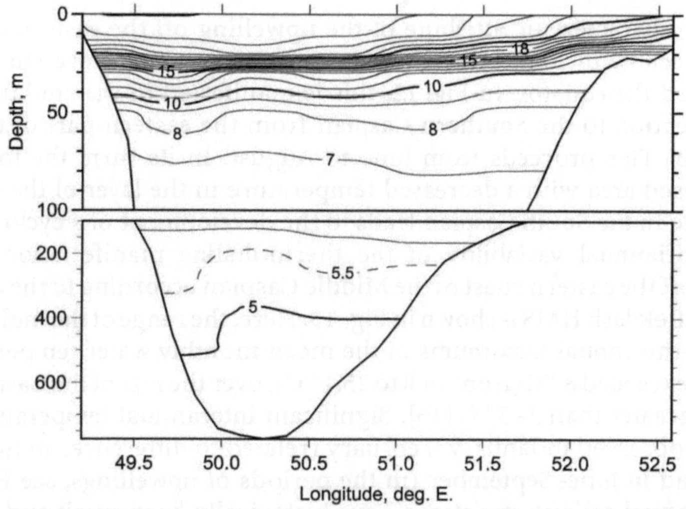

Fig. 11 Climatic distribution of the water temperature (degrees celsius) in the standard section Y (Divichi–Kenderli) in August

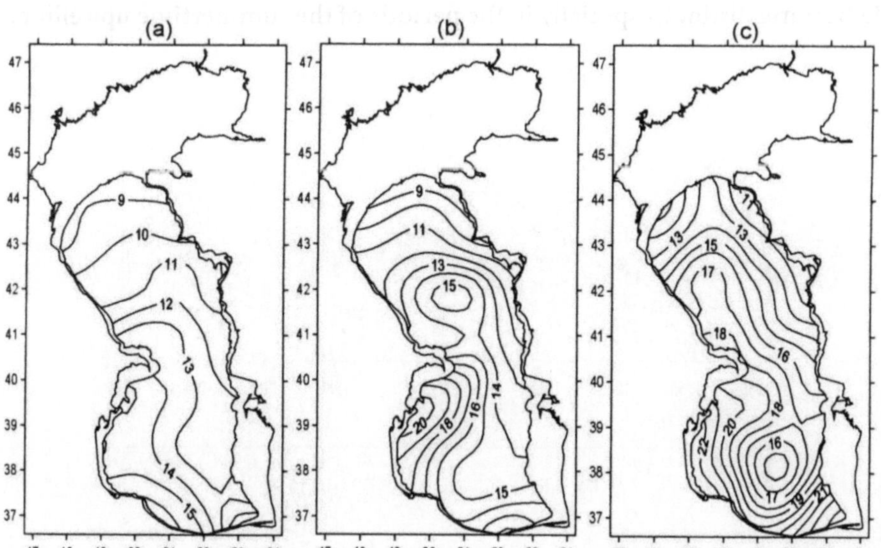

Fig. 12 Climatic fields of the water temperature (degrees celsius) in the Caspian Sea according to the ship observation data at a depth of 30 m in **a** June, **b** July, and **c** August

its western part. Over its entire lifetime, the area of the water upwelling travels northward along the eastern coast of the Caspian Sea, i.e., in the direction of the coastal trapped waves propagation, which represents a characteristic feature of the upwelling dynamics [12].

Another characteristic attribute of the upwelling off the eastern coasts of seas and oceans, including the Caspian Sea, is the alongshore current directed toward the equator. In Fig. 12, this is manifested in the cold upwelling waters advection to the Southern Caspian from the eastern part of the Middle Caspian. This proceeds from June to August. In its turn, the formation of an enclosed area with a decreased temperature in the layer of the seasonal thermocline in the South Caspian leads to the development of a cyclonic gyre.

The multiannual variability of the thermohaline manifestations of the upwelling off the eastern coast of the Middle Caspian according to the observations at the Bekdash HMS is shown in Fig. 13. Here, the range of the multiannual changes in the annual maximums of the mean monthly water temperature in 1949–1995 exceeded 8 °C (from 16.8 to 25.1 °C); over the rest of the Caspian Sea, it was not greater than 2–3 °C [16]. Significant interannual temperature variability was observed in January–February (related to differences in the winter severity) and in June–September (in the periods of upwellings, see Fig. 13a). The interannual salinity variations have historically been weak and chaotic, (see Fig. 13b) and only after 1980, with the formation of vertical salinity stratification in the Caspian Sea, did the interannual salinity variations at the Bekdash HMS became distinct, especially in the periods of the summertime upwelling.

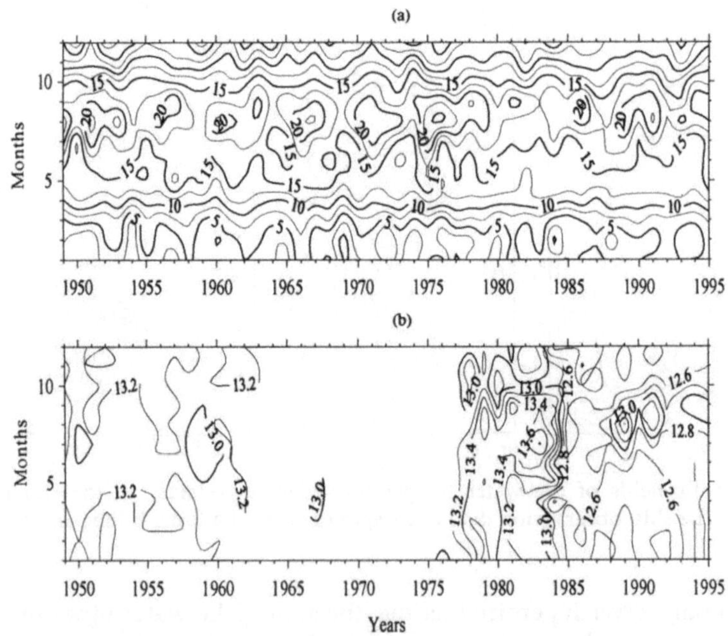

Fig. 13 Seasonal-interannual variability of **a** temperature (degrees celsius), and **b** salinity (practical salinity units), at the Bekdash HMS. The temperature and salinity contour intervals are 2.5 °C and 0.25 , respectively

During 1949–1995, at the Bekdash HMS, ten groups of years with a maximum mean monthly water temperature of less than 20 °C were distinguished (see Fig. 13a). The mean interval between their appearances was approximately 4–5 years. Taking into account the wind-induced character of the summertime Caspian upwelling, it is natural to seek a connection between its observed interannual variability and the corresponding variations in the atmospheric processes. The most prominent process in the World Ocean-Global Atmosphere System featuring a 4–5-year periodicity is the El Niño-Southern Oscillation (ENSO) cycle [17]. The teleconnections of the freshwater

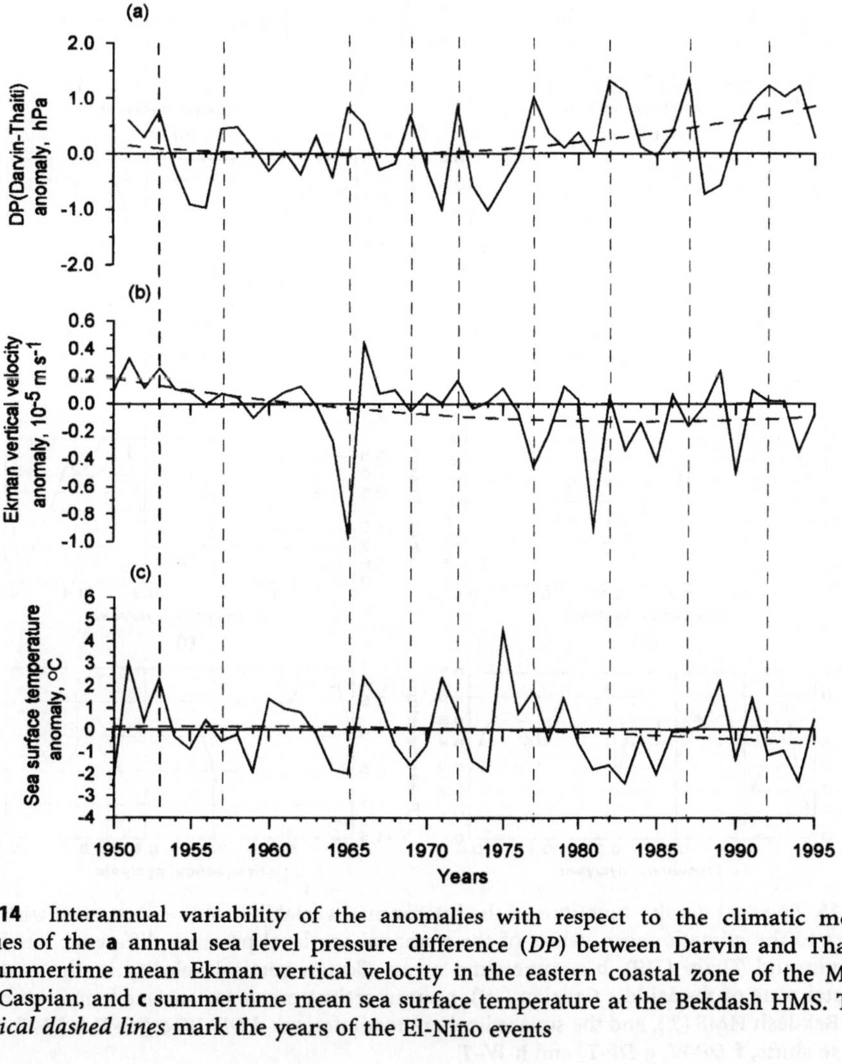

Fig. 14 Interannual variability of the anomalies with respect to the climatic mean values of the **a** annual sea level pressure difference (*DP*) between Darvin and Thaiti, **b** summertime mean Ekman vertical velocity in the eastern coastal zone of the Middle Caspian, and **c** summertime mean sea surface temperature at the Bekdash HMS. The *vertical dashed lines* mark the years of the El-Niño events

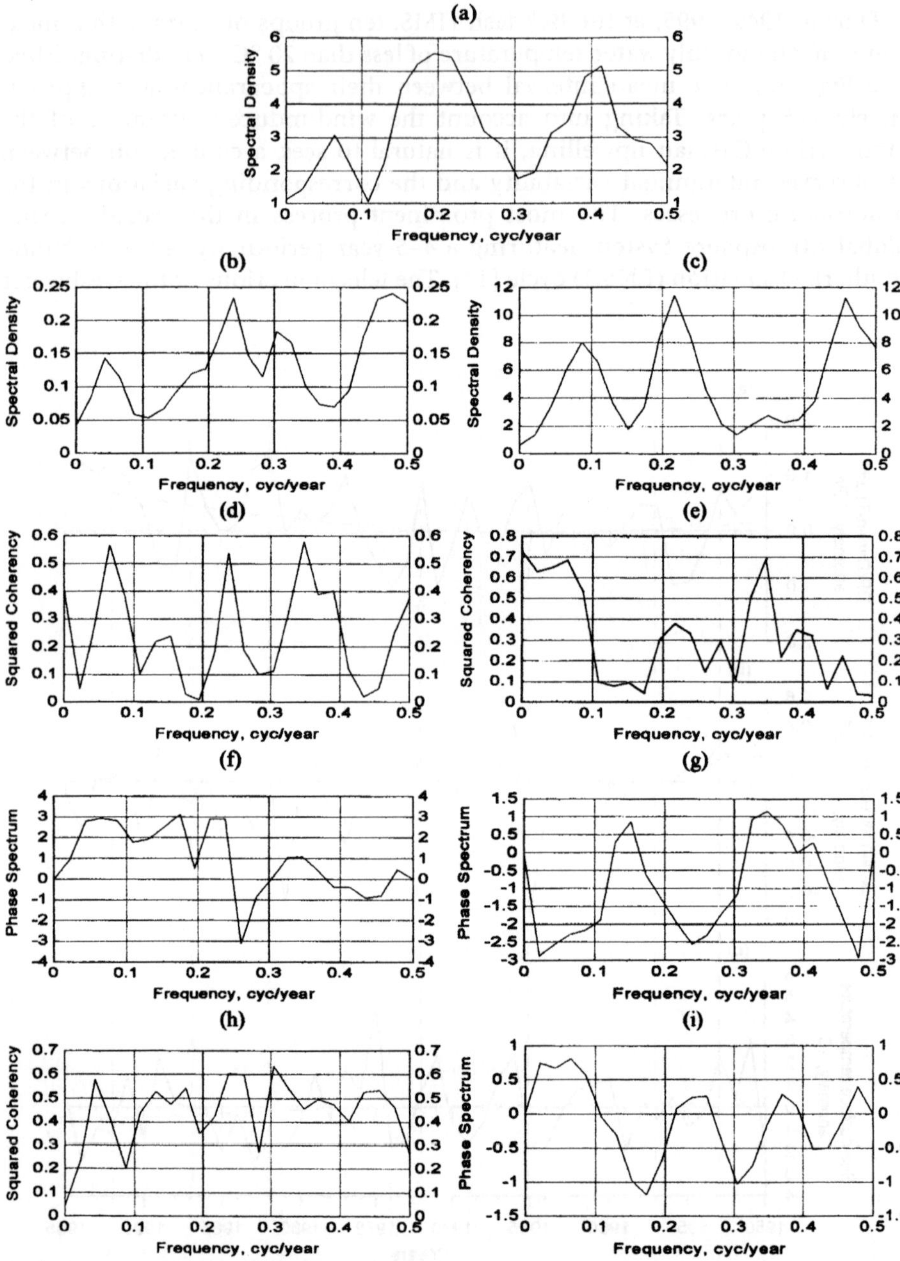

Fig. 15 Spectral density functions of the interannual variability of the anomalies with respect to the climatic mean values of the **a** annual sea level pressure difference between Darvin and Thaiti (*DP*), **b** summertime mean Ekman vertical velocity in the eastern coastal zone of the Middle Caspian (*W*), **c** summertime mean sea surface temperature at the Bekdash HMS (*T*), and the squared coherence functions **d** *DP–W*, **e** *DP–T*, **h** *W–T*; and phase shifts, **f** *DP-W*, **g** *DP-T*, and **h** *W–T*

balance components in the Caspian Sea catchment area with the ENSO had already been recognized somewhat earlier [18].

In order to uncover these kinds of teleconnections, we analyzed multiannual series of anomalies with respect to climatic mean of summertime mean temperature at the Bekdash HMS and Ekman vertical velocities (calculated for the region of the Bekdash HMS from the wind fields obtained as a result of the NCAR/NCEP reanalysis [19]), together with the annual ENSO baric index (annual sea-level pressure difference between Darvin and Thaiti) [17]. These series are presented in Fig. 14, where the vertical dashed lines mark the years of the active phase of the ENSO (El Niño events). From a visual analysis of these series, one can note selected combinations of their extremes. In particular, 6 of the 10 years of El Niño in the period 1950–1995 were mostly coupled with local temperature minimums (see Fig. 14c). However, there was only one case (1965) when this kind of combination was confirmed by a maximum of the negative (i.e., upwelling-favorable) Ekman vertical velocity (see Fig. 14b).

Note that the absence of evident links between complicated natural processes in the temporal domain is a rather common and explicable phenomenon. In order to recognize the links in the frequency domain, the spectral density functions, the squared coherence, and the phase differences for all the processes under consideration were estimated, using the fast Fourier transform algorithm. The results are presented in Fig. 15.

In all the spectra shown in Fig. 15, maximums dominate in the frequency range from 0.20 to 0.25 cycle year^{-1} corresponding to the period range 4–5 years. In the same range, local peaks of the squared coherence between all the processes considered are observed. The phase spectra suggest a counterphase character of the 4–5-year oscillations of the ENSO indexes with respect to similar oscillations of the water temperature and the Ekman vertical velocity, resulting in the phase shifts between the latter two processes being close to zero. From the physical point of view, this means that, with a certain degree of probability, the maximums of the Ekman upwelling and the temperature minimums in the region of the Bekdash HMS correspond to the El Niño events.

The results obtained may be regarded as an additional example of the teleconnections in the World Ocean–Global Atmosphere System, in which the Caspian Sea is one of the active areas [15].

6
Conclusions

The results of the statistical and physical analyses of the data of long-term ship and coastal observations of the thermohaline water structure presented in this chapter, together with the diagnostic calculations of the general water circulation of the Caspian Sea allow us to draw the following conclusions:

- The thermohaline structure of the Caspian Sea waters is characterized by a significant horizontal and vertical inhomogeneity, especially in the near-mouth areas of the Caspian Sea (off the Volga, Kura, and other rivers) and in the area of the summertime upwelling off the eastern coast of the Middle Caspian.
- The seasonal variability of water temperature and salinity is confined to the upper 100-m and 20-m layers, respectively.
- The horizontal thermal structure of the upper layer of the Caspian Sea is characterized by significant seasonal transformations, while the main features of the surface horizontal salinity structure throughout the year are far more stable; this is related to the different spatial and temporal distributions of the external thermal and freshwater impacts.
- The thermohaline structure of the entire water column of the Caspian Sea is subjected to a significant interannual variability, including multiannual trends, which dominate over other kinds of variability (seasonal and synoptic) in the intermediate and abyssal layers of the sea.
- Owing to enhanced river discharge in the 1980s–1990s, a hydrostatically stable vertical salinity stratification was formed in the Caspian Sea deep-water areas, which replaced the previous uniform salinity over the entire water column and caused changes in the summertime vertical thermal structure; the main features of the horizontal thermohaline structure of the Caspian Sea were, however, retained.
- The possible teleconnections of the 4–5-year variability of the summertime upwelling off the eastern coast of the Middle Caspian with the El Niño–Southern Oscillation cycle were traced.
- The interannual cycles and multiannual trends distinguished in the variability of the thermohaline structure of the Caspian Sea waters may exert significant influence upon other components (abiotic and live) of its ecosystem.

Acknowledgements This work was supported by the Russian Foundation for Basic Research Grant No. 03-05-96630.

References

1. Kosarev AN (1975) Hydrology of the Caspian and Aral Seas. Mosk Gos Univ, Moscow (in Russian)
2. Terziev FS, Kosarev AN, Aliev AA (1992) (eds) Hydrometeorology and hydrochemistry of the seas, vol 6. The Caspian Sea, issue 1. Hydrometeorological conditions. Gidrometeoizdat, St. Petersburg (in Russian)
3. Kosarev AN, Yablonskaya EA (1994) The Caspian Sea. SPB, The Hague
4. Samoilenko VS (1963) (ed) Multidisciplinary hydrometeorological atlases of the Caspian and Aral seas. Gidrometeoizdat, Leningrad (in Russian)

5. Kosarev AN, Tuzhilkin VS (1995) Climatic thermohaline fields of the Caspian Sea. Sorbis, Moscow (in Russian)
6. Levitus S (1982) Climatological atlas of the World Ocean. NOAA professional paper 13. NOAA, Washington
7. Kosarev AN, Tuzhilkin VS (1997) Vodn Resur 2:104
8. Baidin SS, Kosarev AN (1986) (eds) The Caspian Sea. Hydrology and hydrochemistry. Nauka, Moscow (in Russian)
9. Trukhchev DI, Kosarev AN, Ivanova D, Tuzhilkin VS (1995) C R Acad Bulg Sci 48:31
10. Tuzhilkin VS, Kosarev AN, Trukhchev DI, Ivanova DP (1997) Meteorol Gidrol 1:91 (in Russian)
11. Kosarev AN (ed) (1990) The Caspian Sea. Water structure and dynamics. Nauka, Moscow (in Russian)
12. Brink KH (1983) Progr Oceanogr 12:223
13. Kosarev AN, Tuzhilkin VS, Kostianoy AG (2004) Main features of the Caspian Sea hydrology. In: Nihoul JCJ (ed) Dying and dead seas. Kluwer, Dordrecht, p 159
14. Tuzhilkin VS, Berlinski NA, Kosarev AN, Nalbandov YR (2004) Ekol Morya 65:75 (in Russian)
15. Rodionov SN (1994) Global and regional climate interaction: the Caspian Sea experience. Kluwer, Dordrecht
16. Kosarev AN, Tuzhilkin VS (2000) Sci Bull Caspian Floating Inst 1:26
17. Harrisson DE, Larkin NK (1998) Rev Geophys 36:353
18. Arpe K, Bengtsson L, Golitsin GS, Mokhov II, Semenov VA, Sporyshev PV (1999) Dokl Ross Akad Nauk 366:248 (in Russian)
19. Kalnay E, Kanamitsu M, Kistler R, et al. (1996) Bull Am Meteorol Soc 77:437

Hdb Env Chem Vol. 5, Part P (2005): 59–81
DOI 10.1007/698_5_004

Published online: 13 October 2005

Sea Surface Temperature Variability

Anna I. Ginzburg · Andrey G. Kostianoy · Nickolay A. Sheremet (✉)

P.P. Shirshov Institute of Oceanology, Russian Academy of Sciences, 36 Nakhimovsky Pr., 117997 Moscow, Russia
sheremet@ocean.ru

Abstract A weekly mean Multi-Channel Sea Surface Temperature (MCSST) data set was used to study seasonal and interannual variability of SSTs averaged over four regions of the Caspian Sea individually (Northern, Middle, and Southern Caspian, and Kara-Bogaz-Gol Bay) and SST trends during 1982–2000. The SST fields averaged for individual months of four hydrological seasons (February, April, August, and October) were calculated and compared with the corresponding climatic SST fields based on in situ measurements in the previous period. The positive SST trends of about 0.05 and 0.10 °C year^{-1} in the Middle and Southern Caspian, respectively, during 1982–2000 were revealed and it was supposed that these trends, which appeared to be several times higher than those in 1940–1980, resulted from the global climate warming. The time correlation of the marked anomalies of the winter and summer SSTs observed in 1950–2000 (satellite and published field data) with the phases of the El Niño-Southern Oscillation and North-Atlantic Oscillation events demonstrated an influence of these atmospheric oscillations on interannual and decadal variability of the Caspian SST. It is emphasized that temperature changes in the Caspian and Black seas during 1982–2000 were similar in general features (positive and close in value trends of SST; a character of changing the annual mean SSTs in 1989–1998, with minimum in 1992–1993; the maximum SSTs in the winter and in the summer of 1998–1999).

Keywords Atmospheric forcing · Interannual variability · Caspian Sea · Sea surface temperature · Seasonal variability

Abbreviations

SST	Sea surface temperature
KBG	Kara-Bogaz-Gol Bay
ENSO	El Niño-Southern Oscillation
NAO	North-Atlantic Oscillation
EA-Jet	East Atlantic Jet
AS	Asian summer pattern

1 Introduction

A basic understanding of the regional peculiarities of the Caspian thermal regime is presented in Chap. 3. Information on spatio-temporal variability of sea surface temperature (SST) presented there as well as in earlier publications [1–5] is based on ship measurements predominantly from the 1940s to the beginning of the 1980s, that is, during the gradual drop of the Caspian Sea level [6–8]. Not much is known about possible changes in the sea thermal regime caused by changes in its level in the subsequent years, which could have the strongest effect mainly in the shallow-water regions (the Northern Caspian and the Kara-Bogaz-Gol Bay), and by the global warming [9]. Interannual variability of winter severity (ice cover) in the Northern Caspian in 1992–2002 is considered in [10]. Some information on changes in thermohaline structure of the Caspian Sea associated with the sea level increase from the second half of the 1970s to the middle of the 1990s is presented in Chap. 3. However, the only possibility to obtain the comprehensive data on temporal variability of the basin-scale SST from the 1980s to date is by using the regular data of satellite measurements [11].

The problem, which is also poorly studied, is a response of the Caspian hydrological regime to the large-scale atmospheric forcing: the El Niño-Southern Oscillation (ENSO), which determines fluctuations in the ocean–atmosphere climatic system and the particularities in the global atmospheric circulation, and the North-Atlantic Oscillation (NAO) that is extremely important for the European region [6, 12–14]. The influence of the ENSO and NAO on the Caspian level changes was studied in a number of works (see, for example, [6, 7, 13]). However, an influence of these oscillations of atmospheric circulation on the Caspian SST variability generally has not been considered yet, although, based on the results of the recent studies of interannual variability of thermal regime in the semi-enclosed basins adjacent to the Caspian Sea (the Mediterranean [15, 16] and Black [17, 18] seas), we can propose that the influence occurs. A correlation between intensity of upwelling along the eastern Caspian coast and ENSO founded recently (see Chap. 3) supports this proposal.

In this chapter, seasonal and interannual variability of the Caspian Sea SST is considered as well as its long-term trend on the basis of an analysis of satellite information during the 19-year period (1982–2000). The results obtained, together with the information available on the seasonal temperature anomalies in the preceding years [1–3, 10], were used to reveal a response of the Caspian SST to the ENSO and NAO. The question of a possible influence of two another atmospheric oscillations (East Atlantic Jet (EA-Jet) [13] and Asian Summer Pattern (AS) [19]) on SST variability is touched upon as well.

2
Data

SST data were derived, as in the cases of the Black [17] and Aral [20] sea studies, from weekly mean Multi-Channel Sea Surface Temperature (MCSST) data set, which was based on measurements by an Advanced Very High Resolution Radiometer (AVHRR) on board satellites of National Oceanic and Atmospheric Administration (NOAA) and available through the Internet (a description of the data is given at: http://podaac.jpl.nasa.gov:2031/DATASET_DOCS/avhrr_wkly_mcsst.html). These data have a spatial resolution of 1/6° and a temperature resolution of about 0.1 °C. Because the subsurface measurements (mainly drifters data) are widely used as ground truth data to retrieve SST values from AVHRR data, the MCSST data correspond, in fact, to the surface layer [21], which allows us to compare the results obtained on the basis of satellite and hydrological data sets.

The MCSST data set contains both valid and interpolated values (in the latter case, pixels void of measurements are filled with the values from the neighboring pixels using the interpolation technique). For the present analysis, we used the nighttime (to exclude solar heating effect) and interpolated MCSST data for the period from November 1981 to December 2000. The time-average (monthly mean, seasonally mean, and annual mean) and space-average (within the specified regions) characteristics of SST were calculated on this basis.

The analysis of the SST characteristics was carried out for four regions of the Caspian Sea including the Kara-Bogaz-Gol Bay (KBG) (see Chap. 2, Fig. 1). Of 1260 pixels grid size from the global weekly mean MCSST data set corresponded to the Caspian Sea basin, 268 pixels belong to the Northern Caspian, 506 pixels to the Middle Caspian, 444 pixels to the Southern Caspian, and 42 pixels to the KBG. Although the temperature distribution within the every region is not homogeneous [3], we calculated the SST characteristics averaged over each region, which for the Middle and Southern Caspian mainly reflected thermal state of their deep-water parts.

Because of gaps in the MCSST data, both valid and interpolated, during a few weeks and even months, especially before 1992, the calculated SST values with monthly, seasonally, and annual averaging could be overestimated or underestimated (in these cases, corresponding elucidations are given in the text). To judge the anomalous winter and summer SSTs, with consideration of the possibility of incorrect estimating SST due to incomplete data set, we also used information on seasonal variations of air temperature in Astrakhan, Atyrau (Guriev), Makhachkala, and Turkmenbashy (Krasnovodsk) for the period 1982–1995 obtained through the Internet (http://www.meteo.ru.data/st_mdata.htm) as well as data on winter severity indices calculated as a sum of daily mean air temperatures below zero at Makhachkala for the period 1956–2000 ([22], see also Chap. 3). Due to the absence of the corresponding valid MCSST data, we did not analyze interpolated data for the Northern Caspian from November to March and the data for the KBG before 1994.

The data on indices of atmospheric oscillations ENSO, NAO, EA-Jet, and AS were obtained through the Internet (http://www.cpc.NOAA.gov/data/teledoc, http://iri.columbia.edu/).

3
Seasonal Variability of the Caspian SST Field

The SST fields in the Caspian Sea for individual months of four hydrological seasons (winter: January–March, spring: April–June, summer: July–September, autumn: October–December) obtained by averaging the weekly mean satellite data for the main basin over the 19-year period (1982–2000) and for the KBG over 7 years (1994–2000) are shown in Fig. 1. Since the character of these satellite temperature distributions and climatic ones given in [1, 2, 4] (see also Chap. 3) agree well, in general, we shall dwell on only a few significant differences in the SST fields given in Fig. 1 and in [2, 4], which could have been caused both by the changes in the Caspian Sea thermal regime during the period under consideration as compared to the previous period and by the specific character of satellite data set used in this study.

In February, satellite SSTs (Fig. 1a) are mainly 1–1.5 °C higher than in [2, 4], which is related, first of all, to the general warming of the Caspian Sea during the period studied (see below). In addition, the interpolated satellite SSTs north of 43°N can be overestimated due to the absence of valid SSTs for the Northern Caspian in winter. In the April distribution (Fig. 1b), SSTs in the utmost southeastern part of the sea are slightly lower than in [2, 4], which reach 14–15 °C. Water temperature in the major part of the Northern Caspian, especially in the region near the Volga River mouth, is 1–2 °C lower. This can be related to the underestimated interpolated SST values for the Northern Caspian due to the influence of cooler waters of the Middle Caspian in this season, with significant gaps in valid SST values for the Northern Caspian in April (for example, the April data in 1987 and 1989 were entirely absent).

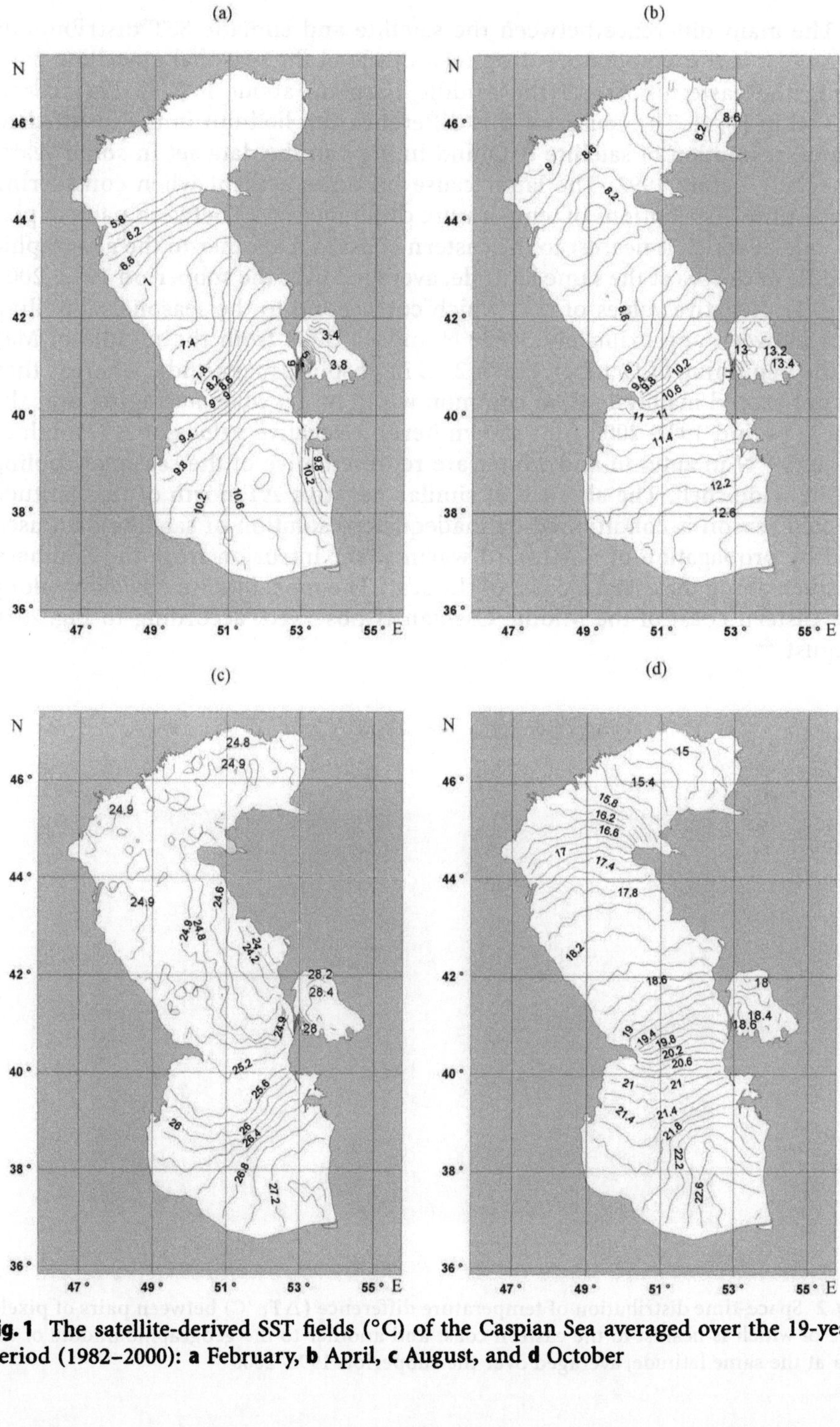

Fig. 1 The satellite-derived SST fields (°C) of the Caspian Sea averaged over the 19-year period (1982–2000): **a** February, **b** April, **c** August, and **d** October

The main difference between the satellite and climatic SST distributions in August is the temperature drop magnitude in the seasonal upwelling zone along the eastern coast of the Middle Caspian: about 1 °C in Fig. 1c and 2–3 °C in [2,4]. The causes of this difference can lie both in the insufficient spatial resolution of satellite data and in gaps in the data set in some years, especially before 1992. The latter cause becomes evident when considering space-time distributions of temperature difference (ΔT) between pairs of pixels, one of which is nearest to the eastern coast and another to the geographic middle of the sea at the same latitude, averaged over the subperiod 1992–2000 (Fig. 2). Negative values of ΔT, which correspond to the seasonal upwelling area between approximately 39°30′N and 44°30′N from the middle of May to the beginning of October, reach 2 °C in absolute magnitude, whereas they do not exceed mainly 1 °C, in common with Fig. 1c, when averaging over the whole period 1982–2000 (not shown here). (Negative values of ΔT south of about 39°N in autumn and winter are representative of the seasonal cooling of the wide shelf. The absence of similar negative ΔT north of this latitude in cold season is conditioned by inadequate resolution of satellite data used and by propagation of northward warm water intrusion from the Southern Caspian along the eastern coast of the sea.) The most intense upwelling along the eastern coast of the Middle Caspian is observed, according to Fig. 2, in August.

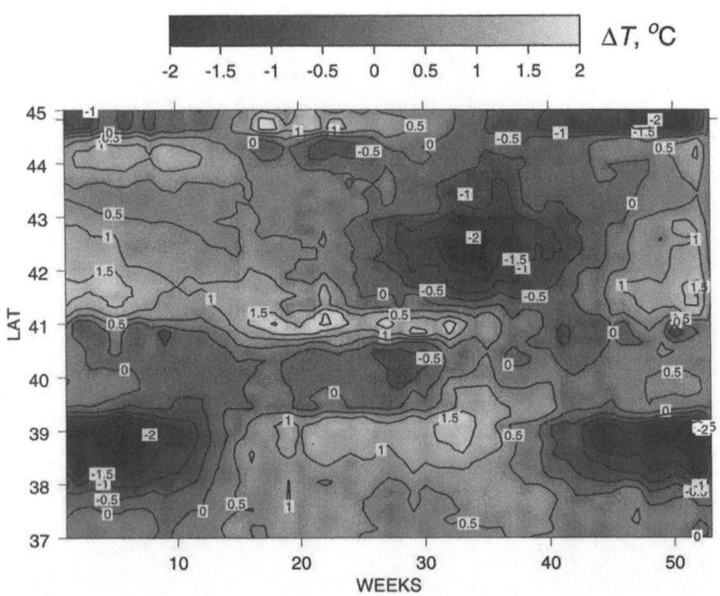

Fig. 2 Space-time distribution of temperature difference (ΔT, °C) between pairs of pixels, one of which is nearest to the eastern coast and another to the geographic middle of the sea at the same latitude, averaged over the subperiod 1992–2000

In October, SSTs in the Middle and Southern Caspian are higher by 1.5–2 °C than in [2,4], which is likely to reflect the fact of the Caspian Sea warming. The obviously overestimated (by 3–4 °C) SST values in the north-eastern part of the Northern Caspian in Fig. 1d can be related to gaps in valid data for the Northern Caspian in October (for example, the SST data for October 1982 were absolutely absent and half of the October SSTs was lacking in 1984, 1985, 1986, and 1987) and, correspondingly, to the influence of higher temperatures in the Middle Caspian on the interpolated SST values.

Figure 1 also demonstrates the seasonal changes in the KBG temperature in 1994–2000. It is seen that, in winter and in autumn (in spring and in summer), the warmest (coolest) waters in this shallow bay are concentrated near the Kara-Bogaz-Gol Strait. In the rest of the bay, SST grows from the north to the south. In the KBG, the lowest (after the Northern Caspian) winter SSTs and the greatest summer SSTs are observed.

4 Seasonal and Interannual Variability of SST Averaged over the Caspian Sea Regions

4.1 Variability of SST and its Long-term Trend

The superposition of 19 (7 for the KBG) annual cycles of the region-averaged SSTs (Fig. 3) gives us an idea about the range of the SST variability in every month (season) and also of the time of the winter and summer extreme temperatures' onset. The summer SST maxima were achieved in the Northern Caspian in July–August, in the Middle, Southern Caspian and KBG they were observed predominantly in August. The winter SST minima in the Middle Caspian appear mainly in February and March, in the Southern Caspian they are achieved in March, and in the KBG – in January–February (the same as in the shallow-water Northern Caspian [2]).

The time series of the weekly (not shown here), monthly (Fig. 4) and seasonally (Fig. 5) mean SST values indicate the same tendencies, on the average, in the long-term temperature variability in four regions of the Caspian Sea. (This correlation of temperature changes in the Caspian Sea basin has repeatedly been reported earlier, see [6]). The observed distinctions in some cases are related to the peculiar features of interpolated satellite data, when valid data are strongly lacking. For example, the SST values in the Middle Caspian in 1988 and 1990 in Fig. 4a were overestimated, in the Middle and Southern Caspian in 1983 in Fig. 5a they were underestimated (in such the cases, the SST values in the corresponding figures are marked with arrows). This, however, does not change the general conclusions about the character of the regional temperature variability.

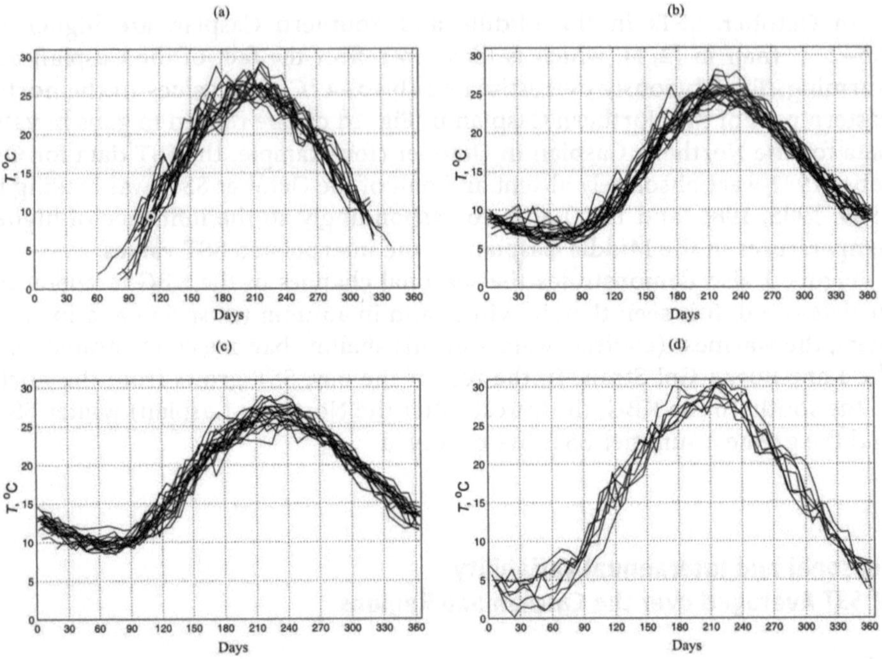

Fig. 3 Nineteen seasonal cycles (1982–2000) of the weekly mean SSTs averaged over the **a** Northern Caspian, **b** Middle Caspian, **c** Southern Caspian, and **d** seven cycles (1994–2000) averaged over the Kara-Bogaz-Gol Bay

The coolest winters (after the Northern Caspian) are typical of the shallow-water KBG, while the warmest ones are characteristic of the Southern Caspian. The same regularity is true for the autumnal period (Fig. 5d). On the contrary, in spring and summer, the highest temperatures are observed in the KBG. In spring and during the maximum warming, SSTs in the shallow-water Northern Caspian are higher than in the Middle Caspian (Figs. 4b and 5b), while in the summer season, on the average, due to the cooling in the Northern Caspian in September, they are close (Fig. 5c). In the period 1982–2000, the lowest winter temperatures in the sea as a whole (Figs. 4 and 5) were observed in 1985 and 1994 (in the Middle Caspian also in 1982, while in the KBG in 1996). The highest winter temperatures fell on 1995, 1997 (except for the Middle Caspian), and 1999–2000. According to [10], the coolest winters in the Northern Caspian were in 1982, 1984, and 1994 (there were no data for 1985), while the warmest winters were observed in 1983, 1995, 1997, 1999, and 2000. Judging from the air temperature data at the locations mentioned above (for example, Fig. 6) and winter severity indices from [22] (see also Chap. 3), the winter of 1983 in the Middle and Southern Caspian was warm as well (the corresponding SST values in Figs. 4 and 5 are underestimated, see above).

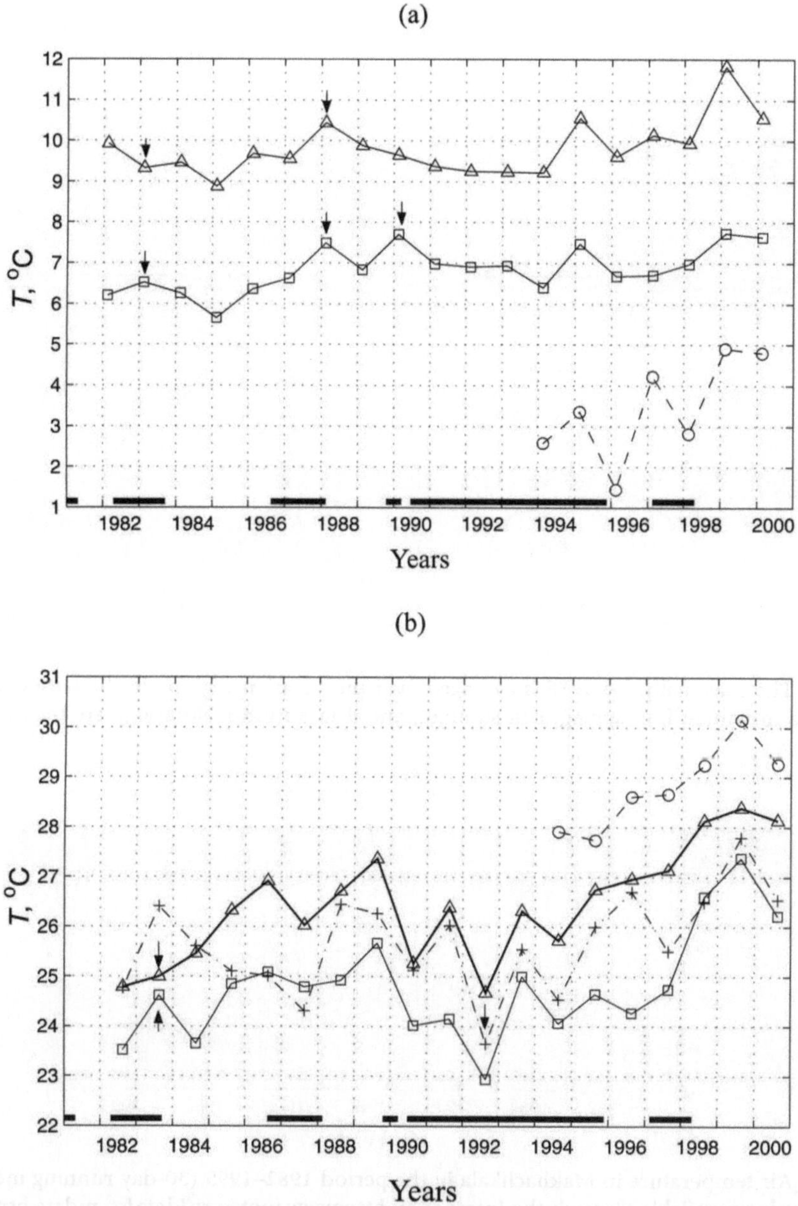

Fig. 4 The extreme monthly mean SSTs averaged over the Caspian regions during the period 1982–2000: **a** winter minima and **b** summer maxima. The *bold solid line with triangles* refers to the Southern Caspian, the *thin solid line with squares* refers to the Middle Caspian, the *dashed-dotted line with crosses* refers to the Northern Caspian, the *dashed line with circles* refers to the Kara-Bogaz-Gol Bay. Hereafter, the *bold horizontal segments* mark time periods of El Niños in accordance with Table 2. The *arrows* indicate underestimated or overestimated values of SST (due to incomplete data set, see the text)

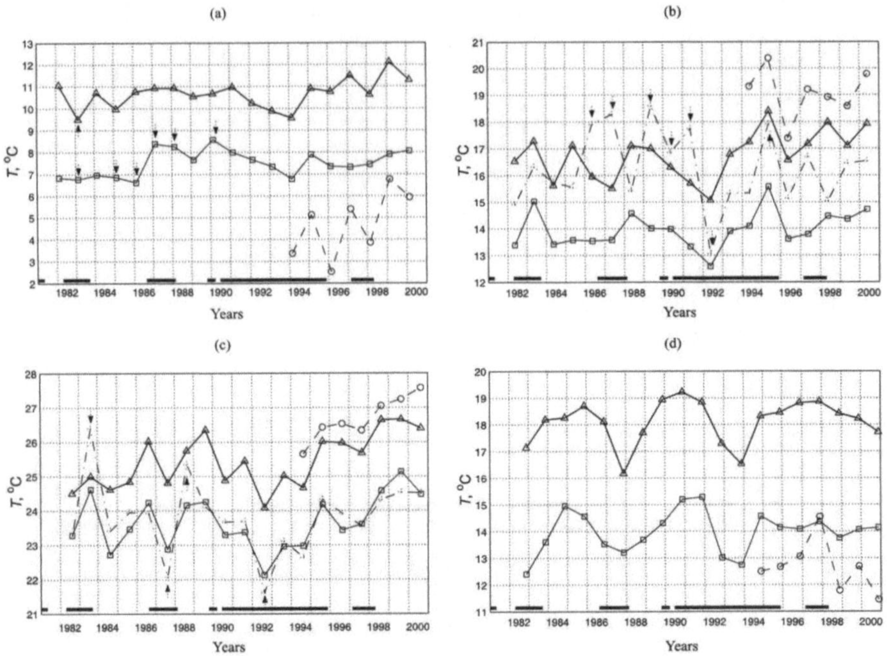

Fig. 5 The seasonally mean SSTs averaged over the Caspian regions in the period 1982–2000: **a** in winter, **b** in spring, **c** in summer, and **d** in autumn. Notations are the same as in Fig. 4

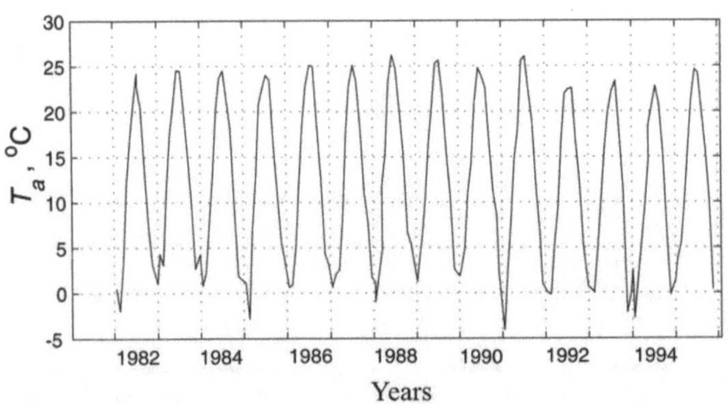

Fig. 6 Air temperature in Makhachkala in the period 1982–1995 (30-day running mean of daily values available through the Internet at: http:www.meteo.ru/data/st_mdata.htm)

The lowest summer SSTs in three regions of the main basin are related to 1982 and 1992. In the summer of 1982, the anomalously low SST values were also observed in the northeastern Atlantic [23], in the Mediterranean Sea [15], in the Black Sea [17], and in the Aral Sea [20], which could be the consequence of the aerosols from the El Chichon (Mexican volcano) eruption in April 1982. The value of the summer SST minimum in 1992 could be influenced by the fact that valid data for the Northern Caspian from mid-June to mid-August 1992 were lacking, while in the Middle and Southern Caspian, the data coverage was very low. Besides, the few in number valid SSTs were related to the region near the eastern coast, and it is not excluded that their low values were the result of coastal upwelling, which was rather intense in 1992 (see Chap. 3). However, in the summers of 1982 and 1992, air temperature over the Caspian Sea was low (Fig. 6), so that the summer SST minima in these years, even underestimated to some extent, were real. The highest summer SSTs were observed in all the regions of the Caspian Sea in 1998–2000.

In 1982–2000, the ranges of variability of the extreme summer SSTs (weekly mean data) were 4.8 °C in the Northern Caspian, 4.0 °C in the Middle Caspian, 3.6 °C in the Southern Caspian, and 2.1 °C in the KBG, while for the winter temperatures the ranges were equal to 2.4 °C in the Middle Caspian, 3.1 °C in the Southern Caspian, and 4.4 °C in the KBG. The maximum annual range of SST (difference between the maximal summer and the minimal winter values) was observed in the Middle Caspian in 1985, 1998, and 1999 (20.5, 21.5, and 20.3 °C, respectively); in the Southern Caspian, it was registered in 1998 and 2000 (19.6 and 18.6 °C, respectively); and in the KBG, it took place in 1996 and 1998 (30.0 and 28.4 °C, respectively).

A statistical processing (linear regression) of the weekly mean SST values allowed us to reveal the SST trends in the Middle and Southern Caspian during the period under study: 0.05 and 0.10 °C $year^{-1}$, respectively. Thus, in the 19-year period (1982–2000), the Caspian SST increased approximately by 1–2 °C, the warming of the Southern Caspian being more intense as compared with the Middle Caspian. Within this period, the trend was not monotonous, however, which is well seen from the long-term variability of the annual mean SSTs in Fig. 7a. The sharpest changes occurred in 1989–1995: a decrease in the SST values in 1989–1992 and their increase in 1992–1995.

The seasonal SST trends in the Caspian Sea, which were obtained also by using the weekly mean SSTs during the 19-year period, differ from the annual ones. In the Middle Caspian, they were equal to 0.04, 0.05, 0.06, and 0.03 °C $year^{-1}$ in winter, spring, summer, and autumn, respectively. In the Southern Caspian for the same seasons they were equal to 0.05, 0.08, 0.09, and 0.03 °C $year^{-1}$, respectively. Thus, during the period 1982–2000, the warming of both regions of the Caspian Sea occurred in all seasons, with the greatest positive trend of SST in summer and the lowest in autumn. In 1994–2000, in the autumn season in the KBG, the temperature trend appeared to be negative (Fig. 5d), whereas during this 7-year period, it was slightly positive on the average (Fig. 7a). It was impossible to obtain the long-term trend in the Northern Caspian SST because of lack of temperature data for the winter season. However, we can suppose that, judging from the increase in the summer SSTs in the recent years (Fig. 4b) and warm winters in the region [10], a positive temperature trend could also be observed here.

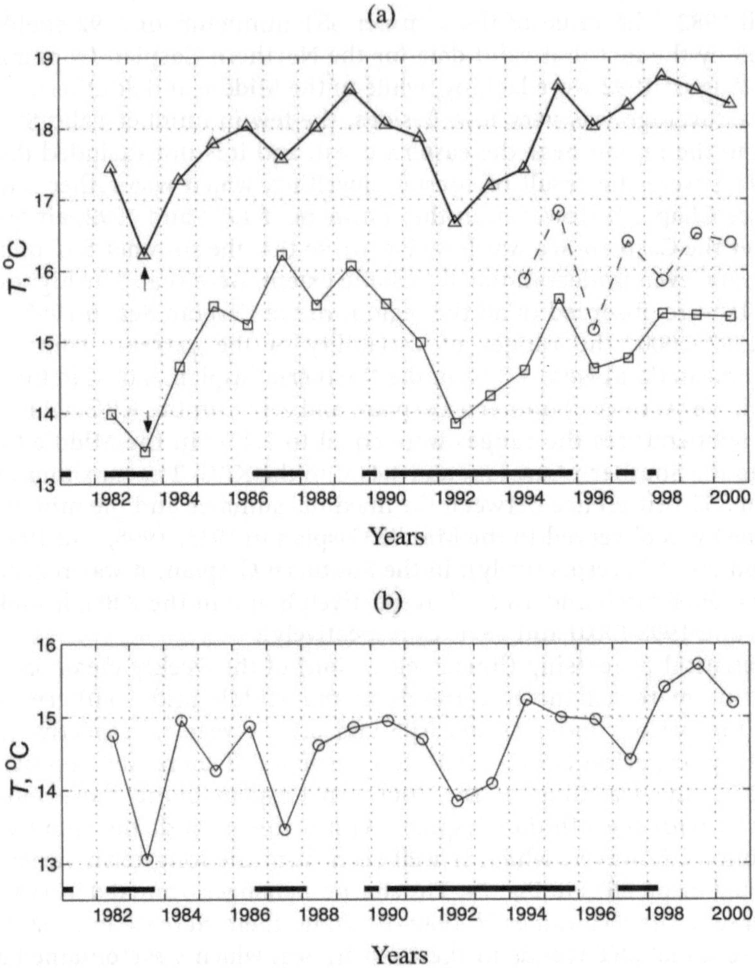

Fig. 7 The annual mean SSTs averaged **a** over the Caspian regions and **b** over the Black Sea in the period 1982–2000. Notations in Fig. 7a are the same as in Figs. 4 and 5. Figure 7b is made on the base of data in [17]. SST minima in 1983 in both basins are conditioned by the absence of data for August and September and gaps in winter data set

4.2 A Comparison of the SST Characteristics in the Modern (1982–2000) and Preceding Periods

The characteristics of SST (annual mean temperature, maximum and minimum values, annual range) for four Caspian regions for both the period 1982–2000, during which the Caspian Sea level increased by about 1.3 m [8], and the period before 1952, with the gradual level decrease [3, 24], are pre-

Table 1 Characteristics of SST on the basis of hydrological (before 1952 [3, 24]) and satellite (1982–2000) data for four regions of the Caspian Sea

SST characteristics, °C	Northern Caspian		Middle Caspian		Southern Caspian		Kara-Bogaz-Gol Bay	
	1	2	1	2	1	2	1	2
Annual mean	12.0	–	13.8	15.0 (13.5*–16.2)	17.1	17.8 (16.2*–18.7)	14.1	16.1 (15.2–16.8)
Maximum	25.2	23.6–27.8 (24.5–29.3)	24.2	23.4–27.4 (24.1–28.1)	26.7	24.8–28.4 (25.3–28.9)	27.3	27.7–30.2 (28.7–30.8)
Minimum	–	–	4.4	5.6–7.7 (5.1–7.5)	8.2	8.9–11.8 (8.4–11.5)	0.1	1.4–4.9 (0.2–4.6)
Annual range	24–26	–	18–20	16.0–19.7 (17.6–21.5)	16–19	14.9–18.2 (16.0–19.6)	25–28	24.4–27.2 (25.8–30.0)

Note. The values of SST in columns 1 and 2 for every region correspond to (1) hydrological and (2) satellite monthly mean data (weekly mean values are given in brackets). Sign * denotes underestimated SST values (see text for description of the table data).

sented in Table 1. For the modern period, these characteristics correspond to the time series of weekly mean SSTs, Figs. 4 and 7a. For the period before 1952, the SST characteristics were derived from data presented in [3, 24].

It follows from the data in Table 1 that the annual mean SST, which practically does not depend on the sea level [3, 25], increased in the modern period as compared to the previous one by approximately 1 °C in the deep-water Middle and Southern Caspian and by 2 °C in the shallow-water KBG (during the period of its level rise, see Chap. 11). The maximum (summer) and minimum (winter) SST values notably increased, while the SST annual range slightly decreased. Since the Caspian Sea level in both the periods did not significantly differ on the average (in any case, its changes could influence only the temperature of shallow-water regions of the sea [3, 25]), the revealed changes in the SST characteristics can be related to the phase of the global climate warming. A similar conclusion that the cause of the water temperature changes in the Northern Caspian in 1939–1973 was climatic changes rather than a sea level drop was made in [25].

A positive trend in the Caspian Sea SST in the period before 1982 has also been mentioned by other authors. In 1960–1983, according to the data of coastal hydrometeorological stations in the Northern Caspian, this trend was equal to 0.2–0.4 °C in 24 years [2], that is, approximately 0.01 $^{\circ}C\,year^{-1}$. The increase in the annual mean water temperature by 0.1–0.5 °C from 1941 to 1982 and the decrease in its annual range by 0.1–0.4 °C in the Northern Caspian were also reported in [26, 27]. The general positive tendency of the long-term changes in water temperature in the Middle and Southern Caspian from 1900 to 1970 was noted in [28]: for example, during this period in Baku, the annual mean temperature increased by 1 °C, that is, the trend was approximately equal to 0.01 $^{\circ}C\,year^{-1}$. Thus, the positive trend in SST in 1982–2000 (0.05–0.10 $^{\circ}C\,year^{-1}$) exceeded several times the values characteristic of the preceding period of the 20th century.

4.3 SST Response to the Large-Scale Atmospheric Forcing

To reveal a possible response of SST in the Caspian regions to the ENSO and NAO, the extreme seasonal SST values were compared in time with the phases of these oscillations and with the character of their decadal variability. For this purpose, the anomalous SST values in the summer and winter periods were chosen from the time series of the weekly mean, monthly mean (Fig. 4), and seasonally mean (Fig. 5) SSTs. The following designations were assigned to the values for their qualitative characterization: warm summer (WS), cold summer (CS), warm winter (WW), and cold winter (CW). These characteristics of the anomalous SST values observed in 1982–2000 were supplemented with the available information on the anomalously warm and cold seasons in the preceding years [1–3, 10]. The summary of information obtained in this way on the anomalies of thermal regime in different regions of the Caspian Sea based on satellite and field data during the period from 1937 to 2000 is given in Table 2 (no data for the Middle and Southern Caspian in 1978–1981, and no data on winter severity in the Northern Caspian in 1985–1992). The time periods of the El Niño and La Niña events registered in 1937–2000 are also indicated there, in accordance with the monthly data on ENSO index. The character of the changes in ENSO and NAO indices in the second half of the 20th century can be judged from Fig. 8.

It follows from Table 2 and Fig. 8 that, from 1950 to 2000, 13 El Niños have occurred (negative values of ENSO index in Fig. 8), from which four the most intense events were observed in 1982–2000. We note that referring the ENSO phase with a low index (in absolute magnitude) either to El Niño or to La Niña suggested in the literature is ambiguous. For example, low ENSO indices with the variable sign in 1985, which are positive after averaging (Fig. 8), are considered as a La Niña event in a number of papers (see [12]) and are not related to this ENSO phase in other papers (for example, [29]); similar disagreement is related to a slightly better manifested phase of the ENSO with a positive index in 1961–1963. The time periods of the El Niño and La Niña events up to 1988 are given in Table 2 according to [29], with the addition of La Niñas in 1961–1963 and 1985. As the time delay of a possible response of the Caspian

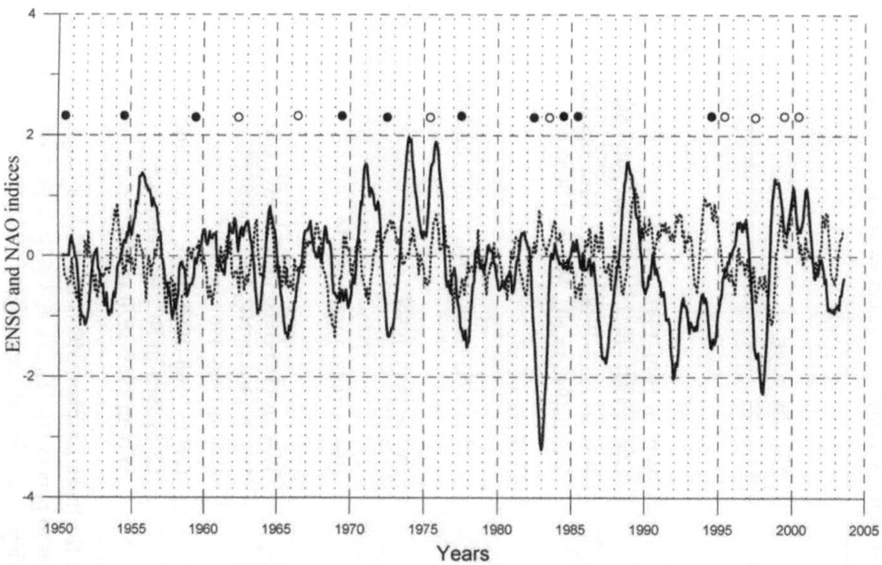

Fig. 8 Six-month running mean of monthly mean ENSO (*solid line*) and NAO (*dashed line*) indices in the period 1950–2003 acquired from the Internet sites (http://www.cpc.NOAA.gov/data/teledoc, http://iri.columbia.edu/). The *circles* indicate years with the anomalous winter SST values: the *black circles* correspond to cold winters, the *open circles* to warm winters

Table 2 Character of the summer and winter extreme SSTs in the years of El Niño and La Niña

El Niño Date	SST character	La Niña Date	SST character
		December 1937–April 1939	WW in 1937–1938 in NC[1]
August 1939–April 1942	WW in 1940–1941 in NC[1] CW in 1941–1942 in NC[1], MC, and SC[2]	May 1942–November 1943	WW in 1943–1944 in NC[1], MC, and SC[2] CW in 1944–1945 in NC[1] and SC[2]
March 1946–February 1947	WW in 1947–1948 in NC[1], MC, and SC[2] WS in 1948 in NC, MC, and SC[2]	December 1949–February 1951	CW in 1949–1950 in NC[1], MC, and SC[3] CS in 1950 in MC and SC[3]
March 1951–April 1952	–		
August 1952–December 1953	CW in 1953–1954 in NC[1,4], MC, and SC[2]	January 1954–January 1957	–
February 1957–February 1959	CW in 1958–1959[2]	November 1961–May 1963	WW in 1961–1962 in NC[1,4], MC, and SC[2]
June 1963–February 1964	CW in 1963–1964 in NC[1,4]	March 1964–October 1964	–
November 1964–December 1966	WW in 1965–1966 in NC[1], MC, and SC[2] WS in 1966 in NC, MC, and SC[2] CW in 1966–1967 in NC[1,4]	January 1967–July 1968	WW in 1967–1968 in NC[4]
August 1968–April 1970	CW in 1968–1969 in NC[1,4], MC, and SC[2] CS in 1969 in NC, MC, and SC[2] WW in 1969–1970 in NC[1,4]	May 1970–March 1972	CW in 1971–1972 in NC[1,4], MC, and SC[2]

Note. NC is the Northern Caspian, MC is the Middle Caspian, SC is the Southern Caspian, KBG is Kara-Bogaz-Gol Bay. Indices 1, 2, 3, and 4 denote references to papers [1], [2], [3], and [10], respectively, from which this information was adopted. Letters W, M, S in brackets correspond to weekly, monthly, and seasonal averaging of SST

Table 2 continued

El Niño Date	SST character	La Niña Date	SST character
April 1972– April 1973	–	May 1973– May 1976	WW in 1974–1975 in NC[4], MC, and SC[2] WS in 1975 in NC, MC, SC, and KBG[2]
June 1976– April 1978	CW in 1976–1977 in NC, MC, and SC[2]	May 1978– July 1979	–
August 1979– April 1981	WW in 1980–1981 in NC[2,4] CW in 1981–1982 in NC[4] and MC (W, M, S)	May 1981– March 1982	–
April 1982– August 1983	CS in 1982 in NC (W, M), MC (W, M), and SC (W, M, S) WW in 1982–1983 in NC[4], MC, and SC WS in 1983 in NC (W, M) CW in 1983–1984 in NC[4] and MC (W, M, S) CS in 1984 in MC (M, S) and SC (S) WS in 1984 in NC (W)	February– May 1985	CW in 1984–1985 in MC (W, M, S) and SC (W, M, S) WS in 1986 in MC (S) and SC (M, S) CS in 1986 in NC (W)
August 1986– February 1988	CS in 1987 in NC (M, S) and MC (S)	March 1988– July 1989	WS in 1989 in NC (M), MC (M, S), and SC (M, S)

Note. NC is the Northern Caspian, MC is the Middle Caspian, SC is the Southern Caspian, KBG is Kara-Bogaz-Gol Bay. Indices 1, 2, 3, and 4 denote references to papers [1], [2], [3], and [10], respectively, from which this information was adopted. Letters W, M, S in brackets correspond to weekly, monthly, and seasonal averaging of SST

Table 2 continued

El Niño Date	SST character	La Niña Date	SST character
November 1989–March 1990, June 1990–November 1995	CS in 1992 in NC (W, M, S), MC (W, M, S), and SC (W, M, S) CW in 1993–1994 in NC[4], MC (W, M, S), SC (W, M, S), and KBG (W) CS in 1994 in NC (M, S), MC (W), and KBG (M, S) WW in 1994–1995 in NC[4], MC (W, M), SC (W, M), and KBG (S) WS in 1995 in NC (S) and SC (S) CW in 1995–1996 in KBG (W, M, S) WS in 1996 in NC (W, M) and KBG (W)	December 1995–December 1996	WW in 1996–1997 in NC[4], SC (S), and KBG (M, S)
March 1997–April 1998	WS in 1998 in NC (W, M, S), MC (W, M, S), SC (W, M, S), and KBG (W, M, S) WW in 1998–1999 in NC[4], MC (W, M), SC (W, M, S), and KBG (W, M, S)	May 1998–December 2000	WS in 1999 in NC (W, M, S), MC (W, M, S), SC (W, M, S), and KBG (W, M, S) WW in 1999–2000 in NC[4], MC (W, M, S), SC (W, M, S), and KBG (W, M, S) WS in 2000 in NC (W, M, S), MC (W, M, S), SC (W, M, S), and KBG (W, M, S)

Note. NC is the Northern Caspian, MC is the Middle Caspian, SC is the Southern Caspian, KBG is Kara-Bogaz-Gol Bay. Indices 1, 2, 3, and 4 denote references to papers [1], [2], [3], and [10], respectively, from which this information was adopted. Letters W, M, S in brackets correspond to weekly, monthly, and seasonal averaging of SST

Sea SST to the ENSO phases is unknown, a conventional referring of one or another temperature anomaly to the nearest El Niño/La Niña event in Table 2 is based on assumption of the significant ENSO influence on atmospheric circulation in the Atlantic-European region during 15 months after its maximum development (see [12]). In the choice of the SST anomalies from those listed in Table 2 for analysis, we preferred the temperature anomalies of monthly and seasonal averaging observed simultaneously in at least two regions of the Caspian Sea. The years with the winter SST anomalies of both signs selected in this way are marked in Fig. 8.

In the period from 1950 to 2000, 11 of 18 winter SST anomalies (Fig. 8) are related to El Niño and seven to La Niña (including the poorly manifested ENSO phase in 1985 between two intense El Niño events and positive ENSO index in 1962 after its weakly negative value in 1961). During these ENSO phases, on the average, cold winters predominated during the El Niño phase, while warm ones prevailed during La Niñas. However, within this period, the relation between cold and warm winters in the ENSO phases was changed: in 1950–1977 (no data for 1978–1981), four of five anomalies related to El Niño were cold, while two of four during La Niñas were warm. Meanwhile, in 1982–2000, only three of six anomalies related to El Niño were cold, whereas in the La Niña phase, only one of three anomalies was cold (in 1985). Hence, by the end of the 20th century, warm winters became more frequent even during the El Niño periods.

The same tendency is observed in the summer SST anomalies. In six cases of nine formally related to El Niño, the summer was cold (1969, 1982, 1984, 1987, 1992, and 1994) and in three cases the summer was warm (1966, 1995, and 1998); five summer anomalies of six related to La Niña were warm (1975, 1986, 1989, 1999, and 2000) (it is not excluded that the 1999 anomaly was actually related to the El Niño event of 1997–1998). The SST minima and the majority of the SST maxima in the transitional seasons - spring and autumn (Figs. 5b and 5d, respectively, see also Table 3.10 in [2]) - corresponded to the El Niño periods. We note that, unlike the winter extreme SSTs, the summer minima of SST in the Middle and Southern Caspian are not always associated with a decrease in air temperature. For example, in the summer of 1950, the negative SST anomalies were the result of intense upwelling near the western and eastern coasts of the Caspian Sea [3].

One can see (Fig. 8) that the winter SST anomalies of both signs related to El Niño are observed in a number of cases during a winter following the maximum development of the phenomenon (1954, 1959, 1966, 1983, and 1995). However, one of the most severe winters in 1969 occurred at the beginning of El Niño (at a large sum of negative ENSO indices in January–February of that year), while the warmest winter of 1999 was observed as the second winter after the peak of El Niño. Anomalies of the summer SSTs frequently follow the winter anomalies in the same year and with the same sign (for example, the warm winter and warm summer in 1966, the cold winter and cold summer in 1969). Starting from 1995, both the summer and the winter marked SST anomalies were only positive. The most marked of them were associated with the 1997–1998 El Niño, one of the most intense in the 20th century. Absolute maxima of the summer (1998–1999) and winter (1999) SSTs corresponded to this El Niño as well as the maximal annual ranges in 1998 in all the regions of the Caspian Sea (see above).

The notably increased warming in the Caspian Sea in 1982–2000, as compared to 1940–1982, as well as a sharp increase in the summer and winter SSTs related to the 1997–1998 El Niño, suggest a change in the character of the global atmospheric processes at the end of the 20th century. This is also seen in Fig. 8. For example, in 1950–1980, El Niño and La Niña were regularly alternating and maximal (by absolute value) ENSO indices during the La Niña periods frequently exceeded those during El Niños. After 1980, however, the intense El Niño events clearly dominated. The NAO character was also notably changed from predominantly negative values of NAO index in the first period (in 1950–1981, the sum of the monthly indices was equal to – 53.2) to predominantly positive values of the index in the second period (in 1982–2000, the sum was equal to + 26.5).

One could expect that an increase in the recurrence and intensity of El Niño, which in the Northern Hemisphere is usually accompanied by a decrease in the zonal western and meridional southern (predominantly in the warm season) forms of atmospheric circulation under intensification of the meridional northern form (predominantly in the cold season) [30], and a domination of positive NAO indices, at which zonal transport increases and the heat flux into Central and Southern Europe decreases [13, 14], will lead to a decrease in the warming rate in the Caspian Sea in 1982–2000 as compared to the previous period. However, the analysis performed above indicates the opposite. It can be supposed that the greater positive temperature trend in the Caspian Sea during the modern period was related to the global warming, which was observed since the 1970s [9], and to the changes in the response of atmospheric circulation to the El Niño events – intensification of the meridional southern form of atmospheric circulation and weakening of the meridional northern form at the end of the 20th century [30]. A change in NAO index against the background of the Caspian Sea warming could lead to variations in SST within this period. It is possible, for example, that the decrease in the annual mean SST in the Caspian Sea in 1989–1992 and its increase in 1992–1998 (Fig. 7a) were caused by high positive values of NAO index in 1989–1992 and its drop (on the average) from 1993 to 1998, respectively. It is also not excluded that the sharp decrease in NAO index in these years and the coincidence of high negative ENSO and NAO indices in 1998 resulted in the significant summer and winter anomalies observed in 1998–1999.

It seems likely that another factors also existed that contributed to the appearance of these marked SST anomalies. First of all is the EA-Jet manifested from April to August [13]. In the years with positive (negative) values of the integrated EA-Jet index (calculated as a sum of indices for these months), wind velocity over the Caspian Sea is lower (higher) than a standard, while air temperature is higher (lower) than a standard. Therefore, it is possible that a change in the dominating negative values of the integrated EA-Jet index in 1963–1984 to positive ones after 1985 served to the high positive trend in the Caspian SST observed in 1982–2000. One of the highest values of this integrated index fell on 1998 (5.5, at a maximal value of 6.4 in 1994). In addition, the integrated (for June–August) positive index of the AS, which characterizes the intensity of the anticyclone centered at about 40–50°N and 40–100°E [19], in 1998, was the highest (7.6) over the period 1950–2000. The year with the second highest value of the integrated positive index of this oscillation (4.5) was 1989, when high summer SSTs were also observed.

5 Conclusions

The analysis of satellite MCSST data has revealed the positive (on the average) trend in the SSTs averaged over the Middle and Southern Caspian – of 0.05 and 0.10 °C $year^{-1}$, respectively, during the period 1982–2000. Interestingly, the general regularities of the changes in thermal regime in the Caspian and Black seas during 1982–2000 were the same: positive and close in values trends of SST (0.08 and 0.11 °C $year^{-1}$ in the western and eastern deep-water regions of the Black Sea [17]), the character of the changes in the annual mean temperature in 1989–1998 with a minimum in 1992–1993 (Fig. 7, see also [17, 18]), the clearly manifested positive temperature anomalies in the summer and winter of 1998 and 1999, and the maximal annual range of SST in 1998.

The large-scale atmospheric oscillations, which determine the fluctuations of the Caspian Sea level (ENSO [7], NAO [6], and EA-Jet [13]), also influence its thermal regime, although there is no unambiguous correlation between them and the character of the SST anomalies. The major part of the winter and summer SST anomalies recorded in the Caspian Sea is related to the El Niño events. In this case, in 1950–1980, with dominating negative NAO indices, negative SST anomalies prevailed, while at the end of the 20th century (in 1995–2000), at a sharp decrease in NAO index after its mainly positive values, positive SST anomalies became prevailed. The extreme values of the annual mean SST also corresponded to the El Niño events: the maximal values in 1966 (the warmest year in the period from 1916 to 1985 [6]), 1995, and 1998, and the minimal values in 1969 [6] and 1992.

It is possible that the time coincidence of large negative ENSO and NAO indices contributes to the appearance of the clearly manifested SST anomalies. However, the sign of the anomalies recorded during such coincidences was different: warm winters in 1966 and 1999 but cold winters in 1969 and 1977. High positive EA-Jet index in the 1998 El Niño could contribute to the appearance of the large summer anomaly in the Caspian Sea, whereas at an even larger index in 1994, also during El Niño but at a positive NAO index and negative index of the AS (– 0.9), the summer was cold. Meanwhile, the warm summer of 1966 was not accompanied by high positive values of EA-Jet (2.0) or AS (0.2) indices. The extremely cold winter of 1953–1954 occurred after a comparatively weak El Niño at a positive NAO index. One can suppose that, in each specific case, various combinations of different factors (ENSO, NAO, EA-Jet, and other oscillations, global warming or cooling, changes in the character of the response of atmospheric circulation to the El Niño events, etc.) determine the value and sign of the SST anomaly in the Caspian Sea and other basins.

Interannual variability in the Caspian Sea SST, especially in the winter seasons, can significantly influence its ecology. (In the Black Sea, this was demonstrated in recent publications [18, 31, 32]; evidence of association of seasonal and interannual variability of chlorophyll concentration with SST variability in the Caspian Sea is given in Chap. 7). Therefore, a continuous monitoring of the Caspian thermal regime is necessary. It is also important to continue the study of the response of the Caspian SST to the global atmospheric forcing, which is frequently manifested in a similar way in the neighboring inland basins (the Caspian and Black seas). As shown in this chapter, satellite information can provide a reliable and effective basis for such research on the scale of the entire sea and its regions.

Acknowledgements This study was supported by the Russian Foundation for Basic Research (Grants NN 04-05-64239, 03-05-64926, 03-05-96630) and the NATO SfP PROJECT No. 981063 "Multi-disciplinary Analysis of the Caspian Sea Ecosystem" (MACE).

References

1. Kosarev AN (1975) Hydrology of the Caspian and Aral seas. Mosk Gos Univ, Moscow (in Russian)
2. Terziev FS, Kosarev AN, Kerimov AA (eds) (1992) The seas of the USSR. Hydrometeorology and hydrochemistry of the seas. Vol VI. The Caspian Sea. Issue 1. Hydrometeorological conditions. Gidrometeoizdat, St. Petersburg (in Russian)
3. Arkhipova EG, Lyubanskii VA, Reznikova LP (1958) Tr Gos Okeanogr Inst 43:53 (in Russian)
4. Kosarev AN, Tuzhilkin VS (1995) Climatic thermohaline fields of the Caspian Sea. Gos Okeanogr Inst Moscow (in Russian)
5. Kosarev AN, Yablonskaya EA (1994) The Caspian Sea. Academic Publishing, The Hague
6. Rodionov SN (1994) Global and regional climate interaction: the Caspian Sea experience. Kluwer, Dordrecht
7. Arpe K, Bengtsson L, Golitsyn GS, Mokhov II, Semenov VA, Sporyshev PV (2000) Geophys Res Lett 27:2693

8. Kostianoy AG, Zavialov PO, Lebedev SA (2004) What do we know about dead, dying and endangered lakes and seas? In: Nihoul JCJ, Zavialov PO, Micklin PP (eds) Dying and dead seas. Climatic versus anthropic causes. Kluwer, Dordrecht, p 1
9. Levitus S, Antonov JI, Boyer TP, Stephens C (2000) Science 287:2225
10. Kouraev AV, Papa F, Buharizin PI, Cazenave A, Cretaux J-F, Dozortseva J, Remy F (2003) Ice Polar Res 22:43
11. Ginzburg AI, Kostianoy AG, Sheremet NA (2004) Oceanology. English Translation 44:605
12. Nesterov ES (2000) Meteorol Gidrol 8:74 (in Russian)
13. Nesterov ES (2001) Meteorol Gidrol 11:27 (in Russian)
14. Stanev EV, Peneva EL (2002) Global Planet Changes 32:33
15. D'Ortenzio F, Marullo S, Santoleri R (2000) Geophys Res Lett 27:241
16. Santoleri R, Bohm E, Schiano ME (1994) Coastal Estuar Studies 46:155
17. Ginzburg AI, Kostianoy AG, Sheremet NA (2004) J Mar Syst 52:33
18. Oguz T, Cokacar T, Malanotte-Rizzoli P, Ducklow HW (2003) Global Biochem Cycles 17:1088
19. Barnston AG, Livezey RE (1987) Mon Weather Rev 115:1083
20. Ginzburg AI, Kostianoy AG, Sheremet NA (2003) J Mar Syst 43:19
21. Fedorov KN, Ginsburg AI (1992) The near-surface layer of the ocean. VSP, The Netherlands
22. Kosarev AN, Tuzhilkin VS, Kostianoy AG (2004) Main features of the Caspian Sea hydrology. In: Nihoul JCJ, Zavialov PO, Micklin PP (eds) Dying and dead seas. Climatic versus anthropic causes. Kluwer, Dordrecht, p 159
23. Djenidi S, Kostianoy AG, Sheremet NA, Elmoussaoni A (2000) Seasonal and interannual SST variability of the North-East Atlantic Ocean. In: Oceanic fronts and related phenomena (Konstantin Fedorov international memorial symposium). GEOS, Moscow, p 99
24. Arkhipova EG (1955) Tr Gos Okeanogr Inst 20:337
25. Potaichuk MS (1978) Tr Gos Okeanogr Inst 139:65
26. Kuksa VI (1994) Southern seas (Aral, Caspian, Azov, and Black) under the conditions of anthropogenic impact. Gidrometeoizdat, St. Petersburg (in Russian)
27. Baidin SS, Kosarev AN (eds) (1986) The Caspian Sea. Hydrology and hydrochemistry. Nauka, Moscow (in Russian)
28. Potaichuk MS (1975) Tr Gos Okeanogr Inst 125:95
29. Sidorenkov NS (1991) Tr Gidrometeorologicheskogo Nauchno-Issled. Tsentra SSSR 316:31
30. Byshev VI (2003) Synoptic and large-scale variability of the ocean and atmosphere. Nauka, Moscow (in Russian)
31. Nezlin NP (2001) Oceanology. English Translation 41:375
32. Nezlin NP, Afanasyev YaD, Ginzburg AI, Kostianoy AG (2002) Adv Space Res 29:99

Hdb Env Chem Vol. 5, Part P (2005): 83–108
DOI 10.1007/698_5_005

Published online: 6 October 2005

Natural Chemistry of Caspian Sea Waters

Valentin S. Tuzhilkin[1] (✉) · Damir N. Katunin[2] · Yury R. Nalbandov[3]

[1]State Oceanographic Institute, 6 Kropotkinsky Per., 119838 Moscow, Russia
valver@orc.ru

[2]Caspian Fisheries Research Institute, 1 Savushkina, 414056 Astrakhan, Russia

[3]P.P. Shirshov Institute of Oceanology, Russian Academy of Sciences,
36 Nakhimovsky Ave., 117997 Moscow, Russia

Abstract Principal features of the natural chemistry of Caspian Sea waters are assessed: the salt(ionic) composition, dissolved oxygen, active reaction (pH), and nutrients. Their climatic regime and interannual variability are considered on the basis of historical data of ship measurements and expeditionary studies of recent years. It is shown that the natural hydrochemical regime of Caspian Sea waters (especially of its near-mouth areas) is characterized by significant spatial inhomogeneity and interannual variability closely related to interannual variability in the volumes and chemical composition of riverine runoff. An alternation of two types of the natural hydrochemical water regime was established manifested in the multiannual variability in all the characteristics considered, especially in the deep and near-bottom layers of the Caspian Sea. One of these types existed under the conditions of large river discharge at the beginning and at the end of the twentieth century. It was characterized by reduced concentrations of dissolved oxygen (down to the values typical of hypoxy) and elevated nutrient contents. The peak development of the other type was confined to the 1960s–1970s under the lowest river discharge. It was characterized by opposite deep and near-bottom anomalies of the dissolved oxygen (positive anomalies) and nutrient contents (negative anomalies).

Keywords Dissolved oxygen · Nutrients · pH · Variability · Vertical structure

1 Introduction

In this chapter, we assess the main large-scale features of the natural chemistry of Caspian Sea waters. The chemical properties of seawaters directly control the condition and functioning of the biotic part of marine ecosystems. In particular, dissolved oxygen and nutrients serve as factors restricting the existence of marine biota.

The first hydrochemical studies of Caspian Sea waters were performed at the beginning of the twentieth century under the supervision of A.A. Lebedintsev, N.M. Knipovich, and S.V. Bruevich. Their principal results were summarized in monographs [1–3]. In particular, it was stated that the Caspian Sea, especially its northern shallow-water part, significantly differs from other regions of the World Ocean in its salt composition and by the high nutrient content; the latter defines the equally high biological productivity of the waters of the sea.

Up to the mid-1950s, chemical observations in the Caspian Sea had been mostly performed by the scientific bodies of the Ministry of Fisheries of the USSR–All-Russia Scientific-Research Institute for Marine Fisheries and Oceanography (VNIRO) and the Caspian Scientific-Research Institute for Fisheries (CaspNIRKh). In the mid-1950s, regular onboard hydrochemical observations over a network of standard cross sections (see Fig. 2 in Tuzhilkin and Kosarev, this volume) under the methodical supervision of the State Oceanographic Institute, Hydrometeorological Service of the USSR were initiated; they lasted until the beginning of the 1990s. The results of the studies performed until the mid-1980s were generalized in Refs [4–9]. The main breakthrough made in this period of the studies was the detection of significant multiannual changes in the nutrient supply to the Caspian Sea with the river discharge that occurred in the second half of the twentieth century manifested not only in the total volume but also in the proportions of mineral and organic forms. Along with the long-term changes in the thermohaline structure of the waters (Tuzhilkin and Kosarev, this volume), this led to a profound transformation of the natural nutrient and oxygen regimes of Caspian Sea waters.

In the first half of the 1990s, onboard hydrochemical observations in the deep-water part of the Caspian Sea were virtually stopped. In the summer of 1995, owing to the efforts of CaspNIRKh and VNIRO, they were resumed with the use of modern facilities for measurement and analysis [10]. Since then, every year, at least one multidisciplinary shipborne survey has been performed in the Caspian Sea including hydrochemical observations over a rather dense network of stations. The results of these recent studies were generalized in a series of publications [11–17] and in other work. They showed that, over the past 2 decades, because of the changes in the multiannual trends, the hydrochemical regime of the Caspian Sea has approached close to the condition observed in the 1920s–1930s, before the significant fall in the mean sea level. In particular, high nutrient contents and low concentrations of dissolved oxygen in the deep layers of the Middle and South Caspian and even in the near-bottom layers of the North Caspian shelf were again observed. In the spring and summer, a significant oxygen oversaturation (up to 130–135%) occurred caused by the active photosynthesis of phytoplankton.

In this chapter, we present the results of a generalization of all the observational data available on the principal natural hydrochemical characteristics of the Caspian Sea: dissolved oxygen, active pH reaction, and nutrients. The total number of ship hydrochemical stations used exceeds 14 000, which is significantly greater than the datasets used in the previous generalizations [1–9]. The proportions between the degrees of knowledge of different hydrochemical characteristics assessed are approximately the same as in other regions of the World Ocean: prevalent are the data on dissolved oxygen (13 800 stations) and pH (11 400 stations), while the number of observations of nutrients is 2–6 times smaller.

The geographical distributions of all the ship hydrochemical stations over the Caspian Sea area and over the years of observations are shown in Figs. 1 and 2, respectively. On the whole, both of these distributions are close to the spatial and temporal structure of the ship thermohaline observations (see Fig. 1 in Tuzhilkin and Kosarev, this volume). Meanwhile, the hydrochemical observations are significantly smaller in number compared with the thermohaline observations (fourfold for dissolved oxygen and 10–20-fold for nutrients). Therefore, the estimates of the statistical parameters for the majority of the hydrochemical characteristics are significantly less detailed and reliable compared with those of temperature and salinity.

Statistical processing of the data was performed following the technique described by Tuzhilkin and Kosarev in this volume. Owing to the changes in the methods for analytical determinations of the hydrochemical characteristics with time, great attention was paid to the expert estimations of the initial data and the results of the calculations. We managed to compile climatic seasonal fields only for dissolved oxygen and pH for the winter (January–March) and summer (July–September) seasons with a horizontal resolution from $1° \times 1°$ to $2° \times 2°$, which is two- to threefold rougher than the resolution of the thermohaline fields. For all of the hydrochemical characteristics considered, only their climatic seasonal and annual values at standard levels averaged over the North, Middle, and South Caspian Sea were obtained.

It should be noted that, owing to the significant multiannual variability of the hydrochemical regime of the Caspian Sea, the climatic (i.e., multiannual mean) values of the hydrochemical characteristics of its waters strongly differ from their actual values in individual years or selected time intervals with steady conditions. Therefore, in this chapter, the main attention is paid to the multiannual variations of natural hydrochemical conditions of Caspian Sea

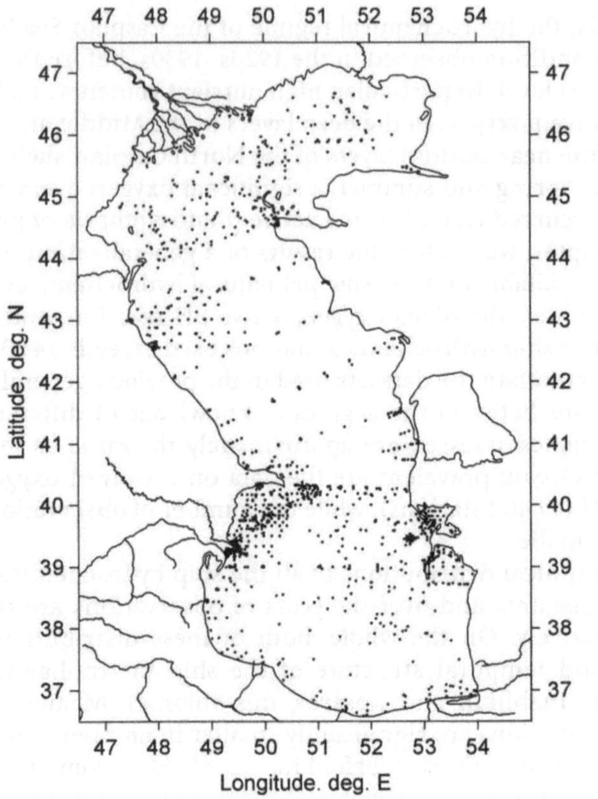

Fig. 1 Location of the historical stations of ship chemical observations in the Caspian Sea

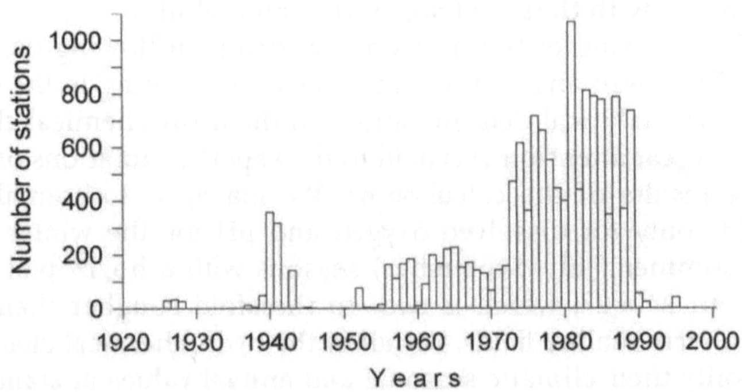

Fig. 2 Distribution of the historical stations of ship chemical observations in the Caspian Sea over a number of years

waters. In so doing, we used the published results of earlier studies [1–9] and the observational data of recent years [10–17].

2
Salt Composition

The present-day salt composition of the waters of the Caspian Sea was formed as a result of transformation of the initial oceanic water under the influence of the continental runoff after the isolation of the Caspian Sea from the World Ocean. This event occurred in the post-Khvalynian epoch of the Quaternary period approximately 10 000 years bp [3]. In Table 1, we present the salt composition of the waters of the Caspian Sea and its individual parts in accordance with Ref [6] as compared with the salt composition of the World Ocean waters after Ref. [18].

Relative to the salt composition of the World Ocean, the Caspian Sea waters feature decreased contents of cations (Na, K, Mg, Ca), whose origin is related to the development of the processes of denudation of rocks.

Owing to the enhanced influence of the continental runoff, the content of chlorine anion in the Caspian Sea is smaller by a factor of 3.5 (up to 7 in its northern part) than that in the oceanic water (Table 1). In the Middle and South Caspian, the content of chlorine anion is approximately the same, being 5.44 and 5.46 $g\,kg^{-1}$, respectively. In the North Caspian, toward the Volga River mouth, it decreases to 0.09 $g\,kg^{-1}$ [6]. For the same reason, the Caspian Sea features increased contents of hydrocarbon and sulfate ions (Table 1).

Since hydrocarbons comprise more than 40% of the ionic river discharge to the Caspian Sea, its waters are characterized by a carbonate oversaturation at a predominance of calcium hydrocarbons. When riverine water is transformed into seawater, calcium hydrocarbons transfer to carbonates, which precipitate from the water to the bottom sediments [3, 6]. This process proceeds especially intensely in the summer in the zone of the waters of the

Table 1 Concentrations of principal ions ($g\,kg^{-1}$) in the waters of the World Ocean and in the main regions of the Caspian Sea

Ions	World Ocean [18]	Parts of the Caspian Sea [6] North	Middle	South
Na + K	10.94	1.62	3.24	3.25
Ca	0.41	0.20	0.35	0.36
Mg	1.27	0.37	0.74	0.75
Cl	18.98	2.65	5.44	5.46
SO_4	2.65	2.19	3.01	3.05
HCO_3	0.14	0.18	0.22	0.22
Total	34.45	7.21	13.00	13.09

eastern shelves of the Middle and South Caspian oversaturated with calcium carbonate. In the summer, the thickness of the layer oversaturated with calcium carbonate in these parts of the Caspian Sea is up to 30 and 60 m and its lower boundary coincides with the seasonal thermocline (Tuzhilkin and Kosarev, this volume).

The ratios of the concentrations of principal ions and of the water salinity to the chlorine ion content according to the data from observations in different years are presented in Table 2. In the Middle and South Caspian, i.e., in the basic volume of the Caspian waters, the ratios of the concentrations of principal ions (except for Na and K) to the chlorine ion content are significantly greater than in the waters of the World Ocean; in particular, the corresponding excesses are twofold for Mg, threefold for Ca, fourfold for SO_4, and 5.5-fold for HCO_3.

The significant spatial and multiannual changes, as compared with those for the World Ocean, in the ratios of salinity to the chlorine content (Table 2, right-hand column) make it difficult to address a common equation of state to the entire Caspian Sea [12, 13]. At present, the equation of state from Ref. [11] is usually applied to the open areas of the sea, though it is not officially recognized. The studies initiated in Refs. [11–13] should be continued in order to compile a practical salinity scale and to derive a refined equation of state of the Caspian Sea waters similar to that available for the World Ocean.

Table 2 Ratios of the concentrations of principal ions and of the water salinity (*S*) to the chlorine ion content in the water in different regions of the Caspian Sea according to the data from observations in different years

Region	Year	Ratios of principal ions and salinity to chlorine ion				
		Ca/Cl	Mg/Cl	SO_4/Cl	HCO_3/Cl	*S*/Cl
Volga near-mouth area	1991 [13]	0.095	0.183	0.828	0.061	2.786
Western part of the North Caspian	1991–1992 [13]	0.071	0.146	0.592	0.063	2.475
Eastern part of the North Caspian	1992 [13]	0.062	0.132	0.534	0.032	2.362
Middle Caspian	1933 [3]	0.068	0.136	0.557	0.039	2.405
	1961–1963 [6]	0.064	0.136	0.553	0.040	2.390
	1992 [12]	0.059	0.135	0.564	0.032	2.415
South Caspian	1933 [3]	0.065	0.137	0.562	0.041	2.415
	1961–1963 [6]	0.066	0.137	0.559	0.040	2.397
World Ocean	1942 [18]	0.022	0.067	0.140	0.007	1.815

3 Dissolved Oxygen

The oxygen regime in the shallow-water North Caspian strongly differs from that in the deep-water parts of the Caspian Sea. In the shallowest zone of the North Caspian with sea depths smaller than 10 m, the vertical increments of the dissolved oxygen as well as those of the water temperature and salinity are small throughout the year (Table 3).

In the North Caspian, the maximum dissolved oxygen concentrations for the entire water column from the surface to the bottom (8–11 $ml\,l^{-1}$) are observed in the cold half of the year from October to April. In the winter, this is favored by the relatively low temperatures, while in the spring, this is caused by the development of phytoplankton and related photosynthetic activity.

In May, with the beginning of intensive heating and desalination of the waters of the North Caspian with the flood riverine waters, a pycnocline is formed in the deepest (10–20-m) outer part of the North Caspian shelf; this hampers the ventilation of the near-bottom water layers. In so doing, the dissolved oxygen concentrations in the near-bottom layer here decrease.

Table 3 Absolute (numerator $ml\,l^{-1}$) and relative (*denominator* %) dissolved oxygen concentrations in the near-bottom layer in the western and eastern parts of the North Caspian in characteristic months of the seasons averaged over intervals of years

Intervals of years, layer	Western part (months) 4	6	8	10	Eastern part (months) 4	6	8	10
Shallow-water zone								
1935–1955 [5] surface layer	7.36/105	6.00/104	5.85/–	8.13/103	8.18/105	5.63/102	5.94/100	8.51/106
1956–2003 surface layer	7.94/103	6.24/104	6.21/106	7.63/98	7.65/103	5.86/101	5.97/101	7.91/98
1956–2003 bottom layer	7.85/102	6.07/101	5.91/100	7.58/98	7.68/103	5.80/100	5.88/99	7.85/98
Deep-water zone								
1935–1955 [5] surface layer	7.41/108	6.22/108	5.92/106	7.11/102	7.58/105	5.91/103	6.12/103	7.49/101
1935–1955 [5] bottom layer	7.20/101	5.53/93	5.23/91	6.82/97	7.52/104	5.62/99	5.72/97	7.27/99
1956–2003 surface layer	7.92/106	6.19/105	5.91/104	7.13/101	8.06/106	5.82/102	5.83/99	7.68/100
1956–2003 bottom layer	7.89/103	5.64/92	4.98/85	6.95/98	8.01/104	5.70/98	5.72/97	7.66/100

By August, the near-bottom concentrations of dissolved oxygen are 15–20% lower than those at the surface (Table 3); meanwhile, the degree of saturation of the near-bottom waters with oxygen commonly exceeds 80% (which marks the hypoxy limit). Only in the western part of the outer North Caspian shelf, almost every year, the degree of saturation of the near-bottom waters with oxygen in the summer is less than 80% [4, 5, 14]. In years with enhanced Volga River runoff, the area of the oxygen-poor regions (with a dissolved oxygen saturation of the near-bottom waters less than 80%) significantly grows; for example, in June of the water-abundant 1991, it reached 27 500 km^2 or approximately 30% of the area of the North Caspian.

The absolute and relative dissolved oxygen values averaged over 1935–1955 and 1956–2003 and generalized over the regions of the North Caspian almost do not differ (Table 3). Only a slight tendency to deterioration of the oxygen regime in the near-bottom layer can be traced in the summer period in the western part of the outer North Caspian shelf. Meanwhile, as shown later, in the southwestern part of this region of the North Caspian, the multiannual tendency to a decrease in the near-bottom dissolved oxygen concentrations is far more clearly manifested.

In the winter, in the upper layer of the entire Caspian Sea, the horizontal distribution of dissolved oxygen (Fig. 3a) is the closest to the value for equilibrium with the atmosphere (about 100% saturation) and, therefore, corresponds well to the water temperature field (Tuzhilkin and Kosarev, this volume). The western and eastern shelves of the Middle Caspian, which suffer stronger impacts of the advection of cold North Caspian waters and of the heat release to the atmosphere, respectively, are characterized by higher oxygen contents compared with those of the deep-water part of the sea, where a northward transport of the relatively warm South Caspian waters is observed. In the Middle Caspian, the winter vertical convective mixing commonly involves an upper layer 100–200-m thick; in extremely cold winters, the mixing reaches the bottom [7, 8]. Therefore, here, the absolute oxygen content in the upper 100-m layer is rather homogeneous over the vertical distribution (7–9 $ml\,l^{-1}$). In the upper layer of the South Caspian, in the winter, the oxygen concentrations are lower (7–7.5 $ml\,l^{-1}$) than in the Middle Caspian owing to the higher water temperatures. The thickness of the upper quasihomogeneous layer caused by the convection here comprises about 50 m.

In the spring, intensive photosynthesis proceeds in the upper 20-m layer of the deep-water areas of the Caspian Sea; therefore, in this season, the absolute oxygen content reaches its maximum value throughout the year, which is 0.2–0.3 $ml\,l^{-1}$ higher than the wintertime values, and the mean oxygen saturation rises to 110% (with absolute maximums up to 130–135%).

In the summer, the greatest absolute (6.0–6.5 $ml\,l^{-1}$) and relative (105–110%) surface concentrations of dissolved oxygen are confined to the areas of the most intensive photosynthetic activity of phytoplankton located in the North Caspian and in the area of the upwelling off the eastern coast

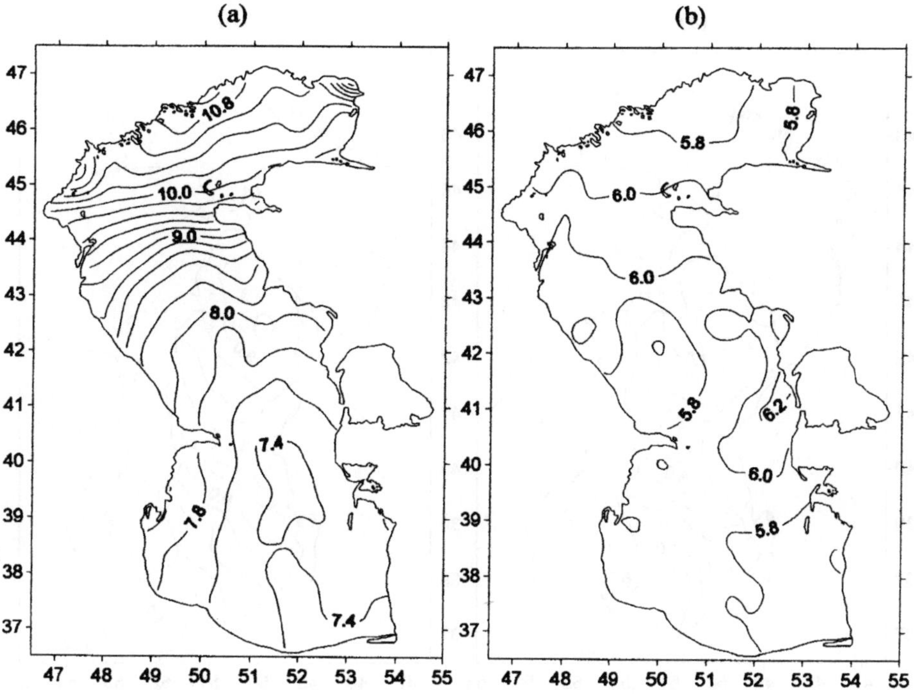

Fig. 3 Climatic fields of dissolved oxygen concentrations (ml l^{-1}) in the surface layer of the Caspian Sea in **a** winter and **b** summer

of the Middle Caspian (Fig. 3b). In the layer from 30 to 100 m, below the seasonal thermocline, the oxygen saturation in the summer becomes minimal (down to 80 and 65% in the Middle and South Caspian, respectively).

In the fall, the oxygen oversaturation of the surface layer in the deep-water areas of the Caspian Sea is retained. The autumn decrease in the temperature of the surface water layers together with the deepening and destruction of the seasonal thermocline result in an increase in the dissolved oxygen concentrations in the upper layers of the Middle and South Caspian up to 6.5–7.5 ml l^{-1}.

In the intermediate and deep layers of the Middle and South Caspian, throughout the year, the absolute and relative concentrations of dissolved oxygen decrease from the northwest to the southeast (Figs. 4, 5). At a depth of 100 m, the amplitude of the intra-annual variations in the absolute oxygen content (Fig. 4) is threefold to fourfold smaller than at the surface of the sea (Fig. 3). Below the 100-m level, virtually no seasonal variability can be found. In the 200–600-m layer of the Middle Caspian, the mean annual absolute (relative) content of dissolved oxygen decreases with depth from 5.5 (70%) to 3.5 ml l^{-1} (50%); in the South Caspian, the corresponding decrease is from 4.0 (60%) to 2.5 ml l^{-1} (40%).

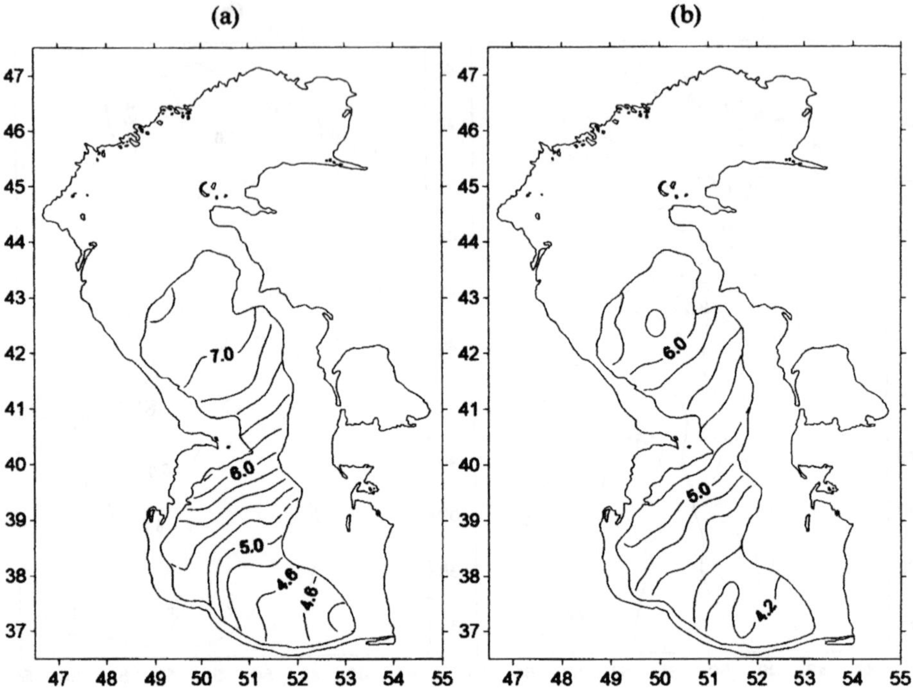

Fig. 4 Climatic fields of dissolved oxygen concentrations (ml l^{-1}) in the Caspian Sea at a depth of 100 m in **a** winter and **b** summer

The interannual changes in the natural and anthropogenic impacts on the Caspian Sea, which is isolated from the World Ocean, cause significant multi-annual variations in the oxygen regime of its waters.

In the North Caspian, the increase in the river discharge that occurred in the 1980s to 1990s (Tuzhilkin and Kosarev, this volume) has lead to a significant enhancement in the stable spring–summer density stratification of the waters over the outer North Caspian shelf; it is presented in Fig. 6a in the form of the buoyancy storage $B = \Delta\rho\Delta z$, where $\Delta\rho$ is the density increment of the water density in a layer with a thickness of Δz. Simultaneously, the river discharge of nutrients to the North Caspian also significantly increased [14]: a fivefold growth in the runoff of mineral phosphorus, a twofold growth in the runoff of organic forms of phosphorus and mineral nitrogen, and a 1.5-fold growth in the silicon compound runoff were observed. The subsequent enhancement of the process of eutrophication of the waters of the North Caspian resulted in a rapid and significant decrease in the near-bottom concentrations of dissolved oxygen in the southwestern part of the outer North Caspian shelf down to the hypoxy threshold (Fig. 6b). In 1996–2003, the summertime oxygen content significantly decreased not only in the near-bottom layer of the southwestern part of the outer North Caspian shelf but also in the east of

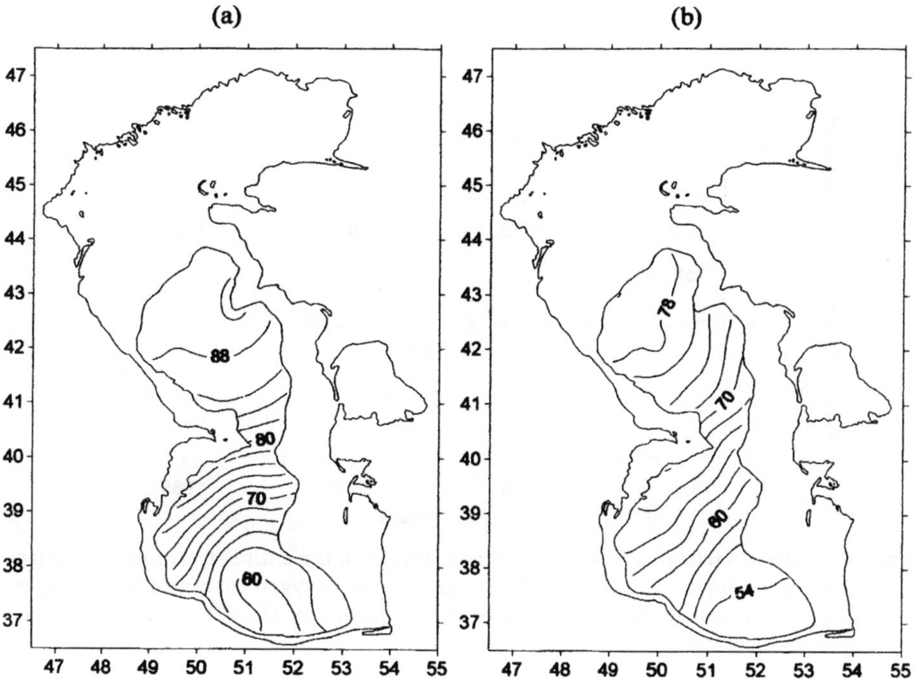

Fig. 5 Climatic fields of dissolved oxygen concentrations (percentage of saturation) in the Caspian Sea at a depth of 100 m in **a** winter and **b** summer

the inner shelf in the region of the Ural'skaya Borozdina trough [14]. On the whole, in the summer–fall period, the areas of the zones with oxygen saturation in the near-bottom layer less than 80% increased almost threefold. The deterioration of the gas regime in the North Caspian had a negative impact on the conditions of formation and dwelling of oxiphilic hydrobionts feeding on benthos.

Over the past few decades, similar tendencies were also observed in the intermediate and near-bottom layers of deep-water basins (Fig. 7). In addition to the intensification of the riverine water inflow and to the related enhancement in the nutrient supply, the decrease in the severity of winter also serves as a strong factor preventing the deeper layers of the Middle and South Caspian from ventilation (Tuzhilkin and Kosarev, this volume).

The conditions of the oxygen regime in the entire water column of the Middle and South Caspian during recent years compared with those for the preceding intervals are presented in Tables 4 and 5, respectively. The greatest (fourfold to sixfold) drops in the degree of dissolved oxygen saturation of the deep waters in 1998–2003 compared with 1964–1980 were observed in the Middle Caspian. In the South Caspian, the corresponding drop was only 1.5-fold to threefold; meanwhile, it started somewhat earlier and has lasted over

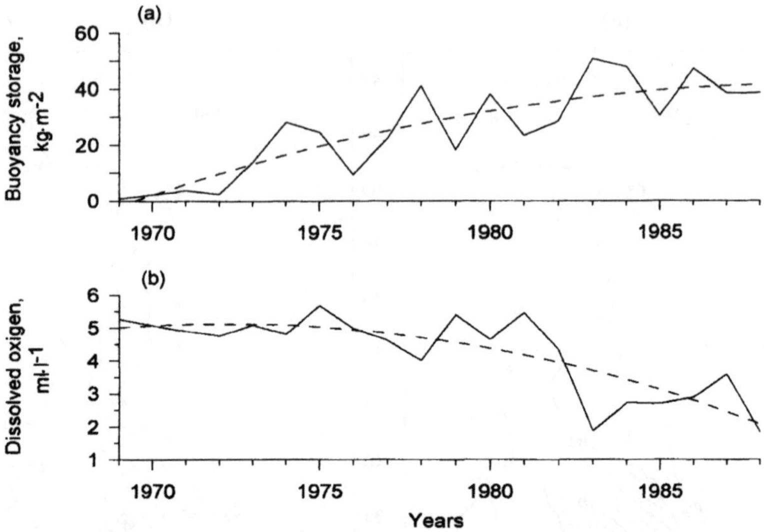

Fig. 6 Long-term variability of **a** the buoyancy storage in the entire layer from the surface to the bottom in August and **b** the near-bottom dissolved oxygen concentration in August according to ship observations in the southwestern part of the North Caspian. The *dashed lines* are the quadratic trends

the recent years as well, while in the Middle Caspian the situation became steady state. On the whole, the present-day oxygen conditions in the waters of both of the deep-water basins of the Caspian Sea are equally unfavorable for marine biota [10, 15–17]. They are caused by the weakening of the wintertime convective ventilation in the deep waters and strengthening of the abyssal destruction processes, that is, the growth in the oxygen consumption required for organic matter oxidation.

In its characteristics, the present-day oxygen regime of the Caspian Sea waters is close to that observed at the beginning of the twentieth century [1–3], when the volumes of the river discharge and the thermohaline structure of the Caspian Sea waters were comparable with the present-day characteristics. One can consider two types of vertical structure of the dissolved oxygen content in the deep-water areas of the Caspian Sea, which correspond to the two types of thermohaline water structure distinguished by Tuzhilkin and Kosarev in this volume. At the beginning of the twentieth century and at present, in the Middle and South Caspian, under the conditions of intensive river discharge and moderate sea level (first type), a rapid drop in the dissolved oxygen concentrations below the upper 100-m layer to values less than 1 ml l^{-1} is observed. The second type of structure is typical of the conditions of a low river discharge and a low sea level. In the Middle Caspian, it is characterized by the near-bottom oxygen concentration values averaged over 1960–1970 of 3.5–4.0 ml l^{-1}; the corresponding values for the South Caspian

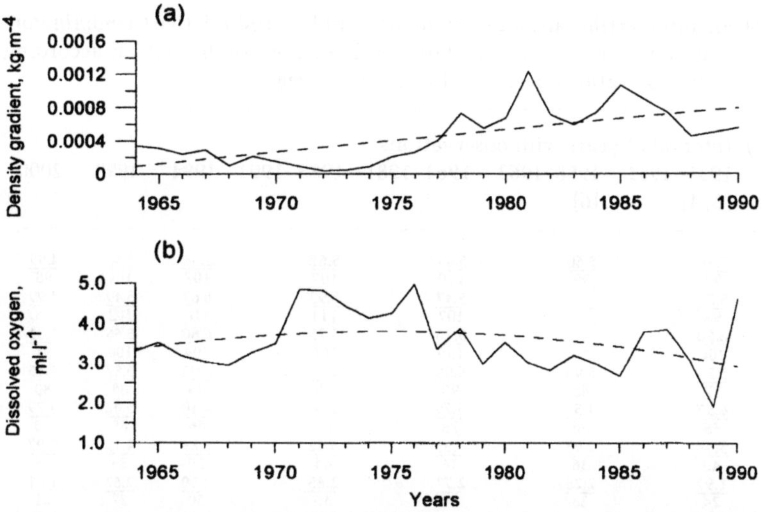

Fig. 7 Long-term variability of **a** the mean vertical density gradient in the 0–600-m layer in February; and **b** the dissolved oxygen concentration at a depth of 600 m in August according to ship observations in the deep-water part of the Middle Caspian. The *dashed lines* are the quadratic trends

Table 4 Mean summertime absolute (*numerator* $ml\,l^{-1}$) and relative (*denominator* %) dissolved oxygen concentrations at standard levels in the Middle Caspian according to the data from ship observations performed in different years

Level (m)	Interval of years with observations 1935–1943 [3, 4]	1958–1962 [6]	1964–1981	1998	2000	2001	2002	2003
0	5.77/101	5.54/97	6.10/100	6.58/109	5.17/98	5.48/102	5.68/103	6.20/99
10	5.84/101	5.77/97	6.14/107	6.26/103	5.27/100	5.32/96	5.96/106	6.17/99
25	6.01/86	5.71/82	5.87/95	5.78/88	5.35/93	5.20/82	5.88/100	5.97/78
50	5.65/77	5.36/70	5.89/86	6.03/77	5.29/75	5.05/67	5.56/74	6.15/78
100	5.42/68	5.29/69	6.22/80	5.94/72	5.08/72	4.26/56	4.53/58	4.38/55
200	4.67/54	3.83/47	5.24/66	5.01/60	2.09/26	1.78/22	2.26/26	1.67/20
400	2.98/37	3.53/42	4.18/53	1.84/22	1.36/17	1.61/20	1.31/16	1.22/15
600	1.86/23	3.49/41	3.97/51	0.98/12	1.33/16	0.77/9	1.23/15	–
750	0.92/11	3.55/42	3.69/46	0.54/7	0.80/10	–	0.65/8	0.70/7

are 1.5–2.0 $ml\,l^{-1}$. In selected years, the maximum oxygen concentrations reached 5.5–6.0 and 3.4–4.5 $ml\,l^{-1}$, respectively (Fig. 7).

If the existing regime of the external impact is retained, it seems quite probable that in future the deterioration of the aeration of the Caspian Sea

Table 5 Mean summertime absolute (*numerator* ml l^{-1}) and relative (*denominator* %) dissolved oxygen concentrations at standard levels in the South Caspian according to the data from ship observations performed in different years

Level (m)	Interval of years with observations 1935–1943 [3, 4]	1958–1962 [6]	1964–1981	1983–1991	1994	1998	2000	2002
0	$\frac{5.38}{97}$	$\frac{5.60}{99}$	$\frac{5.42}{110}$	$\frac{5.62}{107}$	$\frac{5.58}{107}$	$\frac{5.60}{104}$	$\frac{4.99}{98}$	$\frac{5.79}{108}$
10	$\frac{5.57}{98}$	–	$\frac{5.47}{107}$	$\frac{5.93}{111}$	$\frac{6.67}{116}$	$\frac{5.49}{102}$	$\frac{4.99}{98}$	$\frac{5.76}{107}$
25	$\frac{6.50}{106}$	–	$\frac{5.40}{105}$	$\frac{5.75}{100}$	$\frac{6.60}{104}$	$\frac{5.66}{100}$	$\frac{5.31}{91}$	$\frac{5.36}{95}$
50	$\frac{5.21}{77}$	$\frac{5.96}{86}$	$\frac{6.90}{95}$	$\frac{5.75}{85}$	$\frac{6.02}{85}$	$\frac{5.87}{85}$	$\frac{5.45}{80}$	$\frac{5.82}{82}$
100	$\frac{4.57}{58}$	$\frac{4.57}{56}$	$\frac{5.79}{78}$	$\frac{5.04}{66}$	$\frac{4.40}{66}$	$\frac{5.02}{65}$	$\frac{4.79}{60}$	$\frac{4.43}{58}$
200	$\frac{3.44}{42}$	$\frac{2.95}{36}$	$\frac{3.55}{46}$	$\frac{3.42}{44}$	$\frac{3.89}{50}$	$\frac{4.19}{53}$	$\frac{2.97}{48}$	$\frac{3.76}{46}$
400	$\frac{1.97}{24}$	$\frac{2.28}{28}$	$\frac{2.71}{34}$	$\frac{2.85}{35}$	$\frac{2.30}{30}$	$\frac{2.62}{27}$	$\frac{1.61}{21}$	$\frac{1.78}{20}$
600	$\frac{0.47}{6}$	$\frac{2.01}{26}$	$\frac{2.14}{31}$	$\frac{2.28}{28}$	$\frac{1.48}{24}$	$\frac{1.96}{24}$	$\frac{1.71}{21}$	$\frac{0.85}{11}$
800–1000	$\frac{0.26}{3}$	$\frac{1.55}{19}$	$\frac{2.01}{26}$	$\frac{1.93}{24}$	$\frac{1.32}{16}$	$\frac{1.40}{17}$	$\frac{0.91}{11}$	$\frac{0.37}{5}$

waters will be continued [17] up to the appearance of hydrogen sulfide in the near-bottom waters. Trace concentrations of the latter of 0.2–0.3 ml l^{-1} were registered repeatedly during the first decades of the past century [1–3].

4 Active Reaction (pH)

The first pH determinations performed by Bruevich [3] in the early 1930s showed that, in the Caspian Sea, the values were significantly greater than in the waters of other regions of the World Ocean. The principal reason for the enhanced pH values lies in the great contribution of the riverine waters characterized by a high content of anions of weak acids, primarily, carbonic acid, to the formation of the hydrochemical regime of the Caspian Sea. They were also favored by the intensive production processes in the Caspian Sea. Therefore, the greatest pH values are observed in the summertime in the surface layer of the near-mouth areas of the Caspian Sea (Fig. 8).

In the North Caspian, seasonal pH changes are especially great (Table 6). The maximum pH values are observed in August, which is related to the accumulation of the effects of river discharge along with the retention of intensive photosynthesis of phytoplankton. In the fall, when the destructive (oxidation) processes dominate over the productive processes and the North Caspian waters are enriched with free carbon dioxide, pH values decrease, especially in

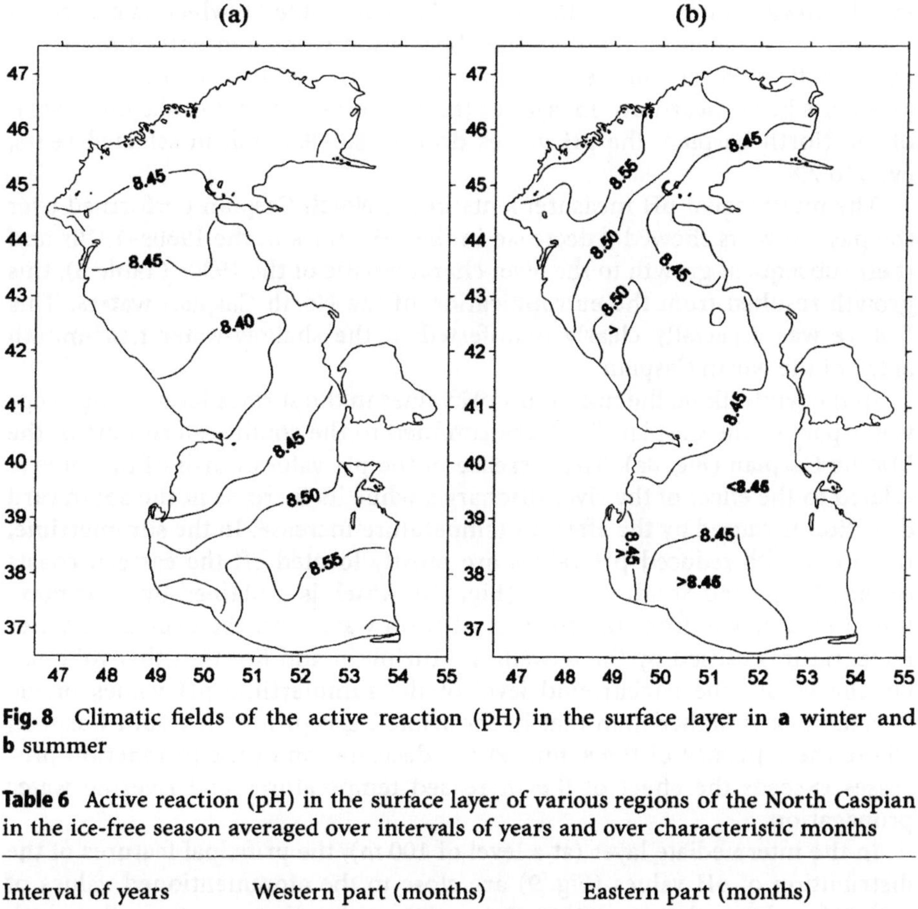

Fig. 8 Climatic fields of the active reaction (pH) in the surface layer in **a** winter and **b** summer

Table 6 Active reaction (pH) in the surface layer of various regions of the North Caspian in the ice-free season averaged over intervals of years and over characteristic months

Interval of years	Western part (months)				Eastern part (months)			
	4	6	8	10–11	4	6	8	10–11
Shallow-water zone								
1935–1955 [5]	8.45	8.45	8.86	8.39	8.36	8.53	8.68	8.28
1961–1963 [6]	8.32	8.43	–	8.34	8.30	8.36	–	8.39
1964–1977	–	8.58	8.74	–	–	8.44	8.48	–
1978–1995	8.40	8.53	8.75	–	8.25	8.41	8.50	–
1996–2003	8.63	8.66	8.79	8.68	–	8.51	8.53	8.61
Deep-water zone								
1935–1955 [5]	8.28	8.46	8.59	8.44	8.27	8.44	8.46	8.37
1961–1963 [6]	8.30	8.28	–	8.37	8.32	8.29	–	8.43
1964–1977	–	8.41	8.45	–	–	8.38	8.40	–
1978–1995	8.43	8.43	8.50	8.42	8.33	8.35	8.41	8.34
1996–2003	8.51	8.53	8.53	8.51	–	8.45	8.51	8.50

the shallowest water zone of the North Caspian (Table 6), which is caused by the low heat content in its waters in this season compared with that for the outer shelf. At the end of the summer, owing to the domination of destructive processes in the near-bottom layer of the southwestern part of the outer shelf of the North Caspian, the pH values drop to 8.0–8.2 and, in selected years, even to 7.8.

The multiannual pH measurements in the North Caspian performed over the past 70 years showed a decrease in the pH values in the 1960s–1970s and their subsequent growth to the level characteristic of the 1930s (Table 6); this growth resulted from the eutrophication of the North Caspian waters. This feature was especially clearly manifested in the shallow-water near-mouth areas of the North Caspian.

In the wintertime, the minimum pH values in the surface layer of the deep-water part of the Caspian Sea were confined to the southwestern part of the Middle Caspian (Fig. 8a). The increase in the pH values north of this area is related to the effect of the river discharge, while its increase in the southward direction is caused by the effect of temperature increase. In the summertime, the areas with reduced pH values are mostly located off the eastern coasts of the Middle and South Caspian (Fig. 8b), which is explained by their maximum remoteness from the sources of fresh water and the decreased water temperatures caused by the upwelling (Tuzhilkin and Kosarev, this volume). On the whole, the background level of the summertime pH values in the Caspian Sea is higher than that in the winter, except for the South Caspian, where the influence of the summertime deceleration of the production processes exceeds the effect of the increased temperatures and riverine water propagation.

In the intermediate layer (at a level of 100 m), the principal features of the distribution of pH values (Fig. 9) are close to the aforementioned values of the surface layer, that is, in the winter, these values increase from the north to the south and in the summer they increase from the east to the west. In both cases, this is related to the water temperature distribution. The summer decrease in the background level of the pH values of 0.1 compared with the wintertime level is caused by the summertime domination of destructive processes in the intermediate and deep waters of the Middle and South Caspian.

The multiannual changes in the pH values and their vertical distribution in the Middle and South Caspian over the past 70 years are characterized by an alternation of two types of structures (Table 7), similar to the thermohaline (Tuzhilkin and Kosarev, this volume) and oxygen (Sect. 3) water structures.

One of these types was observed at the beginning and at the end of the twentieth century under high intensities of river discharge and high mean sea level. It is characterized by enhanced pH values in the 20-m surface layer and their rapid decrease with depth to 7.7–7.9 in the near-bottom layer.

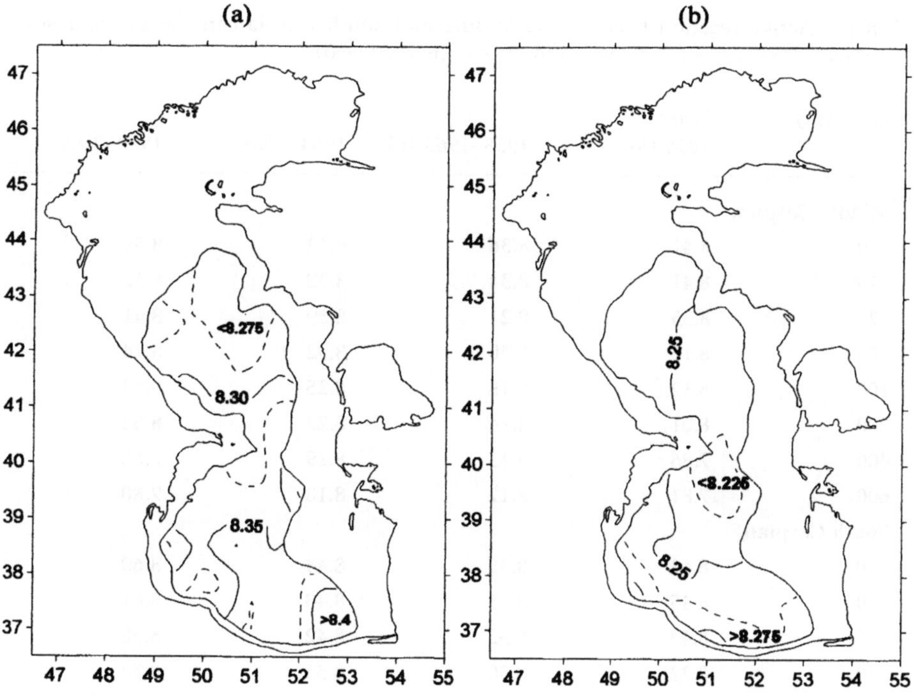

Fig. 9 Climatic fields of the active reaction (pH) at a depth of 100 m in **a** winter and **b** summer

The other type of pH vertical structure observed in the 1960s–1970s under low intensities of river discharge and low mean sea level was distinguished by a significantly greater vertical homogeneity, especially in the Middle Caspian. In these years, the pH values in the surface layer were 0.05–0.15 smaller than at high sea level. Along with this, the decrease in the pH values with depth was significantly slower: down to 8.1–8.2 in the near-bottom layer of the Middle Caspian and down to 7.9–8.0 in the South Caspian.

The previously described particular features of the types of vertical pH structure correspond well to the features of the thermohaline (Tuzhilkin and Kosarev, this volume) and oxygen (Sect. 3) water structures observed simultaneously. They have a common origin related to the multiannual variations in the character of the water ventilation and production–destruction processes in the water column of the Caspian Sea considered earlier. In the Middle Caspian, the multiannual differences in all of the characteristics assessed are manifested better than in the South Caspian. This is caused by the closer position of the former region to the principal sources of the multiannual variations in the condition of the Caspian Sea waters related to the Volga River runoff and to the wintertime heat lost by the water over the northern continental shelf and slope.

Table 7 Active reaction (pH) in the Middle and South Caspian in the summer season averaged over intervals of years and over individual years

Level (m)	Years 1934 [3]	1958–1963 [6]	1964–1981	1998–2004
Middle Caspian				
0	8.41	8.36	8.44	8.54
10	8.41	8.34	8.42	8.52
25	8.35	8.21	8.39	8.41
50	8.13	8.20	8.32	8.24
100	8.12	8.19	8.25	8.13
200	8.01	8.14	8.23	8.02
400	7.86	8.13	8.19	7.88
600	7.81	8.12	8.18	7.80
South Caspian				
0	8.44	8.35	8.44	8.60
10	8.45	8.33	8.43	8.61
25	8.42	8.33	8.42	8.58
50	8.22	8.30	8.35	8.34
100	8.09	8.17	8.28	8.22
200	8.00	8.05	8.21	8.14
400	7.90	7.98	8.14	8.01
600	7.81	7.96	7.97	8.04
800	7.74	7.93	7.93	7.86

5 Nutrients

The principal source of nutrients (phosphorus, nitrogen, silicon) in the Caspian Sea is represented by the river discharge. In Table 8, we present the mean annual sums of main nutrients supplied to the Caspian Sea with the runoff of the largest rivers. The share of the Volga River in the nutrient runoff to the Caspian Sea (80–90%) is even greater than its contribution to the freshwater runoff (Kosarev, this volume). More than 95% of the nitrogen and phosphorus runoff to the North Caspian is provided by the Volga and Ural rivers. For this reason, the northern part of the sea features the highest biological productivity.

In this section, the principal attention is paid to the mineral forms of nutrients, phosphates, nitrates, nitrites, ammonia, and silicic acid, which form

Table 8 Nutrient supply to the Caspian Sea with the runoff of the major rivers: absolute (*numerator*) and relative (*denominator* percentage of the total runoff)

Nutrient	River Volga	Ural	Terek	Sulak	Samur	Kura	Total runoff
Phosphorus (10^3 t)	34.27/89.4	2.72/7.1	0.57/1.5	0.12/0.3	0.06/0.2	0.59/1.5	38.33/100
Nitrogen (10^3 t)	423.6/88.1	35.8/7.4	9.8/2.0	7.5/1.6	1.1/0.2	3.0/0.7	480.8/100
Silicon (10^6 t)	584.3/80.3	30.3/4.2	66.6/9.2	11.2/1.5	6.1/0.8	29.6/4.0	727.9/100

the abiotic basis of the biological productivity of the Caspian Sea, as well as of the entire World Ocean. The climatic (multiannual mean) proportion of mineral forms in the total supply of phosphorus and nitrogen compounds to the Caspian Sea is 10–20% and 30–45%, respectively. In 1986–1999, the proportion of mineral phosphorus exceeded 30%, while that of mineral nitrogen reached almost 50% [14], which is caused by the fivefold and twofold increases in the volumes of their supply to the Caspian, respectively.

These changes in the nutrient runoff to the Caspian Sea in the 1950s are related to the construction of large reservoirs on the Volga and Kama rivers. At the first stage of the formation of the ecosystems of the Volga–Kama reservoirs (before 1985), the delivery of the mineral forms of nitrogen and phosphorus to the North Caspian decreased, while that of their organic forms increased. The subsequent stage of the evolution of the biochemical processes in these reservoirs was accompanied by a domination of organic matter destruction; therefore, the concentrations of the mineral forms of nitrogen and phosphorus sharply increased, while those of their organic forms decreased somewhat.

The bulk of the mineral nutrients supplied to the sea, in particular, 96% of phosphorus and 98% of nitrogen, precipitates to the bottom sediments of the shallow-water near-mouth areas of the Caspian Sea [19]. The rest of the total amount is spread over the sea area and is utilized by phytoplankton. The strongest decrease in the nutrient concentrations with distance from the river mouths is observed in the North Caspian. As shown in Table 9, in the outer part of the North Caspian shelf, the nutrient contents may be 2–3 times smaller as in the shallow-water near-mouth area of the Volga River. A more significant decrease in the nutrient content is observed in the eastern part of the North Caspian compared with its western part. During the period of the enhanced riverine runoff after 1978, these differences became even more distinct.

The greatest seasonal concentrations of nutrients in the North Caspian are observed in May–June, in the period of the spring flood. The minimum content

Table 9 Concentrations of mineral forms of nutrients (μM) in the shallow-water (*numerator*) and deep-water (*denominator %*) zones of the western and eastern parts of the North Caspian in June and August averaged over intervals of years

Months	Years	Western part				Eastern part			
		PO_4 – P	NH_4 – N	NO_2 – N	SiO_3 – Si	PO_4 – P	NH_4 – N	NO_2 – N	SiO_3 – Si
June	1935–1955 [5]	0.53/0.37	9.63/7.32	–	56/33	0.24/0.23	3.42/4.25	–	41/28
	1961–1970	0.20/0.17	5.55/4.12	0.15/0.10	66/41	0.24/0.16	4.27/3.85	0.06/0.04	60/49
	1971–1977	0.20/0.17	5.93/3.28	0.18/0.07	68/43	0.13/0.13	5.64/3.00	0.07/0.04	54/53
	1978–2003	0.52/0.16	6.61/4.44	0.29/0.11	71/30	0.23/0.07	3.47/3.28	0.06/0.04	44/24
August	1935–1955 [5]	0.38/0.30	9.25/10.0	–	42/19	0.17/0.17	3.88/2.98	–	53/30
	1961–1970	0.20/0.20	5.12/4.92	0.06/0.02	73/32	0.17/0.17	3.20/3.85	0.09/0.02	55/65
	1971–1977	0.10/0.17	4.71/3.71	0.07/0.07	69/39	0.13/0.13	4.64/3.57	0.03/0.01	54/62
	1978–2003	0.35/0.12	5.26/2.72	0.10/0.07	53/25	0.26/0.07	3.61/2.42	0.04/0.04	47/25

of their mineral forms is confined to the low-water season (July–September) during the mass development of phytoplankton, i.e., under the domination of the production over the destruction processes. From June to August, the concentrations of the least stable mineral nutrients (phosphorus and nitrite nitrogen) in the North Caspian may decrease manyfold (Table 9).

The significant increase in the supply of the mineral forms of phosphorus and nitrogen to the Caspian Sea that occurred in the 1980s–1990s resulted in an increase of the concentrations in June by a factor of 1.5–3 (Table 9) and in a 1.6-fold increase of the production of new organic matter [14]. In August, owing to the intensive consumption of mineral nutrients by phytoplankton, their concentrations in the North Caspian did not increase so significantly and even decreased (Table 9).

In the Middle and South Caspian, the horizontal distribution, the vertical structure, and the seasonal variability of mineral nutrients are mostly defined by the biochemical (production–destruction) processes as well as by the advective–diffusive water transport by currents and turbulence. The domination of the production processes (photosynthesis) in the upper 30–50-m layer of the Caspian Sea and that of the destruction processes (decomposition of organic matter) in the underlying layers defines the growth in the nutrient concentrations with depth. In the Middle and South Caspian, the highest nutrient concentrations and their horizontal and vertical homogeneity are observed in the winter, when the production biochemical processes are suppressed, while the advective–diffusive processes, first, convective mixing, are most intensive. The greatest inhomogeneity caused by the decrease in the nutrient concentrations in the euphotic layer to zero values is characteristic of the spring and fall under an opposite combination of biochemical and dynamical factors.

The multifold increase in the delivery of the mineral forms of nutrients to the Caspian Sea in the 1980s–1990s has led to the eutrophication of its euphotic layer and to the related changes in its vertical biological structure in deep-water basins; in the following we assess them for the summertime conditions, which are best characterized by the data from observations.

In the period from 1985 to 1991, the concentrations of mineral phosphorus in the upper 100-m layer of the Middle and South Caspian increased approximately twofold compared with the period 1964–1981; in deeper layers, this growth was approximately 1.5-fold (Table 10). In 1998–2004, a slight decrease in the mineral phosphorus concentrations occurred. Deeper, its high concentrations were retained and became more homogeneous over the entire layer from 300 m to the bottom.

Table 10 Mineral phosphorus concentrations (μM) in the summer season at standard levels in the Middle and South Caspian averaged over intervals of years

Level (m)	Years 1934 [3]	1937–1941 [4]	1958–1963 [6]	1964–1981 [9]	1985–1991	1998–2004
Middle Caspian						
0	0.01	0.23	0.21	0.30	0.54	0.32
10	0.01	–	–	0.23	0.57	0.20
25	0.01	0.23	0.11	0.22	0.40	0.29
50	0.12	0.29	0.21	0.22	0.48	0.32
100	0.35	0.55	0.29	0.41	0.83	0.41
200	0.77	0.71	0.57	0.53	0.80	0.43
300–400	1.13	1.12	0.86	0.78	1.08	1.65
500–600	1.42	1.40	1.28	1.08	1.86	1.66
700–800	1.68	–	1.07	–	1.90	1.65
South Caspian						
0	0.04	0.16	0.41	0.28	0.34	0.15
10	0.04	–	–	0.22	0.37	0.23
25	0.05	0.16	0.27	0.20	0.39	0.23
50	0.07	0.19	0.55	0.28	0.41	0.35
100	0.23	0.32	0.70	0.36	0.60	0.42
200	0.84	0.84	1.05	0.60	0.88	1.06
300–400	1.29	1.19	1.30	0.68	0.87	1.22
500–600	1.80	1.52	1.39	1.06	1.27	1.36
700–800	2.16	–	1.79	1.16	1.38	1.26
900–1000	2.26	–	–	–	–	1.07

Nitrates represent the final form of nitrogen mineralization and the principal form consumed by phytoplankton. If their content in the water is small, consumption of intermediate mineral forms such as ammonia and nitrites increases.

In 1985–1991, as compared with 1979–1981, the concentrations of nitrites, similar to those of phosphates, sharply increased, reflecting the conditions at the end of the period with a low river discharge (Table 11). Meanwhile, in the upper water layer of the South Caspian, they were still close to zero. The greatest growth occurred in the depth range from 200 to 600 m in the North Caspian and from 200 to 800 m in the South Caspian. During recent years, a certain decrease in the nitrate concentrations was observed in the entire water column of the Middle and South Caspian.

Table 11 Nitrate nitrogen concentrations (μM) in the summer season at standard levels in the Middle and South Caspian averaged over intervals of years

Level (m)	Years 1934 [3]	1960–1962 [6]	1979–1981	1985–1991	1998–2004
Middle Caspian					
0	0.00	0.71	0.29	0.54	0.18
10	0.00	0.50	0.28	0.79	0.12
25	0.00	0.36	0.30	0.59	0.28
50	0.36	0.50	0.29	1.95	0.98
100	5.78	2.43	0.47	7.53	2.97
200	10.57	5.57	0.64	10.27	8.18
300–400	9.00	6.07	0.81	14.75	10.96
500–600	4.78	4.64	0.63	14.92	12.66
700–800	–	4.29	–	13.68	8.69
South Caspian					
0	0.00	0.01	0.44	0.00	0.05
10	0.00	0.00	0.41	0.22	0.06
25	0.00	0.00	0.29	0.69	0.17
50	0.00	0.09	0.36	0.93	0.50
100	5.93	1.29	0.69	10.41	2.40
200	11.50	3.00	0.74	15.40	6.47
300–400	11.62	2.71	0.94	17.44	9.04
500–600	4.57	1.71	1.08	16.86	9.51
700–800	0.00	0.71	0.71	18.78	8.30
900–1000	–	–	–	–	6.82

Silicic acid is the most conservative kind of nutrient; however, in the middle of the twentieth century, its concentrations in the Caspian Sea decreased significantly (Table 12) caused by the appearance of the large diatom alga *Rhizosolenia calcar-avis*, which was resettled from the Azov–Black Sea basin in the Caspian Sea. Its universal development, except for the areas of desalinated waters, defined the drop in the silicic acid contents over the entire water column of the deep-water part of the Caspian Sea in 1958–1962.

In 1985–1991, with the decrease in the concentrations of *R. calcar-avis* resulting from the development of small cellular algae, the silicic acid contents increased in all of the layers of the Caspian Sea (Table 12). Meanwhile, in recent years, its contents in the upper 30–50-m layer, as well as the concentrations of phosphates and nitrates, have slightly decreased.

Table 12 Silicic acid concentrations (μM) in the summer season at standard levels in the Middle and South Caspian averaged over intervals of years

Level (m)	Years 1934 [3]	1958–1962 [6]	1964–1981	1985–1991	1998–2004
Middle Caspian					
0	12.4	3.9	13.3	11.8	6.2
10	10.9	6.4	10.3	9.1	7.3
25	13.3	8.1	10.4	12.6	9.4
50	18.5	6.8	13.8	11.9	14.1
100	21.2	13.2	21.8	18.5	18.9
200	32.4	26.2	36.9	24.6	38.7
300–400	53.0	25.0	58.9	47.9	59.9
500–600	91.4	48.8	56.6	58.2	114.2
700–800	–	–	–	61.5	128.6
South Caspian					
0	8.1	6.4	14.4	7.8	4.8
10	7.6	3.7	10.7	8.3	3.6
25	8.8	8.9	11.5	9.3	8.3
50	11.8	6.3	11.4	8.2	7.4
100	19.5	17.2	19.7	12.5	16.3
200	26.8	32.5	37.1	17.8	29.2
300–400	46.9	53.6	61.3	25.8	54.8
500–600	77.4	66.2	55.9	36.9	63.7
700–800	97.9	71.4	52.7	38.6	80.8
900–1000	–	–	–	–	78.5

The low concentrations of all the mineral nutrients considered in the upper euphotic layer of the Caspian Sea observed during recent years are related to their intensive consumption by phytoplankton. This was favored by the extraordinary development of the comb jelly *Mnemiopsis*, which appeared in the Caspian Sea in 2000 and consumed up to 80% of zooplankton. This led to a shortening of the trophic web and acceleration of the biochemical cycling of nitrogen and phosphorus in the euphotic layer (Karpinsky, this volume). The drop in the concentrations of mineral forms of nutrients in the surface layer of the Caspian Sea predefined the development of blue-green and pyrophyta small cellular algae capable of successful development under low nutrient contents in the water.

During the past few decades, the strengthening of the dynamical isolation of the euphotic layer from the underlying waters owing to the intensification of the summer thermocline (Tuzhilkin and Kosarev, this volume) might represent an additional factor for the decrease in the surface concentrations of mineral nutrients in the deep-water areas of the Caspian Sea. This favored the weakening of the vertical diffusive flux of nutrients from the near-bottom layer to the surface of the sea.

The lower nitrate concentrations in the near-bottom layers of the Middle and South Caspian compared with the concentrations in the intermediate waters (Table 11) are related to the deficiency of dissolved oxygen required for the oxidation of ammonia and nitrates to nitrites [3]. During the period of the low levels of Caspian Sea (1960s–1970s) and the enhanced ventilation of its deep waters, this effect as well as the related intermediate maximum and near-bottom minimum of nitrate contents were poorly expressed. At present, the vertical distribution of nitrates is close to that characteristic of the 1930s (Table 11). The difference lies in the deeper present-day position of the lower boundary of the nitrate maximum layer (600 m in the Middle Caspian and 800 m in the South Caspian from 1985 to 2004 instead of 400 m in 1934).

On the whole, one can state that the present-day condition of the nutrient regime in the Caspian Sea, the background values, and the vertical distributions are close to those observed in the 1930s [3] and significantly differ from the conditions characteristic of the 1960s–1970s [6–9]. Therefore, one can distinguish two types of multiannual regime of nutrients in line with the previously described characteristics of the natural chemical conditions of the Caspian Sea waters.

6 Conclusion

The generalization of the results of multiannual ship observations of the natural chemical regime of the Caspian Sea waters presented in this chapter allows us to draw the following conclusions:

- The salt composition of the Caspian Sea waters strongly differs from that of the waters of the World Ocean by higher (by a factor of 2–5) ratios of ions of Ca, Mg, SO_4, and HCO_3 to Cl and by significantly smaller homogeneity in space and time.
- The spatial inhomogeneity and multiannual instability of the natural hydrochemical regime of the Caspian Sea waters represent the most prominent distinguishing feature.
- The near-mouth areas of the Caspian Sea essentially differ from its open regions by the enhanced values of all the chemical characteristics considered (dissolved oxygen content, pH, nutrient concentrations) and by the especially strong dependence of their seasonal and multiannual variability on the volumes and chemical composition of the river discharge.
- The chemical regime of the waters of the deep-sea areas of the Caspian Sea is characterized by a greater role of the biochemical (production–destruction) and dynamical (advection–diffusion) processes; meanwhile, even here, the supply of matter with the river discharge is the principal reason of its long-term changes.
- During the twentieth century, an alternation of two types of natural chemical regime was observed in the Caspian Sea; it is manifested in the multiannual variability of all the characteristics assessed, especially in the deep and near-bottom layers.
- One of these types was observed under the conditions of an enhanced river discharge at the beginning and at the end of the twentieth century (before the mid-1930s and after the 1970s); it was characterized by reduced near-bottom concentrations of dissolved oxygen (down to the hypoxy level) and decreased nutrient contents at the surface accompanied by increased contents in the deep and near-bottom layers.
- The other type of hydrochemical regime observed in the middle of the twentieth century reached its peak development in the 1960s–1970s under the conditions of a minimum river discharge; it was characterized by an opposite combination of the anomalies of the aforementioned characteristics: increased dissolved oxygen concentrations and decreased nutrient contents in the near-bottom layer.
- The significant amplitudes and long-term character of the variations in the hydrochemical regime of the Caspian Sea correspond to the multiannual changes in the thermohaline structure of its waters (Tuzhilkin and Kosarev, this volume) and cause obvious aftereffects in the biotic components of the Caspian ecosystem.

References

1. Lebedintsev AA (1913) Transactions of the Caspian expedition of 1904. St Petersburg (in Russian)
2. Knipovich NM (1921) Transactions of the Caspian expedition of 1914–1915. Petrograd (in Russian)
3. Bruevich SV (1937) Hydrochemistry of the Middle and South Caspian. AN SSSR, Moscow (in Russian)
4. Abramov BN (1959) Trans VNIRO 38:117
5. Vinetskaya NI (1962) Trans KaspNIRO 18:4
6. Pakhomova MV, Zatuchnaya BM (1966) Hydrochemistry of the Caspian Sea. Gidrometeoizdat, Leningrad (in Russian)
7. Kosarev AN (1975) Hydrology of the Caspian and Aral seas. Mosk Gos Univ, Moscow (in Russian)
8. Baidin SS, Kosarev AN(eds) (1986) The Caspian Sea. Hydrology and hydrochemistry. Nauka, Moscow (in Russian)
9. Terziev FS, Maksimova MP, Yablonskaya EA(eds) (1996) Hydrometeorology and hydrochemistry of the seas, vol 6. The Caspian Sea, issue 2. Hydrodynamical conditions and oceanological background of the formation of the biological productivity. Gidrometeoizdat, St. Petersburg (in Russian)
10. Sapozhnikov VV (1996) Okeanologiya 36:148 (in Russian)
11. Skorokhod AI, Tsytsarin AG (1995) Trans GOIN 63 (in Russian)
12. Skorokhod AI, Tsytsarin AG (1995) Russ Meteorol Hydrol (1):101 (in Russian)
13. Skorokhod AI, Tsytsarin AG (1996) Russ Meteorol Hydrol (1):76 (in Russian)
14. Katunin DN, Khripunov IA, Bespartochnyi NP, Nikotina LN, Galushkina NV, Radovanov GV (2000) Caspian Float Univ Sci Bull 1:111 (in Russian)
15. Sapozhnikov VV, Katunin DN, Bespartochnyi NP, Kirpichev NB, Luk'yanova ON, Metreveli MP (2002) Okeanologiya 42:634 (in Russian)
16. Sapozhnikov VV, Katunin DN, Kirpichev NB, Luk'yanova ON, Muryi GP, Fesenko VI (2003) Okeanologiya 43:529 (in Russian)
17. Sapozhnikov VV, Belov AA (2003) Okeanologiya 43:368
18. Sverdrup H, Johnson M, Fleming R (1942) The Oceans. Their physics, chemistry and general biology. Wiley, New York
19. Maksimova MP, Katunin DN, Eletskii BD (1978) Okeanologiya 18:454 (in Russian)

Hdb Env Chem Vol. 5, Part P (2005): 109–142
DOI 10.1007/698_5_006

Published online: 13 October 2005

Pollution of the Caspian Sea

Alexander Korshenko[1] (✉) · Alvin Gasim Gul[2]

[1]State Oceanographic Institute, Kropotkinsky per. 6, 119034 Moscow, Russia
korshenko@mail.ru

[2]Institute of Space Research of Natural Resources of Azerbaijan, Azadlig ave., AZ1116 Baku, Azerbaijan

Abstract The pollution of Caspian Sea waters and bottom sediments was described on the basis of long-term monitoring data from 1978 to 2004. It was shown that in the 1980s total petroleum hydrocarbons were in high concentration in the estuarine waters of the Ural River, in the western part of the North Caspian, and near the town of Izberbash on the Dagestan coast, but later they reduced drastically everywhere. In contrast to water in the bottom sediments, zones of high concentrations of petroleum hydrocarbons were located in shallow areas adjacent to the Volga delta and in the western part of the North Caspian, where fine sediments occurred. The other part of the North Caspian, with mainly coarse sediments, had rather clean bottom material. The high concentrations of phenols and detergents in the water occurred rather often at the beginning of the period but later decreased significantly. No seasonal trends and spatial features were observed in their distribution. The average concentration of ammonium was high in the estuarine areas of the Volga, Terek, and Sulak rivers. In water, among the chlorinated pesticides DDT dominated. All pesticides were much more abundant in the 1980s than in the 1990s. In the bottom sediments, pesticides accumulated near the Volga River delta and were practically absent in the central part of the North Caspian. The heavy metal concentration was in the range of regional background levels. In general, for all kinds of pollution studied for this area, there were marked small-scale patches in time and space of very high concentrations.

Keywords Detergents · Heavy metals · Monitoring · Pesticides · Petroleum hydrocarbons · Phenols · Pollution · Spatial and temporal variation

1 Introduction

Investigations into the pollution of the Caspian Sea, mainly of the petroleum hydrocarbons and their influence on the sea and coastal communities, began in the middle of the last century. But the systematic observations of dynamics of pollutants both in coastal areas of the sea, mostly subjected to anthropogenic influence, and in the open sea began only at the end of the 1970s after organization of a marine environment monitoring system, conducted by Hydrometeoservice of the Soviet Union. All chemical analysis was conducted in accordance to the standard guidance manuals [1, 2]. On the basis of the data from the State System of Monitoring (OGSN), assessments of the level of water pollution and description of seasonal and interannual variation were conducted each year [3–5]. The monitoring data became the basis of a general investigation of long-term balance and prediction of pollution of the Caspian Sea by petroleum hydrocarbons [6]. Detailed material of long-term scientific investigations of hydrological and morphological processes and the dynamics of polluting substances and sea level fluctuations on the increase pollution is described in monograph [7]. It gives information about the drainage and concentration of pollutants in the river water and in the suspended matters, their changes in time, and their influence on the condition of the North Caspian. The current data on hydrological processes and water pollution of the North Caspian during the winter period are also described in other articles [8–13].

2 Water Pollution of the North Caspian

In the period from 1978 to 1992 control of the water pollution of the North Caspian was conducted in the framework of the Soviet Union national program of seawater monitoring at the stations of five standard cross-sections (see Tuzhilkin and Kosarev, this volume). Besides temperature, salinity, and the concentration of dissolved oxygen and nutrients, determination of concentrations of ammonium, petroleum hydrocarbons, phenols, detergents, chlorinated pesticides (since 1985), and also heavy metals – mercury (since 1986), zinc and copper (since 1989) and lead (since 1990) was included. On the whole 3174 stations were employed. After stopping of the permanent monitoring investigations of water pollution in the shallow area of the North Caspian in 1992, this work was conducted by different expeditions in 1993–1996 and 2000–2002. The programs of the expeditions and the set of parameters of monitoring varied significantly, however, the data make it possible to describe the time dynamics and space distribution of pollution in water and bottom sediments. In the first expeditions and in 2000 the work was conducted in the summer period adjacent to the delta of the Volga River. The expeditions in 2001 and 2002 covered both the estuarine part and the central part of the North Caspian. On the whole 444 stations were employed.

Total Petroleum Hydrocarbons

According to the monitoring data before 1992 the total petroleum hydrocarbons (TPHs) varied over a wide range and reached 3.78 mg/L in October 1984 at cross-section IIIa in the western part of the shallow area of the North Caspian. The average concentration of TPHs in the water column during this period was rather high and equal to 0.19 mg/L. The particularly high average concentration was observed before 1985. The later level of TPH pollution decreased considerably and mainly varied in the range 0.1–0.2 mg/L (Fig. 1). The spatial distribution of TPHs was patchy. The region in the northeastern part of North Caspian near the estuarine of the River Ural was the most polluted. The average values reached 1.2 mg/L. The level of pollution in the southwestern part of the investigated area near the border with coastal waters of Dagestan also increased. The pattern of seasonal variations of TPHs is not revealed (Table 1).

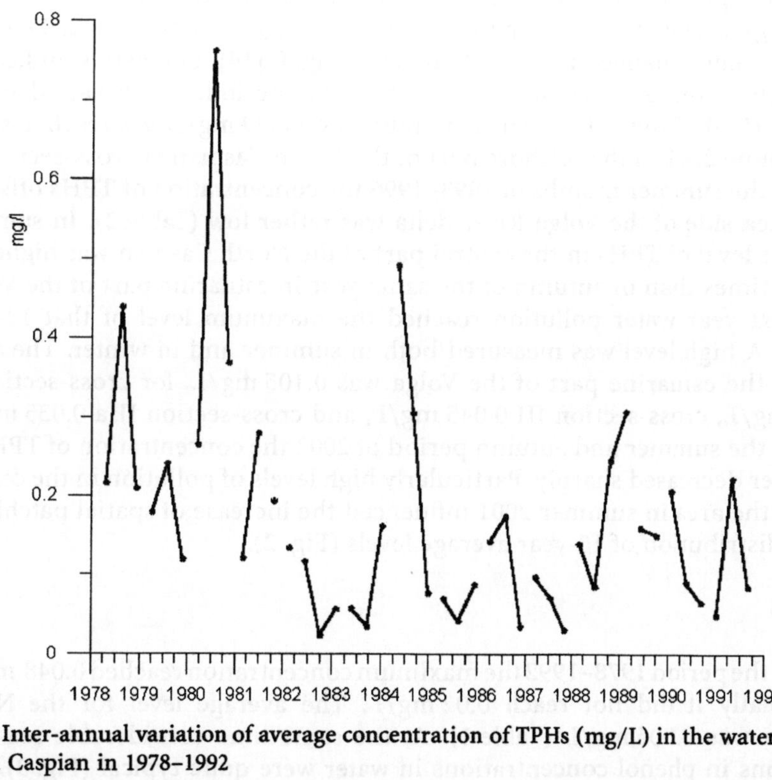

Fig. 1 Inter-annual variation of average concentration of TPHs (mg/L) in the waters of the North Caspian in 1978–1992

Table 1 Average monthly concentration of pollutants in the waters of the North Caspian in 1978–1992

Month	TPHs (mg/L)	Phenols (mg/L)	Detergents (mg/L)	NH_4 (µg/L)
February	0.11	0.002	0.041	54.8
March	0.31	–	0.016	21.4
April	0.28	0.005	0.049	43.8
May	0.05	0.007	0.039	21.9
June	0.14	0.007	0.066	35.6
July	0.13	0.002	0.032	14.3
August	0.25	0.005	0.071	53.1
September	0.29	0.009	0.063	44.3
October	0.13	0.005	0.042	25.8
November	0.08	0.005	0.028	8.8
December	0.35	0.001	0.026	–
Average annual concentration	0.19	0.005	0.054	37.2

In the period of 1993–2002 the average concentration of TPHs was 0.067 mg/L. This data corresponds to the investigations of 2001–2003 where average values changed from 0.029 to 0.073 mg/L [14], but were significantly less than KaspNIRKh estimation for the same period, which varied in the range 0.15–0.35 mg/L [15]. The maximum level 0.29 mg/L was reached at the end of June 2001 in the offshore part of the North Caspian at cross-section II. During the summer months of 1993–1996 the concentration of TPHs offshore of the sea side of the Volga River delta was rather low (Table 2). In summer 2000 the level of TPHs in the central part of the North Caspian was higher by several times than in autumn of the same year in estuarine part of the Volga. The next year water pollution reached the maximum level of that 10-year period. A high level was measured both in summer and in winter. The average for the estuarine part of the Volga was 0.105 mg/L, for cross-section II 0.267 mg/L, cross-section III 0.045 mg/L, and cross-section IIIa 0.035 mg/L. During the summer and autumn period of 2002 the concentration of TPHs in the water decreased sharply. Particularly high levels of pollution in the central part of the area in summer 2001 influenced the increase of spatial patchiness of the distribution of 10-year average levels (Fig. 2).

Phenols

During the period 1978–1992 the maximum concentration reached 0.048 mg/L, but usually it did not reach 0.02 mg/L. The average level for the North Caspian was 0.005 mg/L (Table 1). At the same time considerable seasonal variations in phenol concentrations in water were quite typical (Fig. 3). The

Table 2 Average concentration of pollutants in the waters of the North Caspian in 1993–2002

Pollutant (unit of concentration)	Year, Month, Region									
	1993 June Seaside of the Volga delta	1994 June Seaside of the Volga delta	1995 July–August Seaside of the Volga delta	1996 July–August Seaside of the Volga delta	2000 July North Caspian	2000 November Seaside of the Volga delta	2001 June North Caspian	2001 December Seaside of the Volga delta	2002 August Seaside of the Volga delta	2002 October North Caspian
TPHs (mg/L)	0.049	0.050	0.021	0.020	0.121	0.027	0.151	0.109	0.010	0.008
Phenols (mg/L)	–	–	–	–	–	0.003	–	0.003	–	0.008
Detergents (mg/L)	–	–	–	–	–	0.032	–	0.033	–	Less than DL*
NH_4, (µg/L)	490.3	203.7	204.8	–	192.6	–	701.2	74.7	30.8	14.0
DDT (ng/L)	0.300	0.353	0.175	0.110	–	–	–	–	–	0.244
DDE (ng/L)	0.050	0.073	0.043	0.025	–	–	–	–	–	0.010
DDD (ng/L)	0.069	0.090	0.020	0.007	–	–	–	–	–	less than DL**

*Detection limit 0.025 mg/L
**Detection limit 0.05 ng/L

Table 2 continued

Pollutant (unit of concentration)	Year, Month, Region									
	1993 June Seaside of the Volga delta	1994 June Seaside of the Volga delta	1995 July–August Seaside of the Volga delta	1996 July–August Seaside of the Volga delta	2000 July North Caspian	2000 November Seaside of the Volga delta	2001 June North Caspian	2001 December Seaside of the Volga delta	2002 August Seaside of the Volga delta	2002 October North Caspian
α-HCH (ng/l)	0.070	0.108	0.080	0.029	–	–	–	–	–	3.460
γ-HCH (ng/L)	0.030	0.034	0.025	0.006	–	–	–	–	–	1.394
Fe (μg/L)	–	–	–	–	–	35.6	–	32.2	–	24.0
Mn (μg/L)	–	–	–	–	–	3.6	–	3.0	–	1.3
Zn (μg/L)	–	–	–	1.1	–	5.6	–	4.5	–	7.1
Ni (μg/L)	1.3	–	0	0	–	2.1	–	2.1	–	2.4
Cu (μg/L)	1.2	1.2	0.1	0.1	–	3.1	–	2.8	–	4.3
Pb (μg/L)	0.15	0.1	0.1	0.1	–	4.9	–	4.1	–	0.5
Cd (μg/L)	0.1	0.1	0.1	0.1	–	0.7	–	0.5	–	0.2

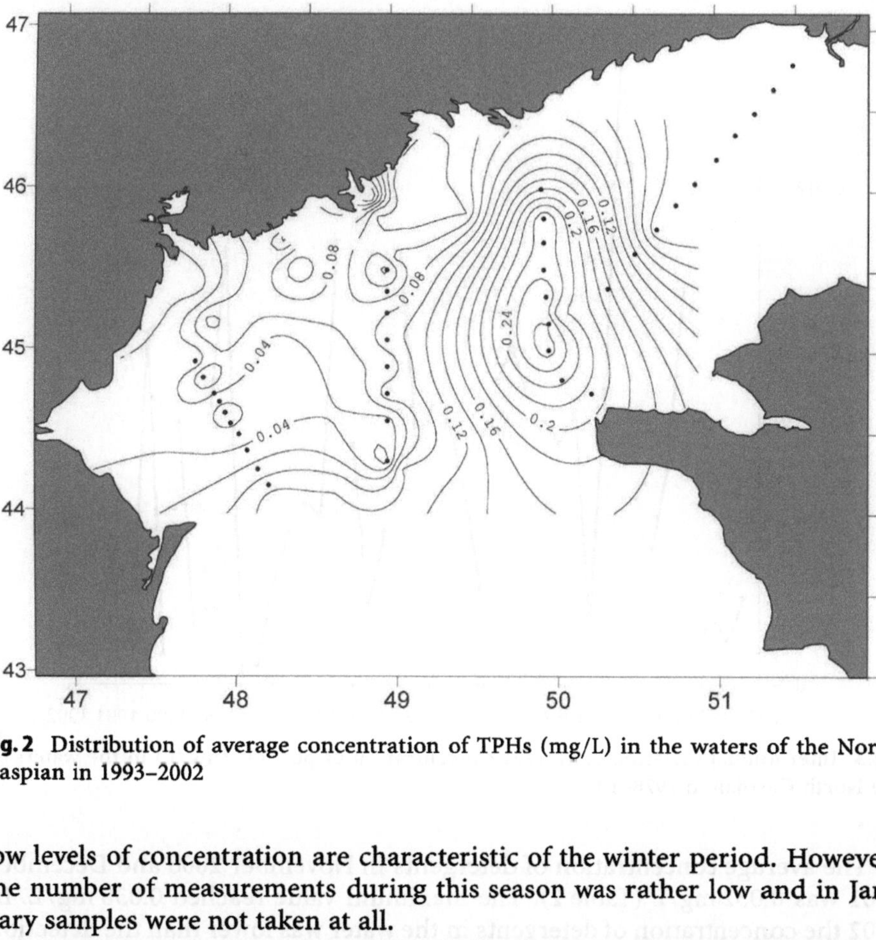

Fig. 2 Distribution of average concentration of TPHs (mg/L) in the waters of the North Caspian in 1993–2002

low levels of concentration are characteristic of the winter period. However, the number of measurements during this season was rather low and in January samples were not taken at all.

In autumn and winter 2000–2002 (Table 2) the average concentration of phenols in water was 0.005 mg/L. This coincides with the results of long-term monitoring. The maximum levels (0.017–0.018 mg/L) were measured in 2002 in the western part of the North Caspian in the zone of influence of the Volga discharge. On the whole, the level of the water pollution by phenols in the North Caspian is rather high.

Detergents

From 1978 till 1992 detergents was observed in 1745 samples taken in the frame of the monitoring programme in the waters of the North Caspian. The average concentration in the explored territory was 0.054 mg/L (Table 1). The higher values were measured in July–September, and the lower values in the cold period of the year. The highest values are more typical for the western part of the North Caspian.

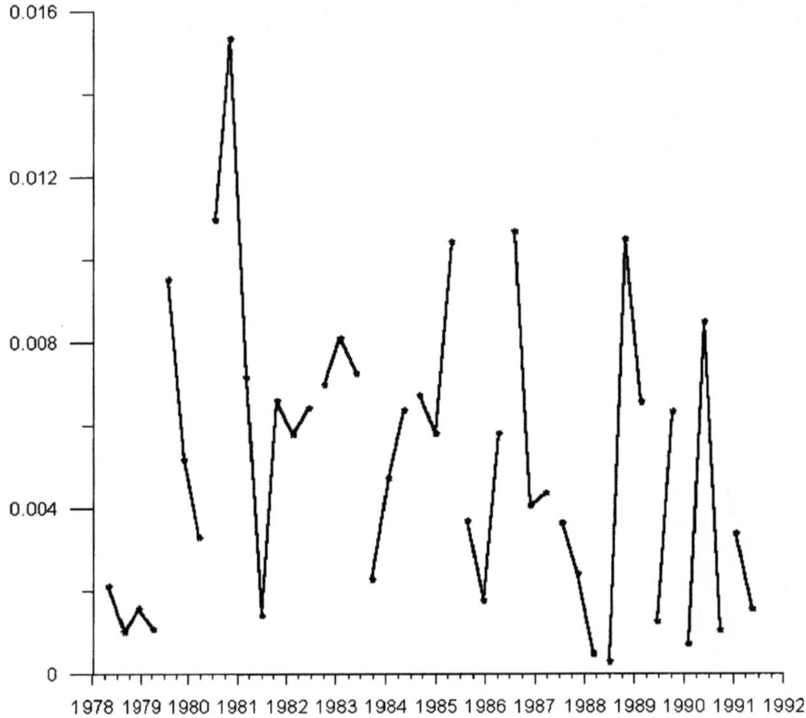

Fig. 3 Inter-annual variation of average concentration of phenols (mg/L) in the waters of the North Caspian in 1978–1992

The average concentration of detergents in November 2000 and December 2001 was 0.032 mg/L (Table 2). The maximum value reached 0.038 mg/L. In 2002 the concentration of detergents in the water was lower than the detection limit of the analytical method.

Ammonium

Until 1992 the average concentration of NH_4 was 37.2 μg/L and the maximum value was 897 μg/L. The higher values are typical for waters adjacent to the Volga delta (avandelta) in its western part. In the seasonal and interannual dynamics there is no clear regularity (Fig. 4, Table 1).

After 1992 the average concentration of NH_4 was 180.1 μg/L and was much higher than in the previous two decades (Table 2). The maximum value (2680 μg/L) was registered in August 2000 in the near-bottom layer in the western part of the North Caspian at cross-section IIIa. At the same time the average concentration of NH_4 in the surface layer (201.5 μg/L) was almost the same as in the near-bottom layer (233.5 μg/L). A slight difference in the vertical distribution of ammonia in the North Caspian was

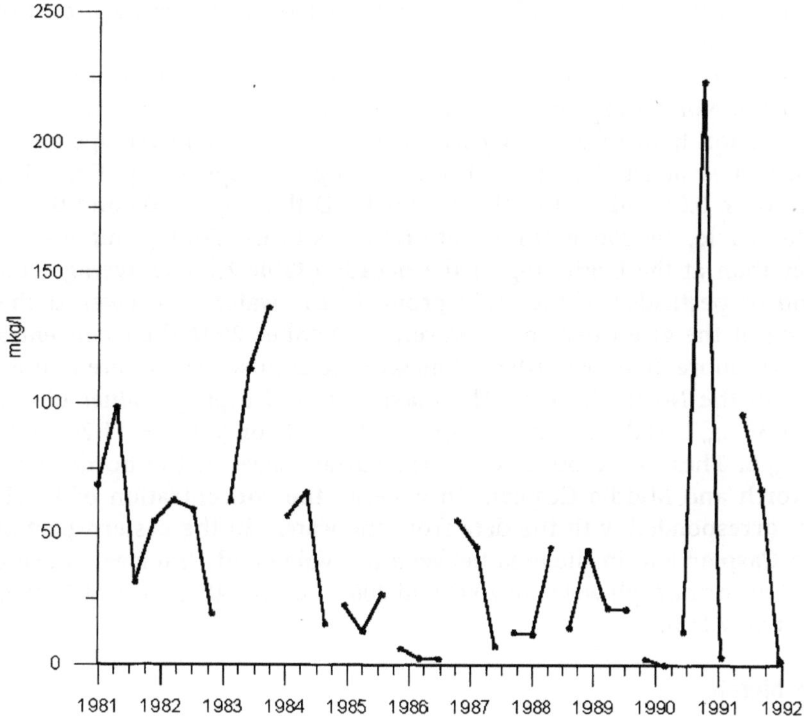

Fig. 4 Inter-annual variation of average concentration of ammonia (μg/L) in the waters of the North Caspian in 1978–1992

also observed in other years: in 2001 the average concentration in the surface layer was 510.5 μg/L and in the near-bottom layer 429.9 μg/L, in 2002 the values were 17.6 and 20.7 μg/L, respectively. This suggests that intensive mixture by wind in the shallow waters prevents stable vertical stratification of ammonia distribution. In general, the highest values occurred more often near the Volga delta and in the western part of the North Caspian.

Pesticides

From 1978 till 1992, of the chlorinated pesticides, DDT gave the highest concentration in water. The average value was 31 ng/L and the maximum value was 1950 ng/L in April 1985 in the eastern part of the North Caspian. DDE concentration was a great deal less than DDT and its average concentration was 1 ng/L. The maximum value (27 ng/L) was registered in June 1985 in the middle part of the area. The average concentration of α-HCH was 2 ng/L and the maximum value (85 ng/L) was measured in June 1991 in the eastern part of the North Caspian. For γ-HCH the average was 1 ng/L and the maximum (264 ng/L) was observed in April 1985 in the eastern part of the North Caspian.

After 1992 the concentration of pesticides from the DDT group in the waters of the North Caspian was rather low and sometimes it was lower than the detection limit of the analytical method. The decade average and maximum values of DDT were 0.23 and 0.92 ng/L, respectively. For DDE the values were 0.03 and 0.21 ng/L, and for DDD they were 0.03 and 0.32 ng/L. On the whole, the concentration of pesticides of the DDT group was a little higher than at the beginning of the decade (Table 2). The average concentration of pesticides of the HCH group in the water was lower at the beginning of the given period. However, in October 2002 their concentration increased more than two orders of magnitude almost everywhere in the open waters of the North Caspian. The maximum and average values of α-HCH were 1.38 ng/L and 0.55 ng/L, respectively, and for γ-HCH 11.20 ng/L and 4.85 ng/L. These were observed in the surface layer at the border between the North and Middle Caspian. In general, the concentration of HCH isomers corresponded with the data from the waters in the eastern part of the North Caspian and in the area between the Volga and Ural rivers during the period of spring high water in 2003 and 2004, i.e., 10–40 ng/L and 7–15 ng/L, respectively [16].

Heavy Metals

Concentrations of metals in the water of the North Caspian was determined episodically in the period 1986–1991. The average value of mercury in the water was practically the same in the central and western parts of the area: 0.15–0.18 μg/L. The maximum value of 0.80 μg/L was registered in the middle of October 1986 at the most northerly station of cross-section IIIa. The average concentration of zinc and copper in the northwestern part of the North Caspian in 1989–1990 was 0.033 and 0.005 μg/L and the maximum value was 0.186–0.019 μg/L. The average concentration of lead in 1990 was 3.43 μg/L.

During the next decade (Table 2) the maximum concentration of heavy metals in the waters of the North Caspian were: iron 52 μg/L, manganese 5 μg/L, zinc 10.6 μg/L, nickel 4 μg/L, copper 6.2 μg/L, lead 8 μg/L, and cadmium 1 μg/L. On the whole, pollution of the waters of the central part of the North Caspian by heavy metals was much lower than in the northeastern part of the area in 2003–2004 [16]. Patches with high concentrations of heavy metals were not observed and the highest values of metals occurred in shallow waters both adjacent to the Volga delta and offshore. Temporal variation was not marked.

3 Bottom Sediments Pollution of the North Caspian

Total Petroleum Hydrocarbons

In bottom sediments near the delta of the Volga and shallow areas of the North Caspian the average concentration of TPHs in 1993–2002 was 13.6 μg/g and corresponded to background concentrations (from 2.81 to 18.56 μg/g) for the different natural complexes of the Russian part of the North Caspian in 2003 [17]. Single high values of TPHs in bottom sediments (higher than 35 μg/g) were registered repeatedly in navigable and fish-passing channels in avandelta of the Volga. It was here that in 1994 the maximum concentration (76.0 μg/g) was registered and the average value of this part was 15.1 μg/g (Table 3). The high level of pollution of bottom sediments by TPHs near the Volga delta and low background concentration in the rest of the vast middle part of the North Caspian caused considerable spatial heterogeneity in the distribution of average values and the appearance of local patches

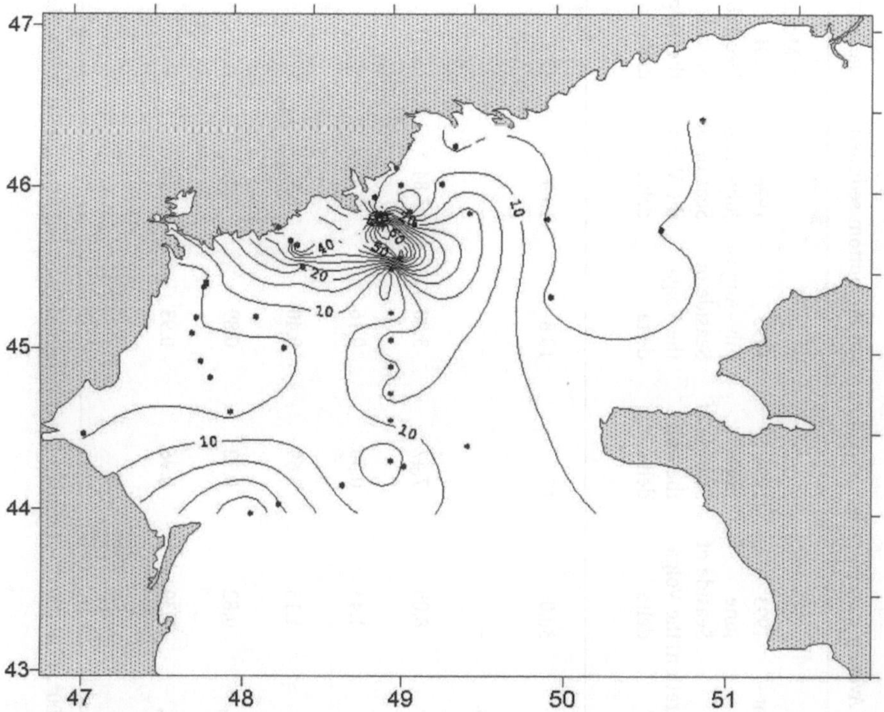

Fig. 5 Distribution of average concentration of TPHs (mg/g) in the bottom sediments of the North Caspian in 1993–2002

Table 3 Average concentration of pollutants in the bottom sediments of the North Caspian in 1993–2002

Pollutant (unit of concentration)	Year, Month, Region								
	1993 June Seaside of the Volga delta	1994 June Seaside of the Volga delta	1995 July–August Seaside of the Volga delta	1996 July–August Seaside of the Volga delta	2001 December Seaside of the Volga delta	2002 August Seaside of the Volga delta	2002 October North Caspian	CEP-2000* October Russian part of the North Caspian	CEP-2001* September Kazakhstan part of the North Caspian
TPHs, (μg/g)	30.0	17.9	10.8	10.0	15.4	14.0	4.8	1.57	11.56
Phenols, (μg/g)	–	–	–	–	–	–	0.49	–	–
DDT, (ng/g)	8.05	7.47	3.80	3.84	–	–	0.07	0.317	4.274 (pg/g)
DDE, (ng/g)	1.43	0.36	0.16	0.25	–	–	0.02	0.246	4.147 (pg/g)
DDD, (ng/g)	1.15	2.69	0.40	0.52	–	–	0.04	0.258	4.718 (pg/g)
α-HCH, (ng/g)	0.82	0.22	0.89	0.29	–	–	0.01	0.016	1.903 (pg/g)
γ-HCH, (ng/g)	5.30	0.96	0.93	0.70	–	–	0	0.173	2.268 (pg/g)

CEP-2000*, CEP-2001* – Caspian Environmental Program expeditions in the North Caspian [18].

Table 3 continued

Pollutant (unit of concentration)	Year, Month, Region								
	1993 June Seaside of the Volga delta	1994 June Seaside of the Volga delta	1995 July–August Seaside of the Volga delta	1996 July–August Seaside of the Volga delta	2001 December Seaside of the Volga delta	2002 August Seaside of the Volga delta	2002 October North Caspian	CEP-2000* October Russian part of the North Caspian	CEP-2001* September Kazakhstan part of the North Caspian
Fe (μg/g)	–	–	–	–	2723	–	3541	5873	6087
Mn (μg/g)	–	–	–	–	44.6	640.4	31.1	–	192.3
Zn (μg/g)	–	–	–	–	4.85	–	8.71	19.57	9.64
Ni (μg/g)	–	–	–	–	2.71	34.6	5.52	16.01	9.37
Cu (μg/g)	9.0	10.3	13.8	13.6	–	56.9	10.3	9.19	5.13
Pb (μg/g)	1.45	1.47	1.84	1.95	0.90	1.60	1.90	4.62	5.60
Cd, (μg/g)	0.05	0.49	0.09	0.09	0	0	0.081	0.060	0.047

*CEP-2000 and CEP-2001 Caspian Environmental Program expeditions in the North Caspian [18].

of high pollution (Fig. 5). Besides, slightly increased values were found in bottom sediments near the Dagestan shelf and in the eastern part of the North Caspian, according to the data from expeditions of the Caspian Environmental Program (Table 3). In the rest of the area the bottom sediments were not polluted by TPHs considerably. This coincides with the data showing decreasing pollutant concentrations in the sediments with predominating coarse fractions, which occupy the central part of the North Caspian [9]. The increased level of pollution in the shallow area adjacent to the Volga delta was evidently connected with the sinking to the bottom of thin fractions of suspended matter with adsorbed petroleum hydrocarbons. During storm mixing of the water, the thin fractions of bottom sediments are stirred-up and carried away by dominating western near-shore currents to the border with the Dagestan shelf. The spatial distribution of the zones with high concentrations of TPHs does not coincide in the water and in the bottom sediments.

Phenols

In Autumn 2002 the concentration of phenols in the bottom sediments varied from 0.183 to 0.888 μg/g and the average number was 0.489 μg/g. The increased values were observed in the central and western parts of the North Caspian.

Pesticides

The concentration of chlorinated pesticides in the bottom sediments in the North Caspian was rather high in 1993–1996. The average values of the DDT group was 5.89 ng/g. However, in autumn 2002 their concentration was rather low and the average value of DDT was only 0.07 ng/g and close to the detection limit of analytical method (Table 3). The average and maximum values of DDT in the North Caspian bottom sediments during the decade were 4.47 and 23.40 ng/g, respectively. Those for DDE were 0.31 and 3.20 ng/g, and those for DDD 1.18 and 11.00 ng/g. On the whole, the concentration of pesticides in the bottom sediments was in the range of values used in the QUASIMEME programme introduced for intercalibration of analytical laboratories [19]. The maximum values of concentration of all pesticides of the DDT group were observed at the end of June 1994, close to the sea side of the Volga delta. In general, the highest values of concentrations were observed in the shallow waters adjacent to delta where the polluted water of the river enters (Fig. 6). Upstream in the Volga delta the average values in 1993–1996 were 12.09 ng/g (DDT), 0.98 ng/g (DDE), 1.67 ng/g (DDD), 1.01 ng/g (α-HCH), and 2.77 ng/g (γ-HCH). These data correspond to the DDT concentrations observed in the water passages of the Volga in 1994, which were 4.4–26.3 ng/g [7].

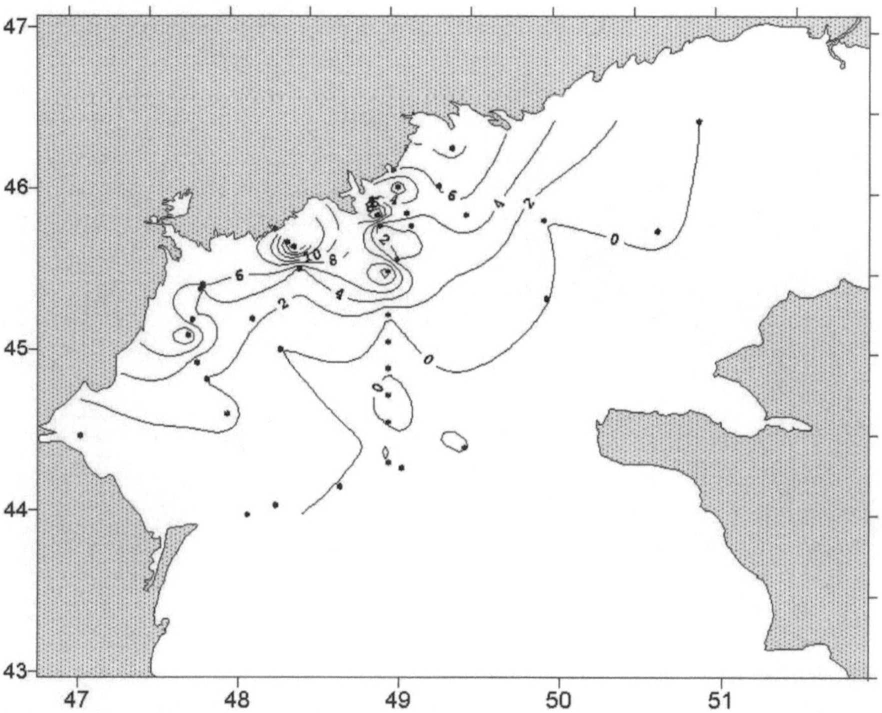

Fig. 6 Distribution of average concentration of DDT pesticides (ng/g) in the bottom sediments of the North Caspian in 1993–2002

The average concentration of the HCH group was rather high in 1993–1996; however, in October 2002 it only reached the detection limit of the analytical method in some samples (Table 3).

The maximum concentration of α-HCH was 3.45 ng/g in August 1995 in the western part of the area at the cross-section IIIa. The highest concentration of γ-HCH (14.20 ng/g) was registered in June 1993 in the part adjacent to the Volga delta. Similarly to DDT pesticides, the patches of high concentration of the HCH group were observed near the Volga delta (Fig. 7).

Heavy Metals

The concentration of heavy metals in the bottom sediments in the North Caspian was determined during the expeditions in 1993–1996 and in 2001–2002 (Table 3). The average concentration of iron was 3207 μg/g and a maximum value (13 900 μg/g) was registered in autumn 2002 in the offshore part of the area at cross-section IV. The main part of the iron data fell in the range from 930 to 5150 μg/g. On the whole, the iron concentration was higher in the offshore part of the area. The average concentration of manganese

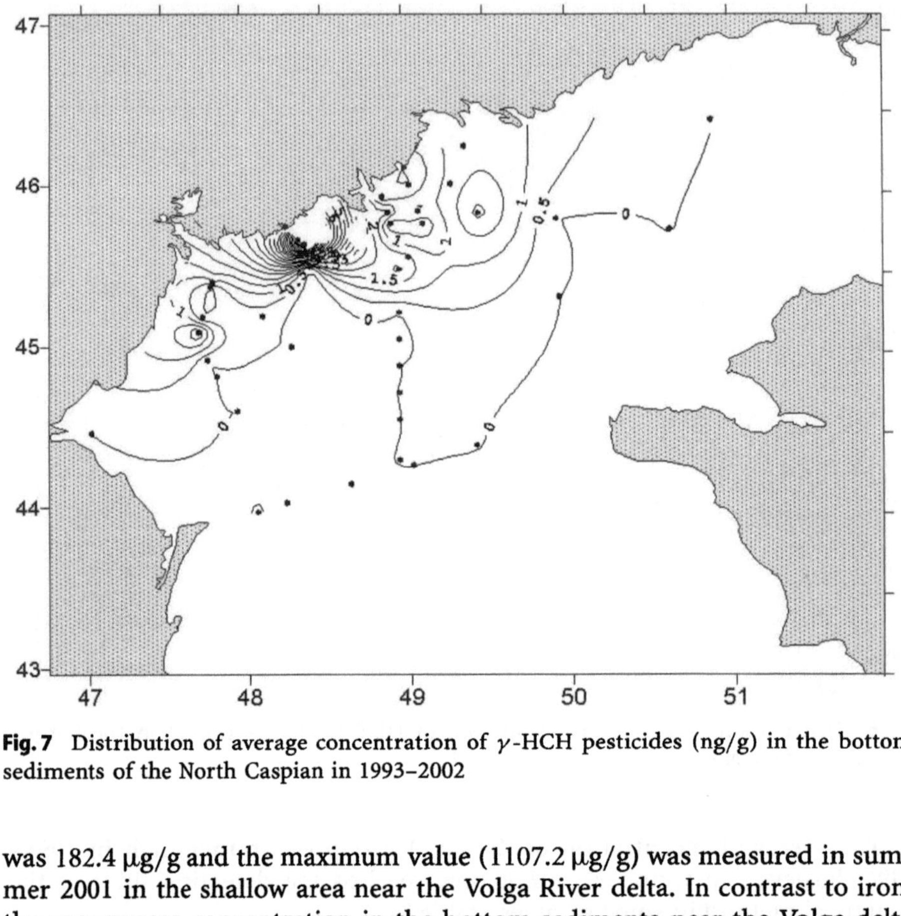

Fig. 7 Distribution of average concentration of γ-HCH pesticides (ng/g) in the bottom sediments of the North Caspian in 1993–2002

was 182.4 μg/g and the maximum value (1107.2 μg/g) was measured in summer 2001 in the shallow area near the Volga River delta. In contrast to iron, the manganese concentration in the bottom sediments near the Volga delta was more than ten times higher than in offshore part of the North Caspian (Fig. 8).

The average concentration of zinc was 7.13 μg/g and the highest value (33.8 μg/g) was registered in the offshore part of the North Caspian.

The average concentration of nickel was 11.7 μg/g and the maximum value (62.1 μg/g) was measured in the shallow area near the Volga River delta. In general, the concentration of nickel with an average of 34.6 μg/g was much higher in bottom sediments close to the delta than in the offshore areas of the North Caspian. The same feature characterizes the copper distribution. The bottom sediments near navigable waterway and fish passages were much more polluted by copper than in offshore areas of the North Caspian. The average value was 16.0 μg/g and maximum was 74.0 μg/g measured in August 2002 near the Volga delta.

The average concentration of lead was 1.61 μg/g and the maximum value (6.18 μg/g) was observed in June 1994 in the vicinity of the island Small

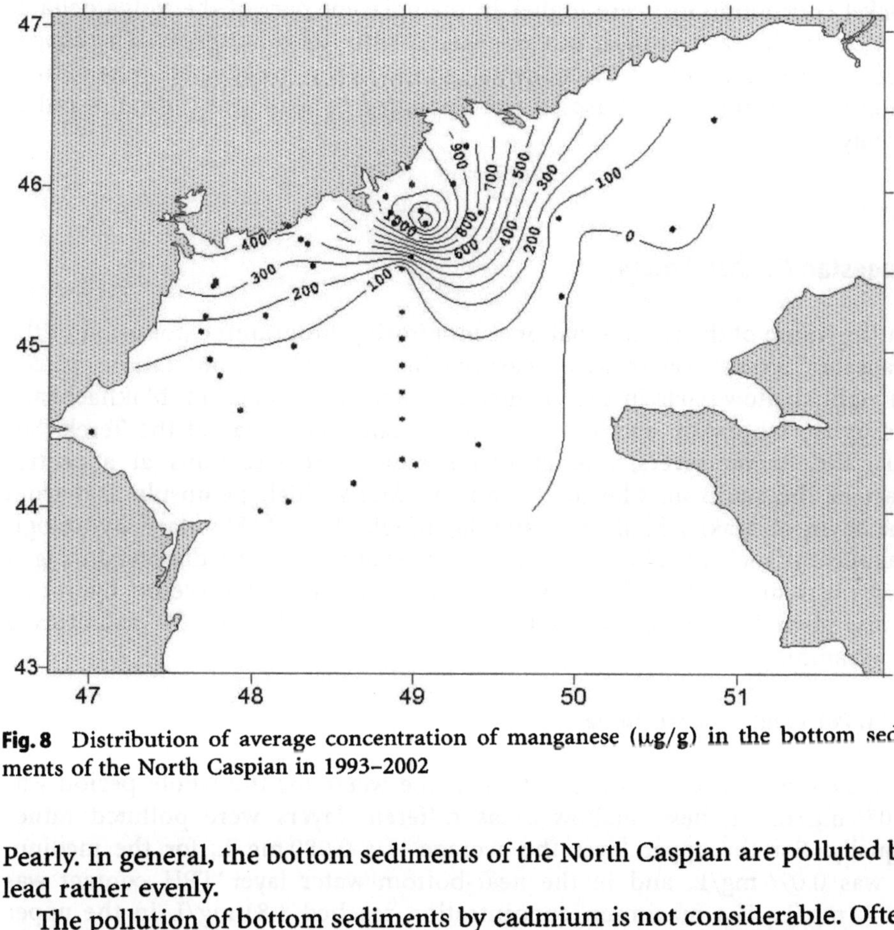

Fig. 8 Distribution of average concentration of manganese (μg/g) in the bottom sediments of the North Caspian in 1993–2002

Pearly. In general, the bottom sediments of the North Caspian are polluted by lead rather evenly.

The pollution of bottom sediments by cadmium is not considerable. Often its concentration was lower than the detection limit of the analytical method. The average concentration of cadmium was 0.061 μg/g and the maximum value was 0.291 μg/g.

The concentration of all heavy metals in the bottom sediments of the North Caspian, including the maximum values, did not exceed the range used for intercalibration in the programme QUASIMEME [19]. The expedition investigations carried out in 2001–2003 in the western part of the North Caspian showed significantly less concentrations of iron, manganese and copper; the same values were found for zinc and nickel, and much higher ones for cadmium and lead, in comparison with our data [14].

The results of monitoring expeditions in 1993–2002 correspond greatly with the data of the expeditions held under the Caspian Environmental Programme (CEP) in 2000–2001 (Table 3) [18]. During those expeditions the increased values of iron concentration in bottom sediments were registered both in deep and eastern parts of the North Caspian. The manganese and nickel concentrations were higher in the adjacent part of the Volga delta as well as in the eastern part, in comparison with the other areas. The copper content was high near the Volga delta and low in the eastern part. The bottom sediments of the North Caspian were polluted by lead and cadmium rather evenly.

4 Dagestan Coastal Waters

In the frame of the routine national monitoring programme pollution of the Dagestan coastal waters was measured in 1978–2004 at 36 stations placed in eight shallow parts in the vicinity of the towns of Lopatin, Makhachkala, Kaspyisk, Izberbash, and Derbent and in estuarine areas of the Terek, Sulak, and Samur rivers, and at two standard cross-sections at a central part of the sea from Chechen' Island to Mangyshlak peninsula and from the town of Makhachkala to Cape Sagunduk (Fig. 1). Standard hydrological and hydrochemical parameters – temperature, salinity, dissolved oxygen and nutrients, and concentrations of ammonia, total petroleum hydrocarbons, phenols and detergents (till 1991) were studied in the monitoring programme.

Total Petroleum Hydrocarbons

The average concentration of TPHs in the water for the whole period was 0.076 mg/L. In these shallow areas different layers were polluted rather equally. For the upper layer the average was 0.080 mg/L, for the medium it was 0.076 mg/L, and in the near-bottom water layer TPH content was 0.072 mg/L. The maximum concentration reached 1.81 mg/L in the upper layer, in mid-March 1989 in an area near the town of Izberbash. In general, this area was most polluted (Table 4). The increased average TPH concentrations were also measured in estuarine areas of the Terek River and near the town of Derbent. In contrast, open waters of the Middle Caspian, the estuarine area of the Samur River, and waters near the town of Kaspyisk were rather clean. Interannual variation of water pollution by TPHs at the Dagestan shelf showed clear evidence of its reduction during recent years (Figs. 9, 10). Within the last decade the high average concentrations were measured only in the estuarine region of the Terek River and near the village of Lopatin.

The seasonal dynamic of TPH content in waters of the Dagestan shelf is not clear (Table 5). The monthly average values were rather similar. A slight increase could be seen in the first part of the year. It is suggested that decreasing TPH concentrations in the upper layer in the warm part of the year is followed by an increase in the speed of degradation of petroleum hydro-

Table 4 Mean values of pollutants in different regions of the Dagestan coastal waters and at the cross-sections in the Middle Caspian during 1978–2003

Region	TPHs mg/L	NH_4 µg/L	Phenols mg/L	Detergents mg/L
Lopatin town	0.076	110	0.006	0.056
Terek River	0.084	101	0.007	0.064
Sulak River	0.074	100	0.006	0.065
Makhachkala town	0.070	90	0.006	0.062
Kaspyisk town	0.068	86	0.005	0.060
Izberbash town	0.094	89	0.007	0.061
Derbent town	0.086	90	0.006	0.066
Samur River	0.069	90	0.006	0.058
Cross-section Chechen-Mangyshlak	0.063	66	0.004	0.053
Cross-section Makhachkala-Sagunduk	0.080	62	0.005	0.043
Average for all regions	0.076	91	0.006	0.060

carbons due to the high temperature of air and water. This seasonal feature is clear in the most polluted part of the Dagestan shelf near the town of Izberbash. Here the maximum TPH concentration was in December–March and the minimum in June–October. Also, the same seasonal variation was typical for the estuarine regions where the maximum occurred in November and February.

Phenols

The concentration of phenols in the waters of the Dagestan shelf varied from 0 to 0.095 mg/L and the average was 0.006 mg/L (Table 4). It was slightly increased in the Terek River estuarine region and near the town of Izberbash. Some values were ten times or more higher than others. These peaks were noted in monitoring investigations during 1978–2004 in all parts of the Dagestan shelf as well as in all seasons. One could suggest that phenols polluted waters in different parts of the Dagestan shelf rather uniformly (Figs. 11, 12). Interannual variation indicated the reduction of phenols concentrations in recent years down to 0.003–0.004 mg/L. The seasonal dynamic is not clear. The minimum (0.004 mg/L) was noted in March and December and the maximum (0.010 mg/L) was in June, mainly due to very high values in 1978.

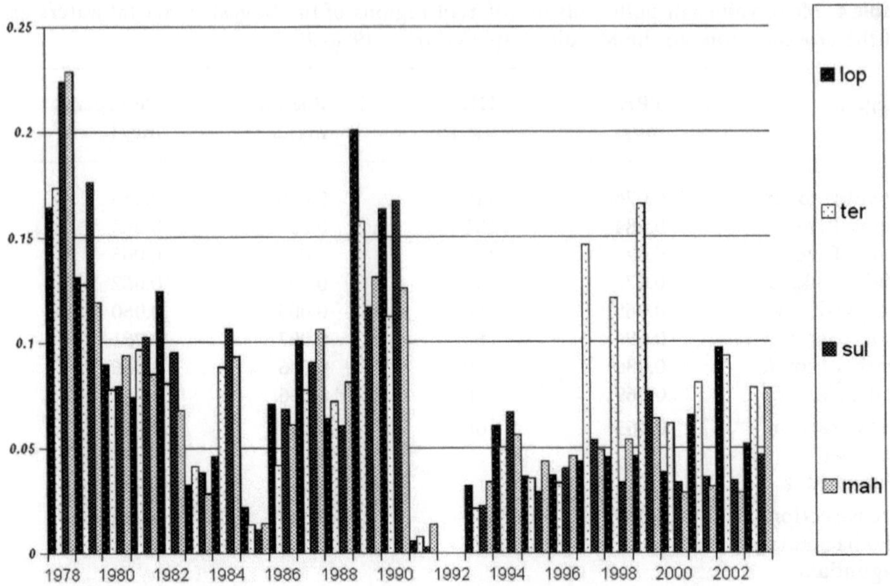

Fig. 9 Inter-annual variation of total petroleum hydrocarbons (mg/L) in the Dagestan coastal waters near Lopatin (*lop*), Makhachkala (*mah*), and estuarine areas of the Terek (*ter*) and Sulak (*sul*) rivers during 1978–2003

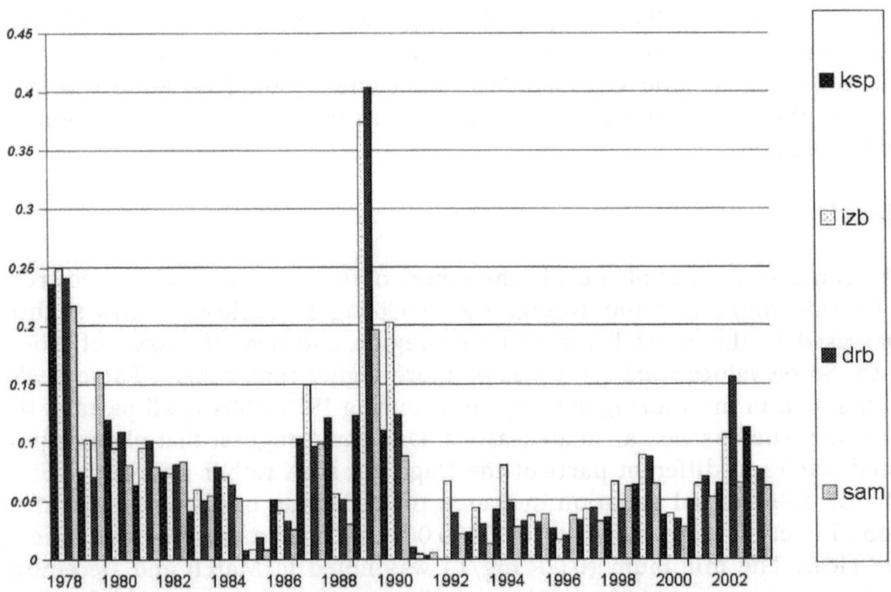

Fig. 10 Inter-annual variation of total petroleum hydrocarbons (mg/L) in the Dagestan coastal waters near Kaspyisk (*ksp*), Izberbash (*izb*), Derbent (*drb*), and estuarine area of the Samur River (*sam*) during 1978–2003

Table 5 Monthly average concentration of total petroleum hydrocarbons (mg/L) in the waters of the Dagestan shelf in 1978–2004

Region	Month											
	I	II	III	IV	V	VI	VII	VIII	IX	X	XI	XII
Lopatin town	0.087	0.082	0.088	0.062	0.072	0.070	0.045	0.069	0.090	0.083	0.111	0.049
Terek River	0.081	0.119	0.081	0.113	0.091	0.068	0.068	0.081	0.079	0.073	0.174	0.113
Sulak River	0.070	0.112	0.065	0.071	0.091	0.061	0.059	0.069	0.076	0.074	0.132	0.072
Makhachkala city	0.079	0.080	0.073	0.093	0.076	0.039	0.065	0.052	0.071	0.062	0.063	0.040
Kaspyisk town	0.068	0.116	0.043	0.064	0.063	0.078	0.067	0.048	0.088	0.066	0.074	0.094
Izberbash town	0.134	0.223	0.137	0.084	0.093	0.045	0.056	0.053	0.105	0.053	0.047	0.118
Derbent town	0.053	0.186	0.099	0.089	0.111	0.141	0.055	0.072	0.130	0.055	0.027	0.050
Samur River	0.056	0.168	0.087	0.099	0.096	0.049	0.049	0.053	0.080	0.036	0.029	0.035
Cross-section Chechen-Mangyshlak	–	0.077	0.036	0.082	–	–	–	0.069	0.064	0.045	0.041	0.040
Cross-section Makhachkala-Sagunduk	–	0.091	–	0.096	0.012	–	–	0.063	0.045	0.042	0.090	0.042
Average for all regions	0.081	0.108	0.080	0.087	0.083	0.061	0.060	0.064	0.085	0.064	0.068	0.061

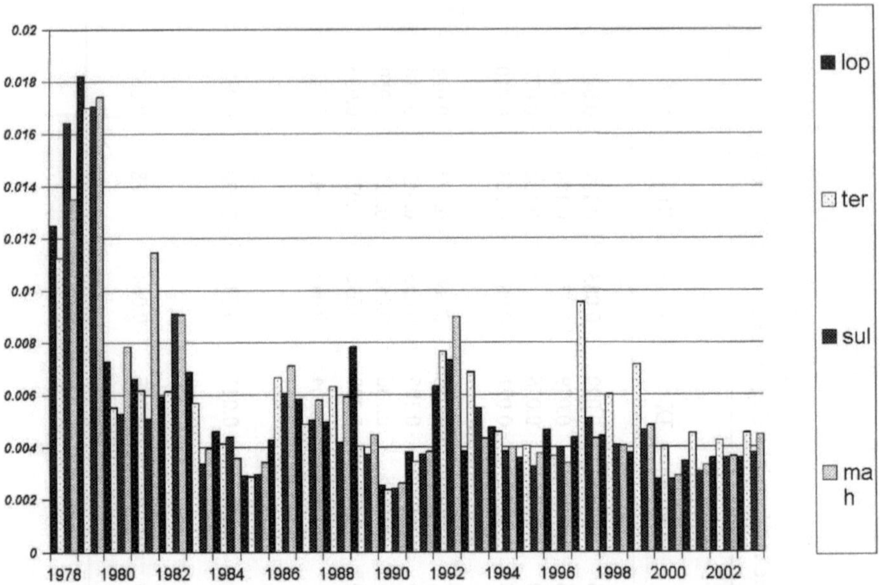

Fig. 11 Inter-annual variation of phenols (mg/L) in the Dagestan coastal waters near Lopatin (*lop*), Makhachkala (*mah*), and estuarine areas of the Terek (*ter*) and Sulak (*sul*) rivers during 1978–2003

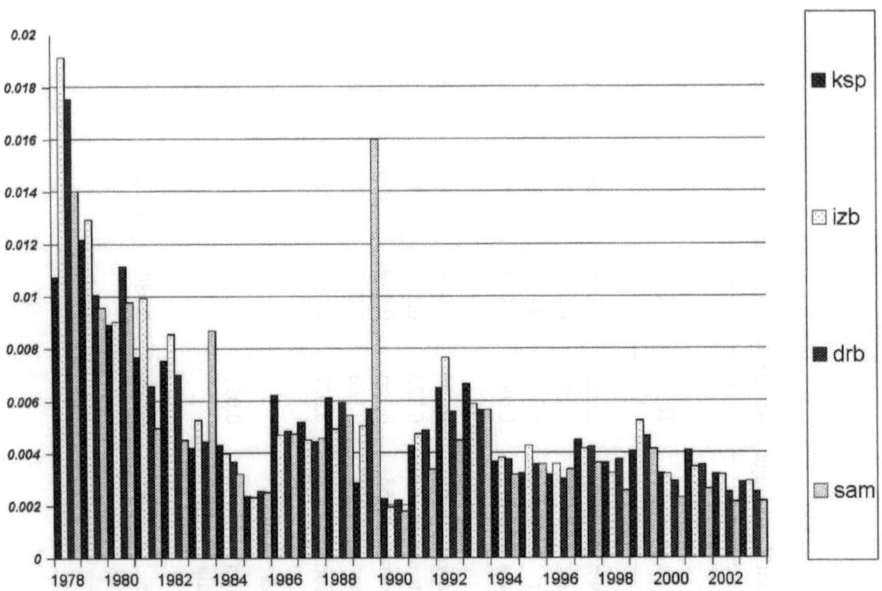

Fig. 12 Inter-annual variation of phenols (mg/L) in the Dagestan coastal waters near Kaspyisk (*ksp*), Izberbash (*izb*), Derbent (*drb*), and estuarine area of the Samur River (*sam*) during 1978–2003

Detergents

The average concentration of detergents from 1978 to 1991 was 0.060 mg/L (Table 4). The maximum value (0.58 mg/L) occurred in waters of the Sulak River estuarine area and near Lopatin in 1979 and 1981. Slightly lower concentrations were measured practically everywhere on the Dagestan shelf. In general, increased content of detergents in the water was typical for estuaries. The open sea water was less polluted in comparison with coastal waters. Monthly average concentrations varied within a narrow range from 0.044 mg/L in June to 0.073 mg/L in February. The seasonal variation was not marked. On the long-term scale, the content of detergents in the water was highest in the 1970s and reduced in the 1990s.

Ammonium

The average concentration of NH_4 was 90.9 μg/L. The maximum content reached 1970 μg/L in September 2001 near Lopatin. The increased content of ammonia was typical for the estuarine waters of the Terek and Sulak rivers and also for the northern part of the Dagestan shelf (Table 4). At the same time the open waters were characterized by low ammonia concentration. The seasonal dynamic of ammonium showed a rather clear increase of average values in January–March (120–130 μg/L) and June (145 μg/L), and a following decrease in August–October (65–77 μg/L). The same seasonal fluctuation was described earlier; but the winter concentration was higher (200–300 μg/L) [20]. In the long-term variation there were two maximum values in 1978 and 1982. Later, the content of NH_4 in waters almost everywhere dropped to about 50 μg/L and slightly increased only in since 2000 (Figs. 13, 14).

5
Water Pollution of the South Caspian

The description of pollution of the South Caspian is based on data from routine monitoring programmes [4, 5] and from investigations carried out by the Academy of Science of Azerbaijan [21]. The main part of the studies was carried out at standard cross-sections of the Hydrometeoservice of the Soviet Union, with some additional stations during 1978–1995. The samples were taken and treated by standard methodology in accordance with the Guidelines [1, 2].

The seasonal dynamic of water pollution in the South Caspian clearly indicated increasing concentrations of petroleum hydrocarbons in the direction from east to west. The average concentrations in winter were 0.15, 0.14, and 0.19 mg/L in eastern, central, and western parts of the region, respec-

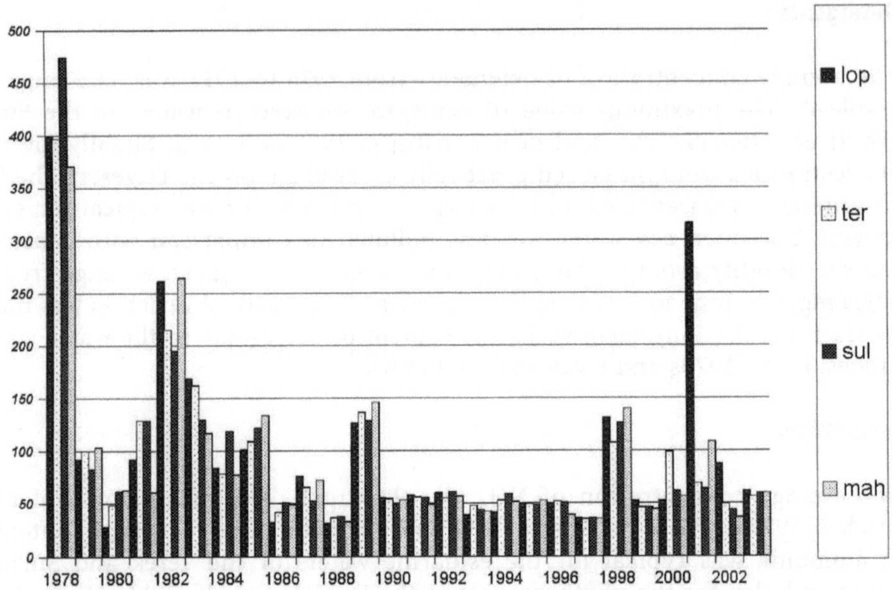

Fig. 13 Inter-annual variation of NH_4 ($\mu g/L$) in the Dagestan coastal waters near Lopatin (*lop*), Makhachkala (*mah*), and estuarine areas of the Terek (*ter*) and Sulak (*sul*) rivers during 1978–2003

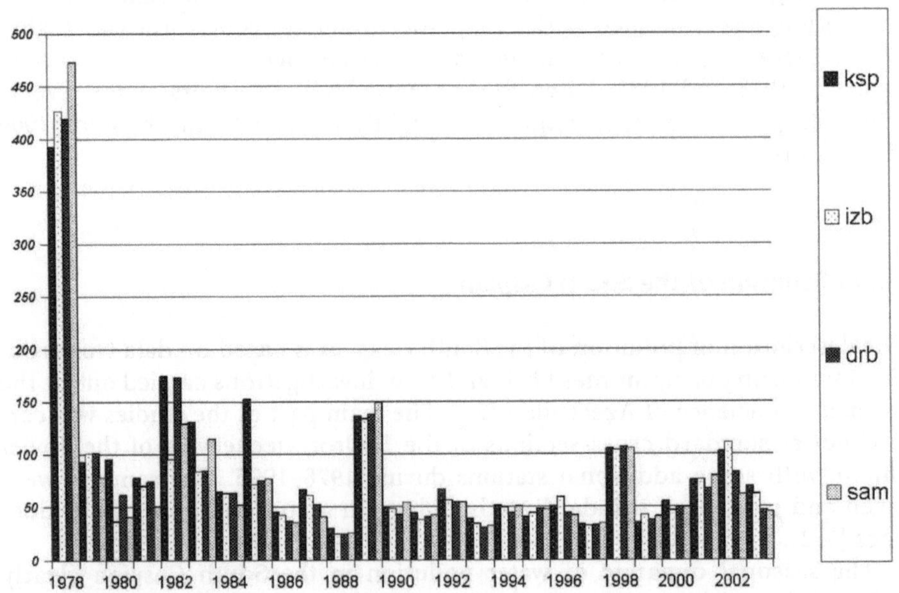

Fig. 14 Inter-annual variation of NH_4 ($\mu g/L$) in the Dagestan coastal waters near Kaspyisk (*ksp*), Izberbash (*izb*), Derbent (*drb*), and estuarine area of the Samur River (*sam*) during 1978–2003

tively. The same features were seen with the spatial distribution of phenols (0.005, 0.005, and 0.008 mg/L) and detergents (0.04, 0.02, and 0.09 mg/L) (Fig. 15).

In spring, concentration of TPHs, phenols and detergents in the upper layer of the South Caspian waters also increased towards the western shelf: 0.08, 0.08, and 0.15 mg/L; 0.006, 0.002, and 0.007 mg/L; and 0.03, 0.01, and 0.08 mg/L, respectively. The pollution was lowest in the central part of the region .

In summer the main features of pollution distribution remained the same. The western part was polluted much more than the others. The average value for this season's concentration was 0.07, 0.07, and 0.16 mg/L for TPHs; 0.006, 0.001, and 0.006 mg/L for phenols; and 0.02, 0.01, and 0.07 mg/L for detergents .

In autumn, along with decreasing temperature, the TPHs concentrations increased (0.12, 0.10, and 0.16 mg/L), but remained practically at the same level for phenols (0.006, 0.003, and 0.006 mg/L) and for detergents (0.035, 0.011, and 0.08 mg/L).

In general, waters of the South Caspian could be considered as heavily polluted. Usually the highest concentrations of petroleum hydrocarbons occurred near the Neftyanye Kamni Bank, Baku Bight (the maximum value reached 1332–1364 mg/L), in the estuarine region of the Kura River, near Turkmenbashy (the average concentration was 0.16–0.27 mg/L) and Cheleken. Relatively clear coastal waters were near the town of Nabran (0.07–0.18 mg/L), in the Lenkoran area (0.08–0.17 mg/L) and in the estuary of the Sefidrud River in Iran (0.05–0.1 mg/L) [4, 22].

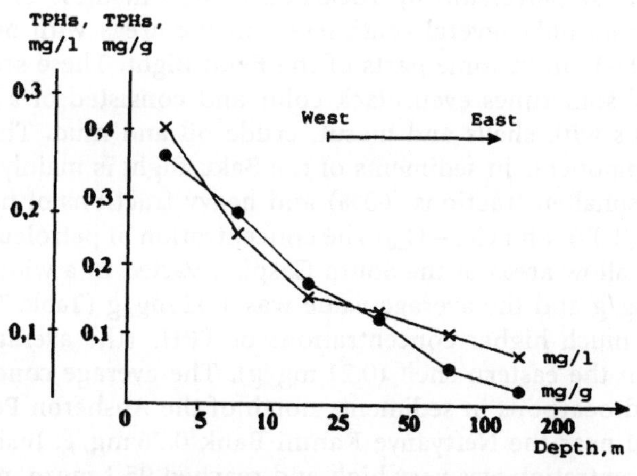

Fig. 15 Reduction in concentration of petroleum hydrocarbons in water (mg/L) and bottom sediments (mg/g) with increasing depth, in the direction from west to east in the Azerbaijanian part of the South Caspian [9]

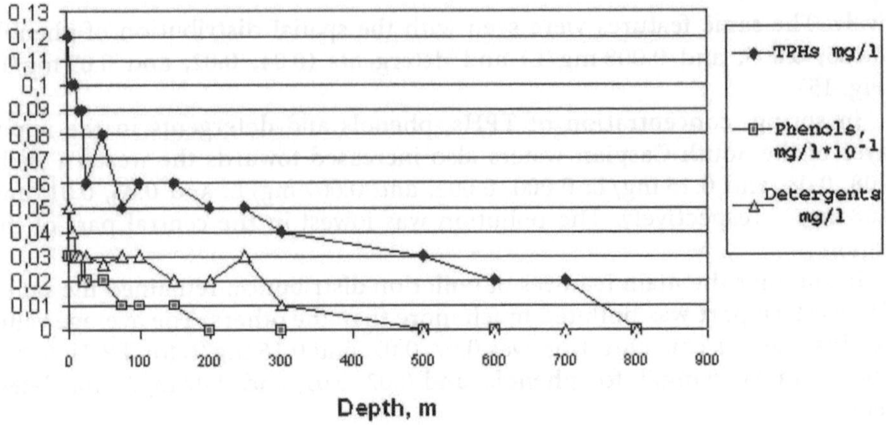

Fig. 16 Decreasing concentration of petroleum hydrocarbons (mg/L), phenols (mg/L $\times 10^{-1}$), and detergents (mg/L) with increasing depth in 1978–1995 [9]

In the Caspian, the concentration of pollutants in waters decreased with depth. The content of petroleum hydrocarbons, phenols, and detergents fell from the maximum value in the surface layer to "zero" in the layer at 500–1000 m (Fig. 16).

6
Bottom Sediments Pollution of the South Caspian

The upper horizons of bottom sediments of the South Caspian were significantly polluted by petroleum hydrocarbons. The thickness of the polluted layer varied from only several centimeters in the areas with negative sedimentation to 8–10 m in some parts of the Baku Bight. These sediments had dark-gray and sometimes even black color and consisted of a friable mixture of detritus with shells and masut, crude oil and mud. The content of petroleum components in sediments of the Baku Bight is mainly represented by resinous-asphalten fractions (60%) and heavy fractions of hydrocarbons like $C_{10} - C_{12}$ (2.0%) and $C_{13} - C_{20}$. The concentration of petroleum hydrocarbons in the shallow areas of the South Caspian varied in a wide range from zero to 226 mg/g and the average value was 1.52 mg/g (Table 7). The western shelf has much higher concentrations of TPHs (the average value was 2.5 mg/g) than the eastern shelf (0.21 mg/g). The average concentration of petroleum hydrocarbons in sediments north of the Apsheron Peninsula was 0.71 mg/g and near the Neftyanye Kamni Bank 0.76 mg/g. Inside the Baku Bight the concentration was very high and reached 95.1 mg/g, near the Sanchagal it was 0.94 mg/g, and in the shallow Makarov Bank 0.81 mg/g. Other parts of the western shelf, like areas near the Nabran, Zuria, Andreev Bank,

Table 6 Concentration of petroleum hydrocarbons (mg/g) in different layers of the bottom sediments in the South Caspian [9]

Region	Depth of sampling place (m)	Hydrocarbon concentration (mg/L) Range of sampling					Type of sediment
		0.0 (mg/g)	0–0.3 m (mg/g)	0.3–0.5 m (mg/g)	0.5–0.7 m (mg/g)	0.7–1.0 m (mg/g)	
Aktau	5.0	0.21	0.15	0.11	0.08	0.03	Mud
Turkmen-bashy	4.0	2.4	1.94	1.85	1.77	1.6	Mud
Cheleken	3.0	0.3	0.20	0.16	0.13	0.11	Sand
Okarem	5.0	0.13	0.11	0.09	0.07	0.03	Mud

and Kurinsky Kamen were much cleaner. TPH average concentrations in bottom sediments in these regions were only 0.03, 0.04, 0.15, and 0.07 mg/g, respectively. At the eastern shelf the level of pollution by petroleum hydrocarbons was rather low and reached 0.19 mg/g in Aktau harbor and 0.35 mg/g in Turkmenbashy Bay [4].

In deep parts of the South Caspian the bottom sediments were rather clean. The shallow areas in the Azerbaijanian part were more polluted and decreasing TPH concentrations corresponded to increasing depth (Fig. 15).

Over the whole Caspian Sea the place most polluted by petroleum hydrocarbon sediments is the Baku Bight. The limited water exchanges with the open sea and long-term dumping of different kinds of wastes, industrial and municipal garbage are the main reasons for appearance the of "anthropogenic" sediments in the bight. Due to stirring of sediments during the stormy weather, or dredging of the bottom materials in harbor, or dumping of sediments from the bight into the sea, the secondary pollution of water occurs rather often. In contrast to the Baku Bight, in shallow areas influenced by waves, and in places with strong near-bottom currents, the sedimentation of thin particles with adsorbed pollutants was strongly limited and consequently the bottom sediments were rather clean. For instance, such areas were noted near the Apsheron threshold, in the vicinity of the Zhiloi Island, and the Neftyanye Kamni Bank (Neft Dashlaru) [4, 23].

In the special places for dumping of materials dredged for industrial and shipping interests the concentration of petroleum hydrocarbons increased. In such a place north of the Apsheron Peninsula the average TPH concentration was 0.14 mg/g. Near Apsheron it was 0.15 mg/g, not far from city Baku 0.36 mg/g, near Karadag 0.24 mg/g, near Aktay 0.12 mg/g, near Bekdash 0.16 mg/g, near Kianly 0.17 mg/g, near Turkmenbashy 0.16 mg/g, near Cheleken 0.15 mg/g, and near Okarem it was 0.30 mg/g [9].

In all polluted areas of the sea the concentration of petroleum hydrocarbons quickly decreased towards the inside of the bottom sediments (Table 6).

Table 7 Size fraction of bottom sediments and concentrations of pollutants in the South Caspian in 1978–1994 [5]

Depth (m)	TPHs, (mg/g)	Phenols (μg/g)	Hg, (μg/g)	Fraction % < 0.005 mm	> 2.0 mm	> 0.5 mm	> 0.25 mm	> 0.1 mm	< 0.005 mm	Type of bottom sediments
Western shelf										
0.0	0.01					53.6				Gravely sand
0–5.0	0.04						72.4			Coarse sand
0.0–1.5	0.09							75.0		Medium sand
0.5–10.0	0.22							82.1		Small sand
0.5–10.0	0	0.32	1.4							Dusty sand
0.5–10.0	0.06	LDL*	1.0	46.0						Clay
0.5–10.0	0.2	0.25	1.2	18.0						Clayey soil
0.5–10.0	0.14	0.09	0.7	8.8						Sandy soil
0.5–10.0	19.0	0.68	0.6	8.2						Sandy mud
0.5–0.25	20.2	1.1	0.7	22.3						Clayey soil mud
0.5–50.0	24.0	1.5	0.74	54.0						Clayey mud
Eastern shelf										
1.0	0.15	0.2	0.32					78.1		Small sand
3.0–10.0	0.20	0.45	0.10						52.5	Dusty sand
10.0	0.93	0.34	0.20	7.6						Sandy mud
5.0–25.0	0.50	0.65	0.32	26.4						Clayey soil mud
5.0–50.0	0.84	1.15	0.44	45.5						Clayey mud
5.0–10.0	0.01	0.0	0.0	43.0						Clay
5.0–10.0	0.12	1.0	–	25.5						Clayey soil

* Less than detection limit

The pollution stopped in clay and clayey soil that was covered with modern sediments consisting of sand and mud.

Among important conditions like distance from the source, depth of place or near-bottom currents, which influence the concentration of pollutants in bottom sediments, the size spectrum of particles in sediments is also of great importance. A predominance of fine fractions leads to increasing concentrations of petroleum hydrocarbons in sediments, e.g. in sand the concentration of TPHs increased from gravel (0.01 mg/g), to coarse (0.04 mg/g), medium (0.09 mg/g), small (0.18 mg/g), and dusty (0.26 mg/g) fractions (Table 7). The same tendency was marked in clay sediments: the petroleum content was 9.6 mg/g in sandy clay, somewhat higher (10.3 mg/g) in clayey soil, and reached the maximum value (12.3 mg/g) in clayey mud. In the Baku Bight a high concentration was measured in the mud (103.8 mg/g) in comparison with the sand (0.75 mg/g) and clay (0.21 mg/g).

The long-term variations clearly indicate a reduction of pollution level (Table 8). It could be the result of increased seawater level and falling industrial activity in the region as a whole.

On the whole, the concentration of heavy metals in bottom sediments did not exceed the regional background value. The most clear exception is Baku Bight where the concentration of mercury in bottom sediments had an average of 1.3 mg/g, for nickel it was 41.3 mg/g, and for strontium 330.3 mg/g.

Table 8 Long-term variations in concentration of petroleum hydrocarbons (mg/g) in bottom sediments of the South Caspian in 1978–1994

Year	Hydrocarbon concentration		
	Western shelf	Eastern shelf	Baku Bight
1978	2.0	1.0	92.5
1979	1.8	1.2	86.4
1980	0.9	1.0	90.0
1981	0.7	0.92	70.5
1982	0.33	0.59	42.5
1983	0.93	0.68	28.4
1984	0.55	0.71	50.3
1985	1.6	0.57	98.0
1986	1.3	0.38	40.0
1987	1.0	0.21	20.2
1988	0.79	0.19	36.3
1989	0.37	0.14	24.5
1990	0.67	0.10	12.5
1991	0.42	0.09	5.2
1992	0.28	0.08	4.0
1993	0.18	–	0.84
1994	0.11	–	0.75

Table 9 Average concentration of pollutants in bottom sediments of Baku Bight in 1978–1995

Depth (m)	TPHs (mg/g)	Organic substances (mg/g)	Phenols (mg/g)	Cu (µg/g)	Pb (µg/g)	Ni (µg/g)	V (µg/g	Be (µg/g)	Sr (µg/g)	Cr (µg/g)	Hg (µg/g)
0.0–0.25	68.4	86.5	0.0015	35.7	24.5	43.1	56.5	1.2	350.0	407.0	1.3
0.25–0.5	78.0	90.5	0.001	33.8	22.8	41.3	59.3	1.1	363.0	421.6	1.4
0.5–0.75	36.4	50.6	0.0008	32.5	20.3	38.4	52.2	1.2	183.7	360.3	1.2
0.75–1.0	28.9	44.8	0.002	28.7	18.0	28.0	51.5	1.0	160.0	361.7	1.0
1.0–1.5	17.1	22.0	0.0006	23.0	13.4	24.1	40.7	1.0	207.5	250.7	0.6
1.5–2.0	6.8	7.7	0.0001	17.0	10.1	14.5	30.4	1.0	LDL*	166.7	0.05
2.0–2.5	2.1	3.4	0.00005	12.8	9.4	7.9	28.4	1.0	LDL*	127.8	0.13
2.5–3.0	1.4	1.8	0.0007	12.1	9.4	5.7	24.7	1.0	LDL*	93.4	0.10
3.0–3.5	0.9	1.9	LDL*	9.7	6.6	5.0	17.5	1.0	LDL*	46.2	0.3
3.5–4.0	0.8	1.0	LDL*	9.5	3.0	5.0	15.1	1.0	LDL*	31.8	LDL*
4.0	0.6	0.8	LDL*	8.8	1.7	5.0	14.0	1.0	LDL*	27.1	LDL*
Average	22.0	28.2	0.007	20.3	12.6	20.0	35.5	1.0	253.0	108.6	0.67

*Less than detection limit

The highest concentration was of chromium, which reached 421.6 mg/g (Table 9). In the Turkmenbashy Bay the bottom sediments were polluted by nickel (28.5 mg/g), chromium (120.0 mg/g), and barium (501.0 mg/g). Increased concentrations of pollutants were registered close to the places of industrial discharges, as well as in places of dredged bottom material or dumping. The concentration of pollutants in bottom sediments decreased towards the central part of the sea. The only exception is zinc, which occurred in high quantities in the deep part of the basin due to sedimentation of dead plankton organisms.

The pollution of bottom sediments by mercury in the western part of the region, with an average of 0.34 mg/g, was several times higher than in eastern shelf, where it was 0.16 mg/g. The maximum concentrations were fixed in areas near the town of Sumgait (0.78 mg/g), Neftyanye Kamni Bank (0.58 mg/g), in Baku Bight (1.22 mg/g), near Aktau (0.34 mg/g), near Turkmenbashy (0.17 mg/g) and Cheleken (0.23 mg/g) [4].

7 Conclusions

The operation in 1978–2004 of long-term monitoring of the Caspian Sea environmental conditions allowed nominating the main anthropogenic factors of pollution of waters and sediments. River discharge, extraction of oil and gas at the sea shelf, discharge of industrial and municipal wastewaters and garbage, and the dredge and dumping of bottom material are the most important.

The long-term variation and spatial distribution of petroleum hydrocarbons in the waters of the North Caspian clearly indicate a high level of pollution in the 1980s in the estuarine region of the River Ural and on the border with the Dagestan shelf. During the following few years the level of TPHs sharply decreased everywhere down to 0.1–0.2 mg/L. In the last few years a large interannual variation was registered and the main patches of high concentration of petroleum hydrocarbons occurred in the offshore part of the North Caspian rather far from the Volga delta. One can suppose that the Volga was not the main source of TPHs in the North Caspian.

In the shallow waters of the Dagestan shelf the TPH concentration was lower than in the North Caspian and had a clear tendency to decline. The most polluted area was near the town of Izberbash. Recently the estuarine area of the Terek River and the waters near the village of Lopatin were temporary places of high pollution. No vertical stratification in the concentration of petroleum hydrocarbons occurred and all layers were polluted evenly in these shallow waters. The average data from long-term monitoring clearly indicate a weak seasonal dynamic. Slightly increased concentrations characterized the cold season, and some decreasing was visible at the end of summer and in autumn. Very high TPH content in the waters occasionally occurred everywhere and there could be patches of high concentrations due to short-term oil spills or permanent riverine discharge.

Phenol concentration in the North Caspian and Dagestan waters was rather high and showed significant variability both in space and time. On a long time-scale the decrease of phenols in the waters and stabilization during recent years at the level of 0.003–0.004 mg/L were clear. The single highest value occurred in practically every controlled region and in each season. The averaged data smoothed the maximum values and therefore the seasonal variability was not marked. One can observe the local (in space) and short (in time) appearance of patches with high concentrations of phenols.

Detergents in the waters of the North Caspian and at the Dagestan coastal area were present permanently. Their average concentration varied from 0.038 to 0.060 mg/L and has decreased during recent years. The highest values commonly occurred in the western part of the North Caspian and estuarine regions of Dagestan Rivers. The central part of the Middle Caspian was less polluted. The seasonal dynamic described a slight increase at the end of summer, in contrast to the cold period of the year.

Distribution of ammonium concentration was nearly the same as for detergents. High average values were measured in estuarine waters of the Volga, Terek, and Sulak rivers, as well as in the western part of the North Caspian. According to monitoring data there were no interannual and seasonal variations nor vertical gradients.

Among the chlorinated pesticides DDT was the most important in the North Caspian waters in the 1970s and 1980s. The average concentration was 31 ng/L. The common concentration of other pesticides was about 1–2 ng/L. The spatial distribution of pesticides was very unequal and local patches of high concentration usually occurred on a rather clean background. The high values were often measured in the eastern part of the North Caspian. After 1992 the concentration of pesticides of the DDT group was lower than the detection limit used in the analytical methodology.

The spatial distribution of petroleum hydrocarbons in bottom sediments in the North Caspian was not the same as in the water. Zones of high concentrations were measured adjacent to the Volga delta area and on the border with the Dagestan shelf. The sediments in the central part of the region mainly consisted of coarse fractions and were rather clean.

The same features of spatial distribution were described for chlorinated pesticides. Their content was rather high in the 1990s and patches were located in shallow parts close to the Volga delta. Over the last few years the concentration of pesticides has been low and has not often exceeded the detection limit.

The heavy metals in bottom sediments of the North Caspian in general were in the range of regional background values. The patches of high concentration of manganese, nickel, and copper located adjacent to the Volga delta area were influenced by freshwater inflow. In contrast, the increased levels of iron and zinc were measured in the offshore part of the area far from the delta. Cadmium and lead had rather low concentrations in bottom sediments and were distributed evenly.

In the South Caspian the upper layer of waters were significantly polluted by TPHs, phenols, and detergents. Their concentration slightly decreased from the western shelf towards the east and sharply increased with depth. In the cold season the content of petroleum hydrocarbons in waters was slightly higher. The maximum values were found close to the places of oil and gas extraction on the shelf.

Among pollutants in the bottom sediments of the South Caspian petroleum hydrocarbons were the most important. The western shelf was more polluted than the eastern. The maximum concentrations of TPHs in sediments were measured in the Baku Bight. In general, the highest values occurred in sediments with a fine size spectrum of particles. Long-term dynamics showed stepwise decreasing of TPH pollution in shallow areas. Sediments in the deep part were rather clean.

The concentration of heavy metals in bottom sediments were significantly higher than the regional background level only in vicinity of industrial and municipal waste-water discharge, and in the places of dredge and dumping of bottom sediments. Very high pollution by chromium, mercury, nickel, and strontium was noted in the Baku Bight. In general, the concentration of metals in sediments decreased from the shore to the deep part of the area. Only zinc occurred in high quantity in the deep parts of the basin due to sedimentation of plankton organisms.

Acknowledgements The authors are very grateful to everybody who took samples during cruises and analyzed them in laboratories during the long-term monitoring programmes studying pollution of the Caspian Sea, and also to T.I. Plotnikova and A.V. Udovenko for their great help in preparation of the database and figures.

References

1. Hydrometeoservice (1993) Guideline on chemical analysis of marine waters. RD 52.10.243-92. Hydrometeoizdat, St. Peterburg [in Russian]
2. Hydrometeoservice (1996) Determination of pollution in marine bottom sediments and at suspended matter. Hydrometeoizdat, Moscow [in Russian]
3. Hydrometeoservice (1978–2002) Annual report on the quality of marine waters based on hydrochemical parameters. Hydrometeoizdat, Leningrad-StPeterburg [in Russian]
4. Azcomhydromet (1978–1995) Annual hydrochemical data on the quality of the Caspian Sea. Azcomhydromet, Baku
5. Azcomhydromet (1978–1992) Annual report on the quality of the Caspian Sea waters. Azcomhydromet, Baku [in Russian]
6. Afanasieva NA (1981) Balance and prediction on the pollution of the North Atlantic, Baltic and Caspian Seas by petroleum hydrocarbons. PhD dissertation, Moscow [in Russian]

7. GEOS (1998) Estuarine region of the Volga: hydrological, morphological processes, regime of the pollution and influence of sea level fluctuations. GEOS, Moscow [in Russian]
8. Buharitsin PI (1996) Hydrological processes in winter in the North Caspian. PhD dissertation, IVP RAN, Moscow [in Russian]
9. Gul AK (2003) The problem of the Caspian Sea pollution. Muallim neshriiatu, Baku [in Russian]
10. Heidarov FA (1981) Dynamic of pollution in the Baku area (current and future conditions). PhD dissertation, Moscow [in Russian]
11. Gul AK (2001) Influence of wind and waves on the pollution spreading in the Caspian Sea. News of the Caspian Sea, N6, Moscow [in Russian]
12. Gul AK (2001) The role of rivers in the pollution of the Caspian Sea. Bilgi, N2, Baku [in Russian]
13. Gul AK (2001) Pollution of bottom sediments in the Caspian Sea. Bilgi, N3, Baku [in Russian]
14. Marchenko EN, Dolgov VV, Ostrovskaya EV, Monakhova GA (2005) Condition of marine environment pollution in water exchange area between the North and Middle Caspian based on data of ecological monitoring during seismic prospecting at sea. In: Problems of the Caspian ecosystem conservation under conditions of oil and gas fields development. Proceedings of the 1st international scientific-and-practical conference, 16–18 February 2005, Astrakhan. KaspNIRKh Publishers, Astrakhan [in Russian]
15. Geraskin PP, Metallov GF, Aksenov VP, Galaktionova ML (2005) Oil pollution of the Caspian Sea as a factor of negative impact on the physiological state of sturgeons. In: Problems of the Caspian ecosystem conservation under conditions of oil and gas fields development. Proceedings of the 1st international scientific-and-practical conference, 16–18 February 2005, Astrakhan. KaspNIRKh Publishers, Astrakhan [in Russian]
16. Amirgaliev NA (2005) On the ecologo-toxicological assessment of the Ural-Caspian basin. In: Problems of the Caspian ecosystem conservation under conditions of oil and gas fields development. Proceedings of the 1st international scientific-and-practical conference, 16–18 February, 2005, Astrakhan. KaspNIRKh Publishers, Astrakhan [in Russian]
17. Burkatski ON, Shelting SK, Sheikov AA, Kurganskaya VV, Kuznetsova TI, Chalenko VA (2005) Application of landscape mapping for marine environmental research. In: Problems of the Caspian ecosystem conservation under conditions of oil and gas fields development. Proceedings of the 1st international scientific-and-practical conference, 16–18 February 2005, Astrakhan. KaspNIRKh Publishers, Astrakhan [in Russian]
18. de Mora S, Sheikholeslami MR (2002) ASTP: Contaminant screening programme. Final report: Interpretation of Caspian Sea sediment data. Caspian Environment Programme
19. QUASIMEME (2204) Laboratory performance studies, Year 9: June 2004–May 2005. Issue 1, Aberdeen
20. Ahmedova GA (2002) The features of hydrological and hydrochemical conditions at the Dagestan shelf of the Caspian Sea. PhD dissertation, Moscow [in Russian]
21. Mehtiev ASH, Gul AK (1999) Bottom sediment pollution of Azerbaijan part of the Caspian Sea. ANAKA, Baku [in Russian]
22. Cruise report of the RV Radon works at Sefidrud grid. (1975–1978) Azcomhydromet, Baku [in Russian]
23. Hydrometeoservice (1986) Hydrometeorological conditions of shelf of USSR seas, vol 2. The Caspian Sea. Hydrometeoizdat, Leningrad [in Russian]

Hdb Env Chem Vol. 5, Part P (2005): 143–157
DOI 10.1007/698_5_007

Published online: 6 October 2005

Patterns of Seasonal and Interannual Variability of Remotely Sensed Chlorophyll

Nikolay P. Nezlin

Southern California Coastal Water Research Project, 7171 Fenwick Lane, Westminster, CA 92683-5218, USA
nikolayn@sccwrp.org

Abstract Seasonal and interannual variability of chlorophyll dynamics in the Caspian Sea remotely sensed by the Sea-viewing Wide Field-of-view Sensor (SeaWiFS) radiometer was analyzed from the start of the SeaWiFS mission in 1997 till 2004. The chlorophyll concentration in the surface layer was especially high in the shallow northern Caspian Sea. Seasonal variability in all Caspian Sea regions except the Kara-Bogaz-Gol Bay was characterized by a seasonal maximum in August to September, related to the period of maximum sea surface temperature and wind stress. On an interannual scale, the variations of chlorophyll in the Caspian Sea were related to North Atlantic Oscillation Index, Volga discharge, sea surface temperature, and wind stress anomalies over different regions of the Caspian Sea.

Keywords Caspian Sea · Chlorophyll · Interannual variations · Seasonal variations · Sea-viewing wide-field-of-view sensor

1 Introduction

In this paper, the principal features of seasonal and interannual variability of the surface chlorophyll concentration in the Caspian Sea are derived from the observations of the Sea-viewing Wide Field-of-view Sensor (SeaWiFS) from 1997 to the end of 2004. The principal features of the physical geography of the Caspian Sea are described in detail by Kosarev in this volume. In this study, we divide the sea area into three major regions (the Northern, Middle, and Southern Caspian; see Fig. 1 in Kosarev, this volume), which have different hydrological and ecological conditions. This classification is conventional

(Kosarev, this volume). The Kara-Bogaz-Gol Bay is analyzed separately, because the hydrological conditions in this shallow lagoon are very untypical: an extensive water evaporation results in extremely high salinity (350 g L^{-1}) (Kosarev and Kostianoy, this volume: "Kara-Bogaz-Gol Bay").

Among other environmental factors influencing the seasonal and interannual dynamics of remotely sensed phytoplankton biomass in the Caspian Sea, we focus on the Volga River runoff, because the volume of fresh water discharged into the Caspian Sea (more than 300 km^3 $year^{-1}$) is very high compared with that of other basins, and the Volga River contributes about 80% to this discharge.

The Northern Caspian, on the one hand, and the Middle and Southern Caspian, on the other, are very different in terms of hydrological and ecological conditions of phytoplankton growth. In the shallow (5–6-m depth) northern part of the sea, a huge freshwater input from the Volga River results in very low salinity (0 psu in the mouths of the Volga and Ural rivers). In deep (maximum depth 1025 m) middle and southern parts of the Caspian Sea the salinity is as high as 10–13 psu, with clear vertical stratification and a sharp thermocline, typically at a depth of 20–30 m (Tuzhilkin and Kosarev, this volume). This thermocline is especially pronounced in summer, resulting in nutrient limitation of phytoplankton growth.

2 Methods

The analysis of spatio-temporal variations of phytoplankton biomass in the Caspian Sea was based on the remotely sensed data collected by the SeaWiFS on board the OrbView-2 satellite. The data were obtained from the National Aeronautics and Space Administration Goddard Space Flight Center Distributed Active Archive Center (NASA GSFC DAAC) [1]. We used monthly averaged Level 3 Standard Mapped Image (SMI) data of SeaWiFS surface chlorophyll calculated during reprocessing 4. The format of the level 3 SMI data is a regular grid of equidistant cylindrical projection of 360°/4096 pixels (about 9.28-km resolution). Use of this data was in accord with the SeaWiFS Research Data Use Terms and Conditions Agreement. The algorithm used at GSFC for calculating the surface chlorophyll concentration was described in Ref. [2]. In this study, we used remotely sensed data (i.e., surface chlorophyll concentration derived from water color) to analyze the dynamics of phytoplankton biomass in the water column. The remotely sensed surface pigment concentration and the total pigment concentration in the water column are correlated [3] but are not identical. However, for this study, the absolute values are not as important as spatial and temporal gradients of phytoplankton biomass. We understand that the absolute values of the surface chlorophyll concentration derived from satellite measurements are subject to significant inaccuracy owing to the technical difficulty of remotely sensed observations. The standard SeaWiFS algorithm was developed for clean, open ocean waters (Case 1), where the color of the ocean surface results mainly from chlorophyll concentration [4]. The Caspian Sea is classified as coastal waters (Case 2), where the sea surface color also depends on the dissolved and suspended matter concentrations. The standard SeaWiFS algorithm overestimates the chlorophyll concentration in Case 2 waters. Regional algorithms based on in situ chlorophyll concentration measurements provide more reliable results in Case 2 waters, but these studies have not yet been carried out for the Caspian Sea [5]. In this study, we do not compare the remotely sensed data directly with in situ absolute values of chlorophyll concentration and phytoplankton biomass. Instead, we consider the variations of satellite-measured chlorophyll concentration as an indicator of the dynamics of phytoplankton biomass in the study region.

The monthly climatological maps of SeaWiFS chlorophyll in the Caspian Sea were obtained from GSFC DAAC. On these maps each pixel of the monthly SMI maps was averaged for each calendar month. To reveal the principal features of the spatial distribution, these maps were smoothed with a 100-km 2D cosine filter.

The mean remotely sensed chlorophyll concentrations in each of four regions (Fig. 1 in Kosarev, this volume) were calculated as medians of all pixel values in monthly SMI maps. We used a median instead of an arithmetic mean because the statistical distribution of remotely sensed chlorophyll in the World Ocean has been shown to be asymmetric [6] and medians were much closer to modal values than means.

To estimate the variations of sea surface temperature (SST), we used remotely sensed Advanced Very High Resolution Radiometer (AVHRR) data from the NASA Jet Propulsion Laboratory Physical Oceanography Distributed Active Archive Center (JPL PODAAC). These data were produced at JPL within the scope of the National Oceanic and Atmospheric Administration (NOAA)/NASA AVHRR Oceans Pathfinder Project. The SST data were derived from the AVHRR observations using an enhanced nonlinear algorithm [7]. We used monthly data for the descending pass (i.e., nighttime observations) to avoid the inconsistency resulting from diurnal heating of the sea surface, which can be as high as a few degrees centigrade [8]. The data format of Pathfinder version 5 is a regular grid of equidistant cylindrical projection of 360°/8192 pixels (about 4.5-km resolution). Only the data with quality flags 4–7 (i.e., the "best SST") were used to calculate the arithmetic mean SST in each of the three regions of the Caspian Sea. In Kara-Bogaz-Gol Bay the number of SST pixels was insufficient for statistical analysis.

The influence of wind stress was analyzed on the basis of meteorological data obtained from the National Center for Environmental Prediction (NCEP) and supplied by the NASA GSFC DAAC as ancillary information for SeaWiFS users. These data contain regular grids of zonal and meridional wind speeds at 10 m above sea level interpolated on an equidistant cylindrical projection of 1° spatial resolution and 6-h temporal resolution (12-h resolution during some periods in 1998 and 1999). Wind data were averaged over monthly periods and transformed to wind stress ($kg\,m^{-1}\,s^{-2}$) using the conventional equation $\tau = C_d \rho_a U^2$, where U is the wind speed ($m\,s^{-1}$) at 10 m, ρ_a is the air density ($1.2\,kg\,m^{-3}$), and C_d is is the dimensionless "drag coefficient" (0.0013) [9].

The monthly estimations of Volga discharge (cubic meters per second) were obtained at the Volgograd Power Plant, located 400 km upstream from the Caspian Sea.

To relate the long-scale variations in the Caspian Sea to the global climatic meteorological cycles, we used the North Atlantic Oscillations (NAO) index, which is defined as the monthly averaged difference between the standardized measurements of the sea level atmospheric pressure at the Azores and Iceland. The NAO index time series was obtained from the International Research Institute of Climate Prediction (Columbia University, USA).

To analyze the interannual variability of chlorophyll concentration in three regions of the Caspian Sea and the parameters of the physical environment (i.e., SST, wind stress, and Volga discharge), each parameter was transformed to a seasonal anomaly. For this, climatic seasonal values were subtracted from each time series.

3
Seasonal Variability

During all calendar months, the chlorophyll concentration in the shallow Northern Caspian Sea was substantially higher than in the Middle and Southern Caspian Sea (Fig. 1). Minimum chlorophyll concentrations (less than $1\,mg\,m^{-3}$) were observed in the eastern part of the Southern Caspian Sea in April–June and in the eastern part of the Middle Caspian Sea in July. Even during the summer months in the deep areas of the Caspian Sea the pigment concentration was higher compared with that of other basins of the Mediterranean region; this feature was noticed earlier by Sur et al. [10].

Chlorophyll concentrations along the western coast were higher than in the eastern areas. This feature correlates with in situ observation that the turbidity of the surface water in the Caspian Sea consistently increases from east to west [11]. We can explain the east–west gradient of phytoplankton biomass with cyclonic circulation in the Caspian Sea [10, 12] (see also Tuzhilkin and Kosarev, this volume) transporting high concentrations of nutrients and phytoplankton from the hyperproductive Northern Caspian shelf. In the summer season, the zone of intensive upwelling is often observed along the eastern coast of the Caspian Sea [10, 12]; see also Tuzhilkin and Kosarev, this volume. We expected that this upwelling could result in the enrichment of the surface layer with nutrients, which in turn could stimulate phytoplankton growth along the eastern coast. However, in the SeaWiFS data we did not observe high chlorophyll concentrations in the eastern Caspian. We speculate that in the productive Caspian Sea the upwelling along the eastern coast does not exert a significant influence on the phytoplankton growth rate.

In the Northern Caspian Sea, the seasonal maximum of phytoplankton biomass was observed in August (Fig. 2c), when the mean chlorophyll con-

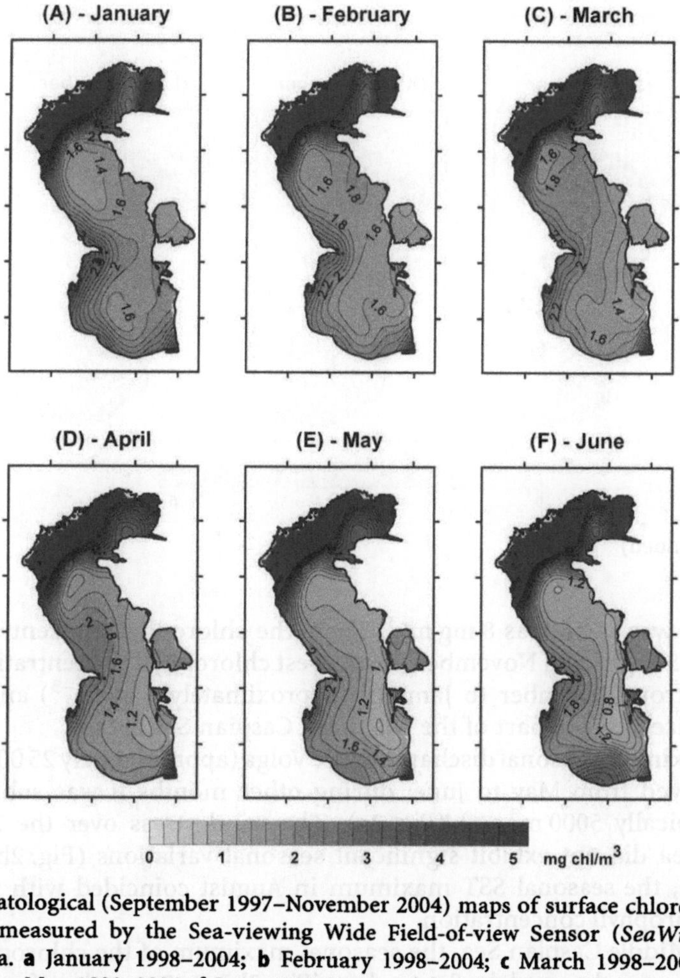

Fig. 1 Climatological (September 1997–November 2004) maps of surface chlorophyll concentration measured by the Sea-viewing Wide Field-of-view Sensor (*SeaWiFS*) in the Caspian Sea. **a** January 1998–2004; **b** February 1998–2004; **c** March 1998–2004; **d** April 1998–2004; **e** May 1998–2004; **f** June 1998–2004; **g** July 1998–2004; **h** August 1998–2004; **i** September 1997–2004; **j** October 1997–2004; **k** November 1997–2004; **l** December 1997–2003. *Isolines* of chlorophyll concentration in milligrams per cubic meter

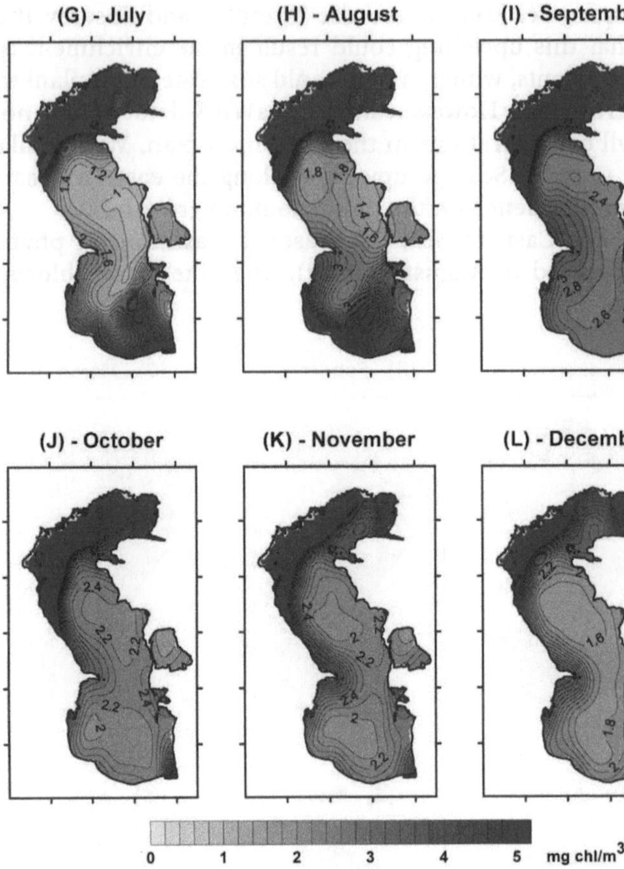

Fig. 1 (continued)

centration was as high as 8 mg m^{-3}. Then, the chlorophyll concentration decreased to 5 mg m^{-3} in November. The lowest chlorophyll concentrations were observed from December to January (approximately 4 mg m^{-3}) and can be related to ice cover of part of the Northern Caspian Sea area.

The maximum seasonal discharge of the Volga (approximately 25 000 $m^3\ s^{-1}$) was observed from May to June; during other months it was substantially lower (typically 5000 $m^3\ s^{-1}$) (Fig. 2a). The wind stress over the Northern Caspian Sea did not exhibit significant seasonal variations (Fig. 2b). At the same time, the seasonal SST maximum in August coincided with the maximum chlorophyll concentration.

In the Middle Caspian Sea, the seasonal maximum of the chlorophyll concentration was observed in September (Fig. 3b), 1–2 months after the SST maximum (Fig. 3a). The seasonal maximum of the wind stress was from July to September (Fig. 3a).

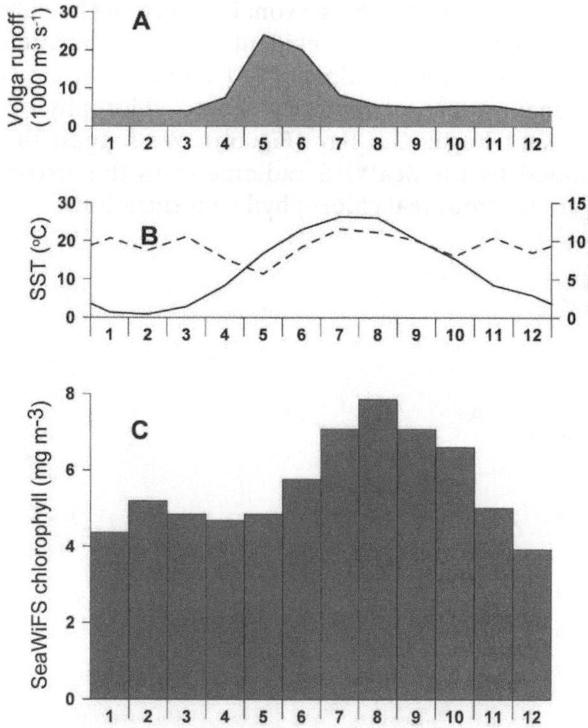

Fig. 2 Climatological seasonal variations of **a** Volga discharge, **b** sea surface temperature (*SST*) (*solid line*) and wind stress (*dashed line*), and **c** SeaWiFS-measured surface chlorophyll concentration in the Northern Caspian Sea

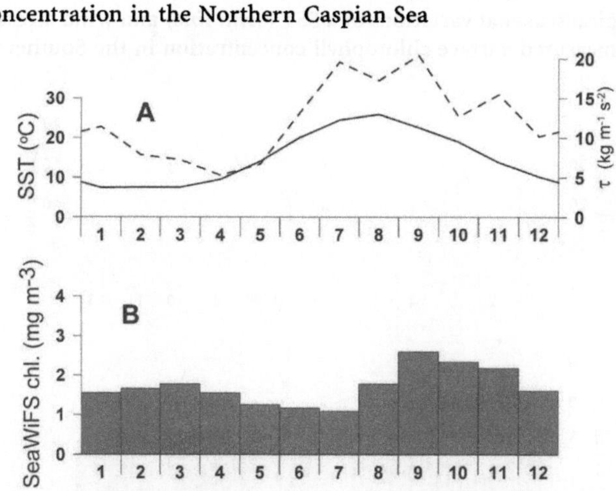

Fig. 3 Climatological seasonal variations of **a** SST (*solid line*) and wind stress (*dashed line*) and **b** SeaWiFS-measured surface chlorophyll concentration in the Middle Caspian Sea

In the Southern Caspian Sea, the seasonal pattern of the chlorophyll concentration was characterized by one evident maximum in August (Fig. 4b), which coincided with seasonal maxima of SST and wind stress (Fig. 4a).

No clear seasonal pattern of remotely sensed chlorophyll concentration was observed in Kara-Bogaz-Gol Bay (Fig. 5b). We suggest that the optical properties measured by the SeaWiFS radiometer in this extremely shallow and salty basin are far from real chlorophyll concentrations.

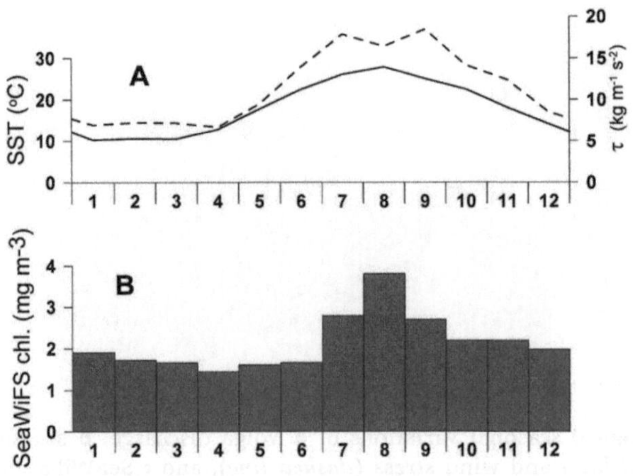

Fig. 4 Climatological seasonal variations of **a** SST (*solid line*) and wind stress (*dashed line*) and **b** SeaWiFS-measured surface chlorophyll concentration in the Southern Caspian Sea

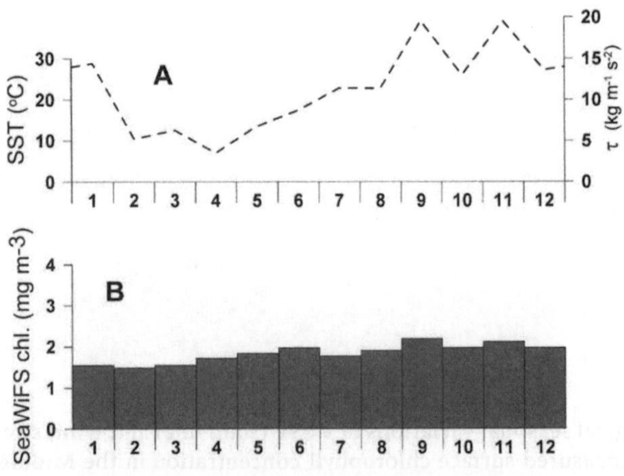

Fig. 5 Climatological seasonal variations of **a** wind stress and **b** SeaWiFS-measured surface chlorophyll concentration in Kara-Bogaz-Gol Bay

4
Interannual Variability

The time series of the NAO index and the seasonal anomalies of Volga discharge, wind stress (Fig. 6), SST, and chlorophyll concentration (Fig. 7) in different regions of the Caspian Sea were correlated at different time lags (Table 1). No prominent interannual trends were detected. In particular, no positive SST trend was revealed in any regions of the Caspian Sea; moreover, substantial negative anomalies were observed in 2003–2004. This observation

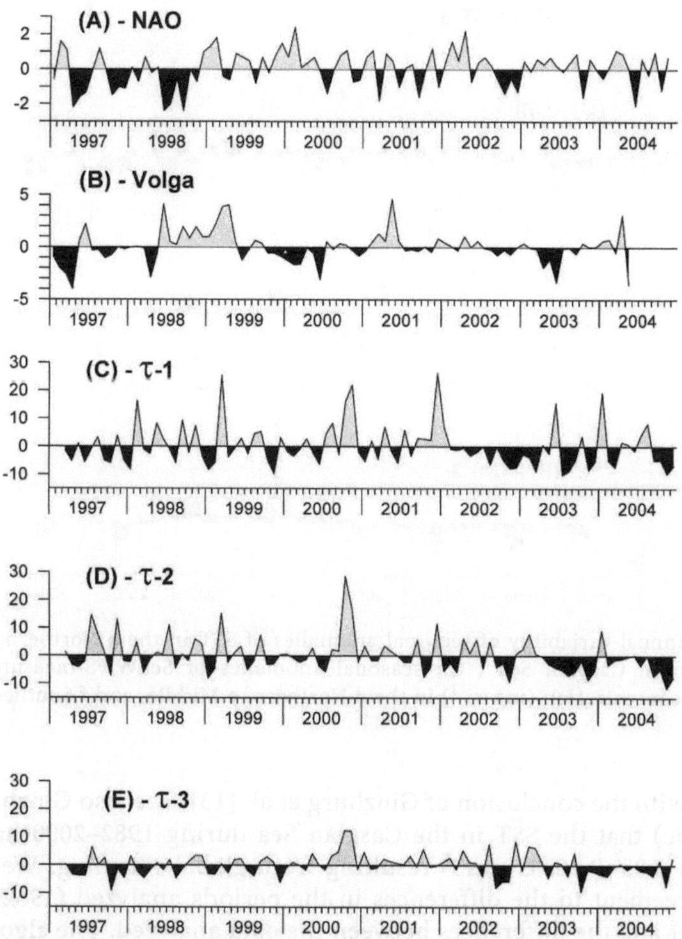

Fig. 6 Interannual variability of **a** the North Atlantic Oscillation (*NAO*) Index, **b** seasonal anomalies of Volga discharge (1000 $m^3 s^{-1}$), seasonal anomalies of wind stress over the **c** Northern, **d** Middle, and **e** Southern Caspian Sea (kg $m^{-1} s^{-2}$)

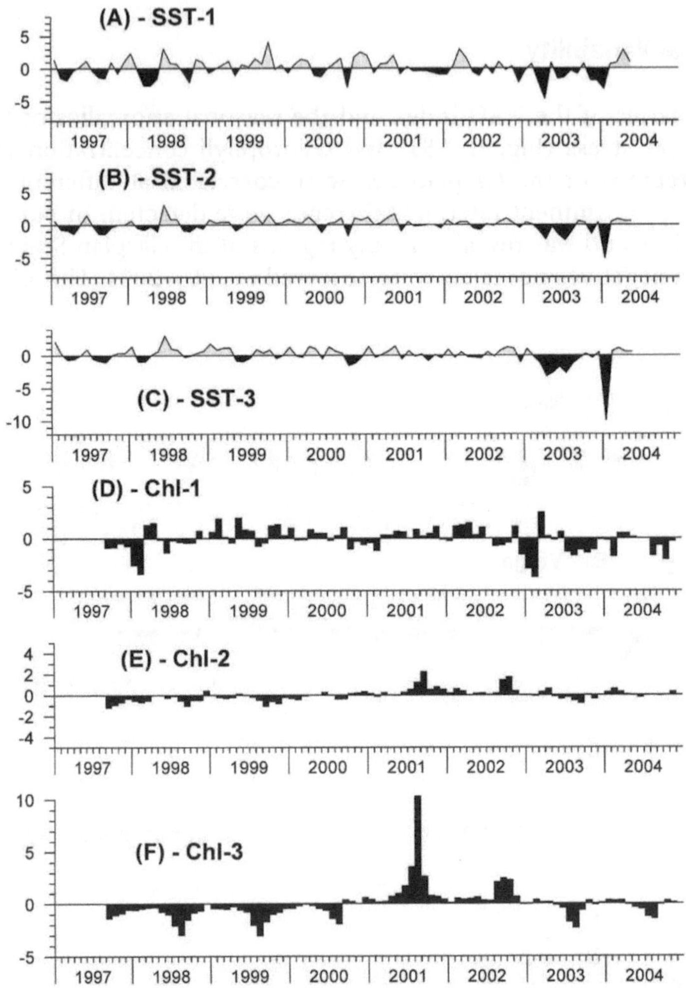

Fig. 7 Interannual variability of seasonal anomalies of SST in the **a** Northern, **b** Middle, and **c** Southern Caspian Sea (°C); seasonal anomalies of SeaWiFS-measured surface chlorophyll concentration (mg m^{-3}) in the **d** Northern, **e** Middle, and **f** Southern Caspian Sea

disagrees with the conclusion of Ginzburg et al. [13] (see also Ginzburg et al., this volume) that the SST in the Caspian Sea during 1982–2000 had a positive trend 0.05–0.10 °C $year^{-1}$ resulting from global warming. We attribute this disagreement to the differences in the periods analyzed (1982–2000 vs. 1997–2004) and the differences between the data analyzed. The algorithms of the estimation of SST from IR satellite radiometry were different in this study and in the study of Ginzburg et al. [13] (Multi-Channel SST, MCSST, algo-

Table 1 Correlation coefficients and the time lags of maximum correlation (in *parentheses*) between North Atlantic Oscillation (*NAO*) and the seasonal anomalies of Volga discharge, sea surface temperature (*SST-1; SST-2; SST-3*), median chlorophyll concentration (*Chl-1; Chl-2; Chl-3*), and wind stress (*τ-1; τ-2; τ-3*) in three regions of the Caspian Sea. The *sign of the time lag* means the variable in the row leads the variable in the column (e.g., maximum correlation between NAO and SST-1 was observed at NAO leading SST-1 by 3 months). Only significant ($p < 0.05$) correlations are given. Correlations of high significance ($p < 0.01$) are in *bold*

	Volga discharge	SST-1	SST-2	SST-3	Chl-1	Chl-2	Chl-3	τ-1	τ-2	τ-3
NAO	–	+0.233 (+3)	**+0.364 (+3)**	+0.262 (+3)	**+0.320 (0)**	–	+0.232 (+10)	–	–	+0.210 (0)
Volga discharge	1.00	**+0.308 (0)**	**+0.308 (0)**	+0.240 (0)	+0.259 (+2)	–	+0.274 (+3)	**+0.288 (−6)**	**+0.310 (−7)**	–
SST-1	X	1.00	**+0.730 (0)**	**+0.524 (0)**	+0.269 (−2)	**+0.283 (+6)**	–	+0.263 (−4)	−0.280 (−8)	−0.267 (−8)
SST-2	X	X	1.00	**+0.827 (0)**	+0.269 (−3)	**−0.367 (−9)**	−0.259 (−9)	**+0.315 (−4)**	+0.241 (1)	–
SST-3	X	X	X	1.00	+0.241 (−3)	**−0.323 (−9)**	–	−0.238 (0)	–	–
Chl-1	X	X	X	X	1.00	–	–	+0.217 (−2)	–	+0.265 (−3)
Chl-2	X	X	X	X	X	1.00	**+0.832 (−1)**	**+0.430 (−10)**	**+0.343 (−10)**	**+0.386 (−11)**
Chl-3	X	X	X	X	X	X	1.00	**+0.434 (−9)**	**+0.415 (−10)**	**+0.482 (−10)**
τ-1	X	X	X	X	X	X	X	1.00	**+0.587 (0)**	–
τ-2	X	X	X	X	X	X	X	X	1.00	**+0.555 (0)**
τ-3	X	X	X	X	X	X	X	X	X	1.00

rithm [14] in the study of Ginzburg et al. [13] vs. Pathfinder SST algorithm [7] in this study), which could influence the results of the quantitative analysis.

The seasonal anomalies of chlorophyll concentration in the Northern Caspian Sea were not correlated with chlorophyll concentrations in the Middle and Southern Caspian Sea. In contrast, the chlorophyll concentrations in the last two regions were highly correlated ($r = 0.832$), with the variations in the Southern Caspian Sea leading the variations in the Middle Caspian Sea by 1 month. The absence of correlation between the shallow Northern Caspian Sea and the deep Middle and Southern Caspian Sea regions can be explained by dramatic differences in the environmental conditions between these two areas. In the deep regions of the Caspian Sea, the process of phytoplankton growth is regulated by vertical stratification of the water column, similar to other deep open regions of the World Ocean [15]. The pycnocline established in summer as a result of the heating of the surface layer by solar radiation works as a natural boundary separating deep layers rich in nutrients from the well-illuminated upper mixed layer where phytoplankton is concentrated. This boundary results in a nutrient limitation of phytoplankton growth. Wind stress and/or the cooling of the sea surface erode the pycnocline, resulting in an increase of nutrient flux into the upper layer, which in turn stimulates phytoplankton growth.

In the shallow Northern Caspian Sea (depth 5–6 m) with high water turbidity and significant freshwater discharge this scheme does not work and other processes regulate the phytoplankton growth rate. In particular, positive anomalies of Volga discharge resulted in positive anomalies of chlorophyll concentration 2 months later. We speculate that more intensive Volga runoff transported more nutrients [16], which resulted in higher phytoplankton biomass. The chlorophyll concentration in the Southern Caspian Sea is also correlated with Volga discharge, but the time lag is 3 months rather than 2 months in the Northern Caspian Sea. Three months look like the period during which the Volga river input reaches the Middle and then the Southern Caspian along the western coast [5].

The chlorophyll concentration in the Northern Caspian Sea was positively correlated with the NAO index and SST. This correlation can result from the correlation between Volga discharge and phytoplankton growth, on the one hand, and the correlation between SST, NAO index, and Volga discharge, on the other. The NAO index was positively correlated with SST over the entire Caspian Sea area, SST trailing NAO by 3 months. The positive NAO index phase shows a stronger than normal subtropical high-pressure center over the North Atlantic and a deeper than normal Icelandic low. The increased pressure difference results in more and stronger winter storms crossing the Atlantic Ocean on a more northerly track, which in turn results in warm and wet winters in Europe [17]. In the Caspian Sea area, these meteorological variations result in higher SST a few months later.

The waters of the shallow Northern Caspian are mixed and turbid; so, light limitation of phytoplankton growth can play a prominent role there. As such, higher SST enhances vertical stratification and retains phytoplankton in the thin well-illuminated upper layer, stimulating primary production, which results in increased chlorophyll concentrations. Wind stress is also positively correlated with chlorophyll concentration anomalies in the Northern Caspian with a time lag of 4 months. We speculate that an intensive wind stress results in resuspension of bottom sediments in this shallow area, enriching the water column with nutrients and stimulating phytoplankton growth.

SST anomalies are negatively correlated with chlorophyll concentration anomalies in the deep Middle and Southern Caspian Sea areas. Indeed, low SST indicates more intensive upwelling or turbulent mixing, resulting in increased nutrient flux to the upper euphotic layer, an increase of the phytoplankton growth rate, and more phytoplankton biomass. However, the time lag between the signal (low SST) and the response (high chlorophyll concentration) is as long as 9 months. We speculate that more intensive convection of the water column in winter results in greater phytoplankton biomass during the period of its maximum in late summer (August to September). Earlier we observed a similar correlation in the western Mediterranean Sea [18]. A similar relationship is observed between the remotely sensed chlorophyll concentration and wind stress. More intensive wind mixing results in higher phytoplankton biomass in the deep parts of the Caspian Sea 10–11 months later.

5 Discussion

Analyzing the time series of the seasonal anomalies of chlorophyll concentrations in the Southern Caspian Sea (Fig. 7f), we focus on the prominent maximum of the chlorophyll concentration (i.e., phytoplankton biomass) in summer 2001. This maximum was not related to the variations in either SST or wind stress. Some authors [5] attribute this phytoplankton maximum to the consequences of the invasion of the carnivorous ctenophore *Mnemiopsis leidyi*, which was observed in the Middle and Southern Caspian Sea in summer 2001 [11], being especially abundant in the southern part of the sea [19]. Increased carnivorous pressure on zooplankton could decrease its biomass and grazing pressure on phytoplankton, resulting in prominent phytoplankton blooms observed on satellite images. A similar effect (the bloom of *M. leidyi* with simultaneous increase of phytoplankton biomass) was observed in the 1990s in the Black Sea [20]. In the second half of the 1990s *M. leidyi* was transported from the Black Sea to the Caspian Sea in the ballast waters of cargo ships traversing the Volga–Don canal [21] and during this short period achieved a high density in the Caspian Sea, resulting in a decline of the biomass of zooplankton and zooplanktivorous fish [11].

The results of this study indicate that seasonal and interannual variability of remotely sensed chlorophyll concentration in the Caspian Sea can be explained with meteorological conditions in that area, in particular SST and wind stress. In the Northern Caspian Sea, Volga discharge also plays a significant role. All these factors are connected to global meteorological cycles, e.g., the NAO index, which regulates the weather conditions in Eurasia, including the Caspian Sea region.

Our study shows that during 1997–2004 Volga discharge was not correlated with the NAO index. At the same time, on an annual time scale over several decades, the periods of the positive phase of the NAO are related to lower Volga discharge and declining Caspian Sea level. Rodionov [17] provides the following explanation. During positive NAO the mid-latitude westerlies over the North Atlantic are strong and the storm tracks over Europe are shifted northward, leaving room for a high-pressure cell which is established over the Volga River basin. This climatic situation results in lower than normal precipitation in southeastern Europe and Volga discharge. During recent years (1983–1990), the pattern of atmospheric circulation changed. Northward shift of the rainstorm tracks was not observed, resulting in positive rather than negative correlation between the NAO index and the precipitation in the Caspian Sea basin [17]. However, the correlation on a monthly time scale can be significantly different from the correlation on an annual time scale. We speculate that on a monthly time scale during 1997–2004 an intensification of zonal atmospheric circulation during positive NAO did not significantly influence the Volga discharge. At the same time, the shift of rainstorms northward resulted in higher SST of the Caspian Sea, which was manifested in positive correlation between SST and the NAO index (Table 1).

The weather in the Caspian Sea region is related to climatic patterns in other regions of the Earth. A teleconnection exists not only between NAO and the Caspian Sea, but also between the SST in the equatorial Pacific and the Volga discharge resulting in the Caspian Sea level variations [22]. The SST anomalies in the equatorial Pacific, called the El Niño southern oscillation (ENSO) cycle influence weather conditions all over the world [23, 24], including the precipitation over the Volga watershed in autumn and the river discharge during the following spring. SST in the Caspian Sea is also influenced by the ENSO cycle [13]. Some authors relate these global meteorological cycles with solar activity [17, 25, 26]. Other authors [27] recognize the North Sea–Caspian pattern (NCP) oscillation, not correlated with NAO, as the most important for the weather in eastern Europe. The influence of these interannual cycles on phytoplankton biomass in different basins, including the Caspian Sea, requires further data collection and analysis.

Acknowledgements The author would like to thank the SeaWiFS project (code 970.2) and the DAAC (code 902) at the NASA GSFC for the production and distribution of remotely sensed images, respectively. These activities are sponsored by NASA's Mission to

Planet Earth program. I also thank the NASA PODAAC at the JPL, California Institute of Technology, for SST data. I thank A.G. Kostianoy and S.A. Lebedev for cooperation and providing me with important information on recent observations in the Caspian Sea and S.N. Rodionov for critical discussion of the results.

References

1. Acker JG, Shen S, Leptoukh G, Serafino G, Feldman G, McClain C (2002) IEEE Trans Geosci Rem Sens 40:90
2. O'Reilly JE, Maritoena S, Mitchell BG, Siegel DA, Carder KL, Garver SA, Kahru M, McClain C (1998) J Geophys Res Ocean 103:24937
3. Smith RC, Baker KS (1978) Limnol Oceanogr 23:247
4. Morel A, Prieur L (1977) Limnol Oceanogr 22:709
5. Kopelevich OV, Burenkov VI, Ershova SV, Sheberstov SV, Evdoshenko MA (2004) Deep-Sea Res II 51:1063
6. Banse K, English DC (1994) J Geophys Res Ocean 99:7323
7. Walton CC (1988) J Appl Meteorol 27:115
8. Deschamps P-Y, Frouin R (1984) J Phys Oceanogr 14:177
9. Large WG, Pond S (1981) J Phys Oceanogr 11:324
10. Sur HI, Ozsoy E, Ibrayev R (2000) Satellite-derived flow characteristics of the Caspian Sea. In: Halpern D (ed) Satellites, oceanography and society. Elsevier, Amsterdam, p 289
11. Kideys AE, Moghim M (2003) Mar Biol 142:163
12. Kosarev AN, Tuzhilkin VS, Kostianoy AG (2004) Main features of the Caspian Sea hydrology. In: Nihoul JCJ, Zavialov PO, Micklin PP (eds) Dying and dead seas: climatic and anthropic causes. Kluwer, Dordrecht, p 159
13. Ginzburg AI, Kostianoy AG, Sheremet NA (2004) Oceanology 44:605 (English translation)
14. McClain EP, Pichel WG, Walton CC (1985) J Geophys Res Ocean 90:11587
15. Longhurst A (1995) Prog Oceanogr 36:77
16. Leonov AV (2002) Oceanology 42:683
17. Rodionov SN (1994) Global and regional climate interaction: The Caspian Sea experience. Kluwer, Dordrecht
18. Nezlin NP, Lacroix G, Kostianoy AG, Djenidi S (2004) J Geophys Res Ocean 109:C07013. DOI 10.1029/2000JC000628
19. Vostokov SV, Ushivtsev VB, Lisitsyn BE, Soloviev DM (2004) Oceanology 44:90
20. Kideys AE (2002) Science 297:1482
21. Ivanov VP, Kamakin AM, Ushivtzev VB, Shiganova T, Zhukova O, Aladin N, Wilson SI, Harbison GR, Dumont HJ (2000) Biol Invas 2:255
22. Arpe K, Bengtsson L, Golitsyn GS, Mokhov II, Semenov VA, Sporyshev PV (2000) Geophys Res Lett 27:2693
23. Diaz HF, Hoerling MP, Eischeid JK (2001) Int J Climatol 21:1845
24. Ropelewski CF, Halpert MS (1986) Mon Weather Rev 114:2352
25. Shermatov E, Nurtayev B, Muhamedgalieva U, Shermatov U (2004) J Mar Syst 47:137
26. Lyatkher VM (2000) Geophys Res Lett 27:3727
27. Kutiel H, Benaroch Y (2002) Theor Appl Climatol 71:17

Hdb Env Chem Vol. 5, Part P (2005): 159–173
DOI 10.1007/698_5_008

Published online: 6 October 2005

Biodiversity

Mikhail G. Karpinsky

Russian Federal Research Institute of Fisheries & Oceanography (VNIRO), 17, V. Krasnoselskaya, 107140 Moscow, Russia
Karpinsky@vniro.ru

Abstract Modern autochthonous Caspian fauna evolved from a few marine species over 1.8 million years in conditions of splendid isolation, existing long-term in brackish water and without competition from marine species. A weak morphological disconnection of the autochthonous Caspian species testifies to the fact that the fauna is young and that the speciation process in the Caspian is very active. The Caspian Sea became one of the largest centers of speciation of brackish-water fauna. Autochthonous brackish-water species, found either only in this sea (46% of the species checklist consisting of macrobenthos, zooplankton and fishes), or in the Ponto-Caspian region (20%) predominate. Some species are of freshwater origin, and a few organisms from the Arctic and the Mediterranean Seas also penetrated the Caspian Sea. In the Caspian, the brackish-water species occupied an area with salinity of 12–13‰. The brackish-water species, having colonized the Northern Caspian, turned out to be able to compete with the freshwater ones. The result of a comparison of Caspian, Azov and Black Seas fauna is that the marine species are more adaptive to environmental conditions, more specialized and more competitive.

Keywords Biodiversity · Brackish water fauna · Caspian Sea · Endemics · Speciation

1 Biodiversity

The flora and fauna of the Caspian Sea have been widely studied, and many new species both for the Caspian and to science have been described. However, there remain poorly studied groups as such as the infusorians, nematodes or ostracods. The total number of species differs according to the author read, but the variation is due more to where the limit between the brackish-water and freshwater species is drawn and whether protozoa and parasites are taken into account, rather than to the description of new species or the appearance of introduced species. The most complete list is given by Kasymov [1]: 733 species and subspecies of plants and 1814 species and subspecies of animals, of which 1069 are free-living invertebrates, 325 are parasites and 415 vertebrates, but freshwater species constitute the bulk of the list. One of the striking peculiarities of the Caspian Sea, caused by its long and splendid isolation from the ocean is the high percentage of the species and genera of animals and plants that are Caspian endemic. That peculiarity, in many respects, determines the uniqueness of the biological diversity of the Caspian.

Unicellate algae, referring to phytoplankton and partially to phytobenthos, constitute the main mass of plants. The largest groups are diatoms (Bacillariophyta, 292 species), blue-green (Cyanophyta, 203 species), green (Chlorophyta, 139 species) and pyrrophytes (Pyrrophyta, 39 species) algae; there are also 5 species of euglenas (Euglenophyta) and 2 species of yellow-green (Chrysophyta) algae. Phytoplankton of the Northern Caspian is considered typically freshwater estuarine, whereas euryhaline neritic marine and brackish-water species predominate in the Middle and Southern Caspian. According to the data for 1962–1976, 449 species (414 species in the Northern, 225 in the Middle and 71 in the Southern Caspian) were discovered in the Caspian Sea phytoplankton. The species diversity of the phytoplankton decreases from the north to the south, which occurs due to the lack of freshwater types. Most of the Caspian phytoplankton algae (210 species) are freshwater, 74 species are brackish-water–freshwater, 66 are brackish-water and only 47 can be considered marine, while another 52 are considered indeterminate [2]. The 129 species of diatoms and 55 species of blue-green algae inhabiting, mainly, sediment substrata and included in the fouling biotope, as well as some species of green algae such as *Enteromorpha* and *Cladophora* that form filamentous colonies or more complicated formations able to attach to the bottom, are referred to as microphytobenthos.

The number of macrophytobenthos plants is not great in the Caspian, comprising 13 representatives of brown (Phaeophyta) and 25 species of red (Rhodophyta) algae. However, only stonewort (Charophyta, 10 species) form appreciable settlements, developing mainly in the shallow water and in silty bays, protected from seaway. Five species of the higher plants inhabit the Caspian Sea, not including lesser reedmace (*Typha angustifolia*), common reed (*Phragmites communis*) and other coastal plants growing in the river deltas. Eelgrass (*Zostera minor*) is found at sandy and sandy–shelly grounds, fennel-leaved pondweed (*Potamogeton pectinatus*) grows in the coastal zone, and naias (*Najas marina*), and sea grass (*Ruppia spiralis, R. maritima*) grow in the bays.

Most species of phytoplankton and phytobenthos are included in the groups of freshwater or marine plants and are found in other reservoirs. However, a large number of species in the brackish-water plants group and to a lesser degree in the brackish-water–freshwater plants group are found only in the Caspian Sea. According to the data of Makarova [3], of the 73 discovered phytoplankton species of diatom algae, 22 are Caspian endemic, and of these 10 are relicts; all of them belong to brackish-water and brackish-water–freshwater groups. The highest content of species endemic is among diatom algae, but they are in other groups as well, and again, mainly in brackish-water groups.

The flora of the Caspian Sea available for study by light microscope is now very well known, and it is likely that any new species of plants discovered will be introduced species. However there is a large group of small, naked flagellate algae referred to as nano- and picophytoplankton, the study of which is in its infancy. It is already known that they make up more than half of the total phytoplankton biomass, and due to their small size and high abundance, their role in production of organic matter is much higher [4]. At the same time the species composition of that group in the Caspian Sea is so far unknown, and advantage should be taken of any opportunity to update the checklist of vegetable organisms in the future.

The composition of the free-living invertebrate fauna of the Caspian Sea is very different from that of the ocean, seas with marine salinity and even from those brackish-water seas of similar salinity and a considerable common history that are united into one Ponto-Caspian (the Black, Azov and Caspian Seas) biogeographic region. This is because many of the high taxa characteristic of the seas and oceans are either absent (Radiolaria, Siphonophora, Brachiopoda, Scaphopoda, Loricata, Cephalopoda, Echiuroidea, Sipunculoidea, Priapulida, Chaetognatha, Echinodermata, Tunicata and others), or are represented by only a small number of species (Porifera, Coelenterata, Nemertini, Kamptozoa, Polychaeta, Bryozoa, and Decapoda from Crustacea) due to the low salinity and long-term isolation of the Caspian.

Protozoa are represented in the Caspian by some taxa referring to four classes. Ten species of naked amoebae (Amoebida) inhabiting water depth and sediment surface, as well as 18 species of benthic foraminifers (Foraminifera) belong to the class of Sarcodina. Amoebae are characteristic for freshwaters, whereas foraminifers are typically marine dwellers; they are all found in the Southern Caspian, while there are 16 species in the Middle Caspian, and this number drops to 12 in the desalted Northern Caspian. In the coastal zone there are 2 species of Heliozoa and Flagellata are represented by 40 species. Of the protozoa of the infusorians class (Ciliata), 439 free-living and 21 parasitic species constitute most of the mass, and are very well studied. Most of them (305) are psammophilous interstitial species dwelling in water among the fine sand grains of sediments of the coastal zone, 172 species are classed as part of the fouling community and 135 as part of the plankton, though some species can be classed in two communities at once. Marine species make up 90% and freshwater 10% of the free-living infusorians of the Caspian Sea.

The only species of endemic Caspian sponge (Porifera) has three forms differing from each other by some peculiarities of the skeleton and constituents of their spines. Five species, two autochthonous and three Azov–Black invaders, of coelenterates (Coelenterata) are found in the Caspian. Practically all 29 species of free-living flatworms (Turbellaria) inhabit the Middle and the

Southern Caspian, mainly near the eastern shore. Cosmopolitan freshwater nemertean (Nemertini) common to the Volga are also found in the Caspian. Free-living nematodes (Nematoda) are classed as meiobenthic organisms, and although 52 species have already been discovered in the Caspian, the abundance of the species in that group along with the widespread marine species new to science testifies not so much to a large number of species endemic, as to the fact that this group has been little studied so far. Rotifers (Rotatoria) are chiefly freshwater organisms; most of the 67 species are found in the pre-mouth space of the Volga, and they enter the sea together with river waters. However, there is a group of euryhaline species in the Caspian with a vast distribution, and there is also a group of endemic (four species and subspecies) brackish-water rotifers. A representative of the class Kamptozoa, having rather recently penetrated into the Caspian Sea, became one of the major species involved in the fouling of ships and hydraulic structures.

All three classes of annelid worms (Annelida) are present in the Caspian Sea. Oligochaetous worms (Oligochaeta) are represented by the most species, 31, of which 24 inhabit the Northern Caspian. These worms being freshwater by origin and being Caspian endemic in most cases, are the most frequently found organisms in the benthos. Of seven species of polychaetous worms (Polychaeta) one is Caspian endemic, four are autochthonous Ponto-Caspians and two are introducers from the Mediterranean (one of which is not acclimatized). Three autochthonous leeches (Hirudinea) are freshwater in origin, and two of them are endemic.

Crustaceans are one of the organisms most characteristic of and at the same time specific to the Caspian Sea; most of them are autochthonous and many of them are endemic. Cladocerans (Cladocera, 55 species) are characteristic for freshwater and most of them are found in the desalted territories near the river mouths. At the same time representatives of family Polyphemidae have inhabited the Caspian brackish-water part with salinity of up to 12–13‰, where all 24 species live; of these 23 are autochthonous and 16 are endemic, 2 species have 12 endemic forms and only one is a cosmopolitan introducer that was discovered for the first time in 1957. Copepods (Copepoda, 19 species) are a group of diverse origins: seven species are endemic, four live also in the Azov and Black Seas and come into rivers, six are widely distributed in other seas and brackish-water waters, one is also known in the Baikal lake and one is of Arctic origin. Furthermore, eight species of parasitic copepods, of which none is marine, have been found in the Caspian Sea. Two species of Mediterranean invaders represent acorn barnacles (Cirripedia); however one of them, *Balanus eburneus*, was discovered in the Turkmenbashy Bay, but it has not been found since. The numerous ostracods (Ostracoda, 46 species) are mostly benthic. There are a lot of newly described species among them, which testifies to insufficient study of them.

Representatives of three orders of crustaceans are the most characteristic of the Caspian. A large proportion of the Caspian mysids (Mysidacea) are nektobenthic, but some are capable of coming up into the water column, and four species are constantly met in plankton. All 20 species of mysids are autochthonous; 13 of them are endemic, and 7 species only to the Ponto-Caspian, while some euryhaline species are entering the deltas and lower rivers. In the opinion of G.O. Sars [5] the diverse 18 species of Caspian Cumacea arose from one progenitor form, an immigrant from the Mediterranean Sea (Fig. 1); thus all are autochthonous, and of them, 3 genera and 7 species are endemic, and a further 11 only to the Ponto-Caspian, although many species come into the deltas and lower rivers and start colonizing them. Amphipods (Amphipoda) are the most abundant of the crustaceans by species (74), and are the most important in the benthos community. Sixty nine species are autochthonous, four species are of Arctic origin, one is of the Mediterranean, 31 species are Caspian endemic, and 38 are also found in the Azov–Black Sea basin, from where many species have penetrated into rivers, some have colonized the Baltic Sea, and one reached the coast of England. Most of the Caspian Amphipoda as well as Cumacea are benthic; only two species, of genera *Pseudalibrotus*, are frequently found at water depth.

Isopods (Isopoda) are present in the Caspian as two endemic subspecies inhabiting the Arctic basin and freshwaters. Finally, in addition to two species of crayfishes from decapods (Decapoda), one of which is endemic, two Azov–Black Sea shrimps and the crab, whose homeland is the Atlantic coast of America, were introduced in the last century.

Bivalve mollusks (Bivalvia) as well as crustaceans are among the most characteristic benthos organisms in the Caspian Sea. Their diversity is not great: 25 species, 8 of which comprise 17 subspecies. Bivalves are the largest benthic organisms; their mass can reach 10 g, as a result of which mussel biomass makes up 90% of the biomass of the total benthos and thus de-

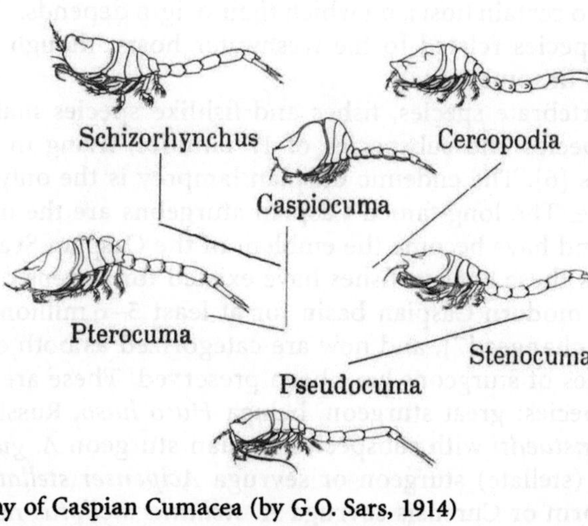

Fig. 1 Phylogeny of Caspian Cumacea (by G.O. Sars, 1914)

termines its quantitative distribution. Three genera, *Dreissena*, *Didacna* and *Hypanis*, to which most of the species belong, are autochthonous and are composed of 16 endemic Caspian species, 4 of which are divided into 10 subspecies, while another 4 Ponto-Caspian endemic species have 6 subspecies. One subspecies (and, correspondingly, the species) is widely distributed, and, finally there is one Azov species introduced into the Caspian. However, two endemic species of *Dreissena* are now completely supplanted by the introducer *Mytilaster*, together with another two Mediterranean species introduced into the Caspian Sea in the twentieth century. Gastropods (Gastropoda) in the pre-delta spaces of rivers are represented by various freshwater species, though two brackish-water species of genus *Theodoxus*, one of which is endemic, 3 endemic species of family Planorbidae, as well as a large group of up to 78 mollusk species of the family Pyrgulidae inhabit the Caspian itself. The mollusks of this group are small and very variable, and, since they are described only by the shell, the number given may be unreliable. Nevertheless, most of the species of that group are undoubtedly Caspian endemic.

The species of the typically freshwater groups of water mites (Hydracarina, Arachnoidea, two species) and the larvae of harlequin flies of families Chironomidae (eight species) and Ceratopogonidae (one species, Insecta, Diptera) mainly inhabit the Northern Caspian and the coast, but two species of chironomids, *Chironomus albidus* and *Clunio marinus*, the first of which is Caspian endemic, have colonized to depths of more than 100 m and start to predominate in the benthos at depths of 200 m and deeper, where other benthos species penetrate poorly. All six species of bryozoans (Bryozoa) are widely spread, though they belong to different groups: two species are freshwater, three are Mediterranean and one is autochthonous.

Concluding a review of the Caspian invertebrates it is necessary to mention 325 parasitic species of different taxa from protozoan to crustaceous. Most of them are related to certain hosts, on which their origin depends. Chiefly, these are widespread species related to the freshwater hosts, though the Caspian endemics formed in some taxa.

Of the 415 vertebrate species, fishes and fishlike species make up about a quarter, 100 species and subspecies of 17 families, living in the sea and in the river deltas [6]. The endemic Caspian lamprey is the only representative of the fishlike. The long-famed Caspian sturgeons are the most popular and important, and have become the emblem of the Caspian Sea. Judging by the fossil remains these ancient fishes have existed for dozens of millions of years and in the modern Caspian basin for at least 5–6 million years without considerable changes [7], and now are categorized as both endemic and relicts. Five species of sturgeons have been preserved. These are the Caspian Sea migratory species: great sturgeon, beluga *Huso huso*, Russian sturgeon *Acipenser gueldenstaedti* with subspecies Persian sturgeon *A. gueldenstaedti persicus*, starred (stellate) sturgeon or sevruga *Acipenser stellatus*, with the South-Caspian form or Curensis sevruga *A. stellatus stellatus* natio *cyrensis*, spiny sturgeon or ship (russian name), *A. nudiventris* and living only in the Volga basin, the sterlet *A. ruthenus*.

Family clupeid (Clupeidae) is represented by 18 species and subspecies found only in the Caspian and in the rivers of its basin: 3 sprats and 15 herrings. All 5 species with 15 subspecies of the Caspian herrings developed from one species [8]. Three Far East species (calico salmon, humpback salmon, hoopid salmon) were acclimatized in addition to two Caspian salmons, Caspian salmon *Salmo trutta caspius* and Caspian inconnu *Stenodus leucichthys*. A large group of freshwater fishes of the families Esocidae, Cyprinidae, Siluriae, Percidae (31 species) inhabit the desalted zones near rivers, mainly the Volga Mouth, although some species have formed brackish-water Caspian subspecies. Pipefish *Syngnathus nigrolineatus caspius* as well as sand smelt *Atherina mochon pontica* natio *caspia*, having penetrated into the Caspian 13 000 years ago from the Azov and Black Seas, formed endemic subspecies. Two species of gray mullet and starry flounder acclimatized in the previous century, and now gray mullets are of commercial importance. The family Gobiidae is very interesting: 26 species of 37 are Caspian endemic; the rest settled in rivers, and one even penetrated into North America. They inhabit both weakly salty regions of the sea and regions of high salinity. Most of the species inhabit the Middle and Southern Caspian, mainly, at depths down to 50 m. They prefer the coastal regions of the sea with sandy soil, stones and rocks, and avoid silty areas.

Among the other vertebrates in the Caspian Sea there are 2 species of grass snake (*Natrix*) and 312 species of waterfowl and near-waterfowl birds. The only mammal species is a seal *Pusa caspica*, an Arctic immigrant to the Caspian Sea in the Glacial period that developed into an independent species.

A general analysis of the described fauna allows some conclusions. The smaller species able to endure unfavorable conditions, spread in part by birds (mainly, infusorians), are chiefly of marine origin and are widespread. Larger brackish-water species are for the most part autochthonous in spite of marine (Amphipoda, Cumacea, Bivalvia and others) or freshwater (Rotatoria, Oligochaeta, Hirudinea) origins and are endemic in most cases; where they penetrated into other reservoirs, it was through freshwaters. Finally, the group of brackish-water–freshwater organisms is widespread and also penetrated the Caspian via freshwaters. The highest number of endemics are benthic, not only among invertebrates, but among fishes as well, insofar as gobies can be referred to as bottom fauna. The proportion of endemics among plankton and nekton organisms is smaller, since freshwater species play an important role in these biotopes.

In the Caspian Sea, autochthonous brackish-water species found either only there (46% of the species checklist, consisting of macrobenthos, zooplankton and fishes) or in the Ponto-Caspian region (20%) predominate (Fig. 2). Autochthonous organisms, being initially Tertiary period marine fauna and flora that have adapted to substantial desalting and have evolved into groups of similar species are considered an ancient relict complex originating from the marine population of the Tethys [9]. Until recently these very species were the major inhabitants of the Caspian; however after the appearance of a relatively small group of species of Mediterranean origin in the previous century, some have been excluded from their biotopes. The highest species diversity is observed among benthos organisms, and the highest number and share of autochthonous species and endemic species are also in that group.

Flora and fauna of the Caspian have a series of peculiarities that distinguish it from other seas and lakes and determine many biological processes occurring in it. In order to understand the reasons for the emergence of these peculiarities and to determine the source of uniqueness of the organisms, it is necessary to analyze the history of the formation of Caspian Sea fauna (it is more difficult with flora, as the paleontological data are poor).

The historical development of the water body known today as the Caspian Sea proceeded rather impetuously by geological standards. The periods when there was contact with the ancient Atlantic and Indian Oceans or with the Mediterranean Sea, gave place to periods of long-term isolation. Together with all these changes, fauna was also replaced. The waters at the junction with the ocean reached oceanic salinity; even in the case of desalting they were desalted insignificantly and fauna was rich, diverse and typically marine. During the periods of isolation, the reservoir became enclosed, the water brought by rivers accumulated in it, and it was desalted considerably; at a lower river discharge the reservoir might become hypersalty. At the same time, few marine species were able to adapt to the new conditions that initiated formation of a new brackish-water fauna, which had the advantage of these conditions hampering the penetration of more competitive marine and freshwater species. At a very strong river discharge, the almost complete desalting resulted in the brackish-water fauna being replaced by a freshwa-

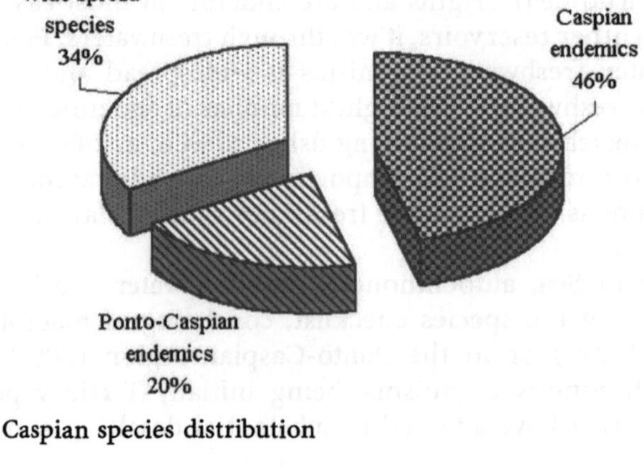

Fig. 2 Types of Caspian species distribution

ter one, and the marine species becoming entirely extinct. However, at the next junction with the ocean, the salinity rose and the marine species appeared again forcing out brackish-water species either completely or to river mouths and estuarine zones, breaking off the evolution of the brackish-water fauna and occupying the main space of the reservoir. The paleontological data mainly concern mollusks; ancestral forms of the modern brackish-water Dreissenidae and Cardiidae that have been preserved, and evolved, perhaps, from the end of the Tethys period in the changing reservoir at the replacement of marine with brackish-water and freshwater fauna. There are much poorer data for some crustaceans and fishes, from which it can be said that sturgeons existed here about 5–6 millions years ago and changed rather too little. In relation to other groups the information is insignificant and fragmentary. As a result it is hard to judge now whether the brackish-water fauna of the Caspian Sea has been formed over and over again or if the fauna is transformed relics of the old fauna; the fact that the overwhelming majority of brackish-water species are of obviously marine origin makes the first theory more probable.

Thus, the modern autochthonous Caspian brackish-water fauna replaced that which existed before the marine one, having preserved a genetic relationship with it. This occurred at the last disconnection from the ocean about 1.8 million years ago, when the Akchagyl basin lost its connection with the Indian Ocean, and its rich marine fauna became almost entirely extinct as the reservoir desalted. The fauna of the enclosed brackish-water Apsheron Sea-Lake originated from the remains of the Akchagyl basin, as well as from the freshwater and Euxin (Chaudine basin, the modern Azov–Black Sea province) basins of the Kumo-Manych depression. That sea and the Baku basin having emerged to replace it were similar in salinity to the modern Caspian and were inhabited by similar fauna consisting of species of the genera *Didacna*, *Adacna* (*Hypanis*), *Dreissena* and some others. In the Quaternary period the salinity variations were not great (from 5‰ to 20‰), and the fauna preserved in the Caspian is not only similar to the modern fauna, but also has species existing even now. In the period of the Early Khvalyn transgression (40–70 thousands years ago) the Caspian level rose greatly [10] and connection with the Black Sea was established through the Kumo-Manych depression, mainly, in the form of Caspian water discharging towards the west. At that time the salinity in both basins was approximately the same 5–7‰ and the Black Sea basin turned out to be inhabited by species of Caspian origin, and some Mediterranean species penetrated into the Caspian [9].

After separation from the ocean, salinity in the reservoirs forerunning the modern Caspian varied from 5–20‰, which corresponds to brackish waters. Brackish-water content as a phenomenon is met in nature rather frequently—in each zone of fresh and marine water mixing, territories with a lower salinity inevitably exist; they are seldom marked out morphologically by terrain features, forming rivermouth areas, lagoons, estuaries, bays, desalted bights, etc. The inland seas are often desalted, sometimes even too strongly, by the rivers outletting into them, as, for instance, in the Baltic, White, Black, and Azov Seas; however, there is always a contact with other seas of marine salinity and ocean and a possibility of marine species incoming from there, with which the brackish-water species have to compete. In the case of the Caspian Sea, there is a unique, splendidly isolated, long-term existing brackish-water reservoir quite commensurate with the seas by scale, where specific fauna was formed over a long period with no competition from marine species. The Caspian thus became one of the largest centers of speciation of brackish-water fauna. The fact that no endemic brackish-water taxon of family rank is known elsewhere, whereas the Caspian amphipod *Caspicola knipovitschi* is an example of a monotypic endemic family, serves as an indicator of the duration of the process of the brackish-water fauna formation in the Caspian.

Only a very small number of marine species were able to adapt to brackish-water conditions, out of which the fauna of the Caspian started to form under isolation. As mentioned, all the diversity of the Caspian Cumacea is descended from one progenitor form, an immigrant from the Mediterranean, as well as 5 species and 15 subspecies of the Caspian herrings. A high morphological similarity of species in other groups as well makes it possible to suppose the speciation to have originated in one or several initial forms. Almost all 76 species of the Caspian amphipods can be classed into several morphologically similar groups, and there are good reasons to believe that they have descended from five to six ancestral forms; the variability of the shells of autochthonous Cardiidae (Bivalvia) allows identification of transition forms among almost all species, though three major groups corresponding to genera are singled out; about 60 species having occasional minimal morphological differences are singled out from a large group of small gastropods of genus *Pyrgula*, but there is no clarity as to whether these species are independent or if it is merely species variability. This checklist can be continued. A weak morphological disconnection of the modern autochthonous Caspian species testifies to the fact that the fauna is young and is still in the process of formation, to which one of the first researchers of the Caspian fauna Grimm [11] paid attention, and the speciation process itself in the Caspian is very active. Besides the marine species, some species of freshwater origin have adapted to new conditions and can be also included in the autochthonous brackish-water species complex, although their share in the total species checklist is about 15–20%, and their role in the biomass is even smaller. Examples are oligochaetes, some species of rotifers, leeches, cladocerans, chironomids and gastropods. A splendid isolation, not only geographical, but also hydrochemical (low salinity and hydrochemical composition differing from ocean and freshwater) made it possible for a new fauna to develop and be formed beyond competition from both marine and freshwater species.

Besides autochthonous species and species of freshwater origin another two groups of organisms—Arctic and Mediterranean—penetrated into the

Caspian (Fig. 3). Representatives of the Arctic fauna (14 species), also currently found in the desalted bays of the Arctic basin, entered the Caspian during the Glacial period (about 13 000 years ago), having preserved species characters or having formed new subspecies remaining thus far similar to the initial forms. The other group is the Mediterranean species. Seven penetrated into the Caspian via the Kumo-Manych depression approximately 40–70 thousand years ago, and insignificant differences developed in some of them, allowing their classification as subspecies. All these species became organically part of the Caspian ecosystem long ago. However, a new group of about 30–50 mainly Mediterranean species introduced by man by chance or intent and united by the general name of introduced species appeared in the Caspian in the twentieth century. They have changed the Caspian ecosystem so greatly that it is necessary to discuss this problem separately in the next chapter [Karpinsky, Shiganova, Katunin, this volume].

The most accurate estimation of the resources of the Caspian fauna is possible by comparison with the fauna of other seas of analogous salinity and in the same climatic zone, such as the Azov and Black Seas. These were connected closely with the Caspian in their historical development and have been relatively recently inhabited by practically the same fauna before discovery of the Bosporus (about 12 000 years ago), which enables us to describe them as united in the Ponto-Caspian biogeographic region.

The Mediterranean species (about 80%), to which species of freshwater and Caspian origin (10% each) are added in the desalted, estuarine regions, are the basis of the Black Sea fauna. Nevertheless, the Black Sea species checklist is approximately 3.5 times less than that of the Mediterranean. In its turn, the Caspian fauna is inferior in diversity to that of the Black Sea. The Caspian species checklist according to the estimation of Zenkevitch [9] is 2.5 times less than that of the Black Sea (41.3%), mainly on account of benthos species, since there are more zooplankton species due to the plenitude of freshwater species in the Caspian. The diversity of the Black Sea benthos species is most striking in such

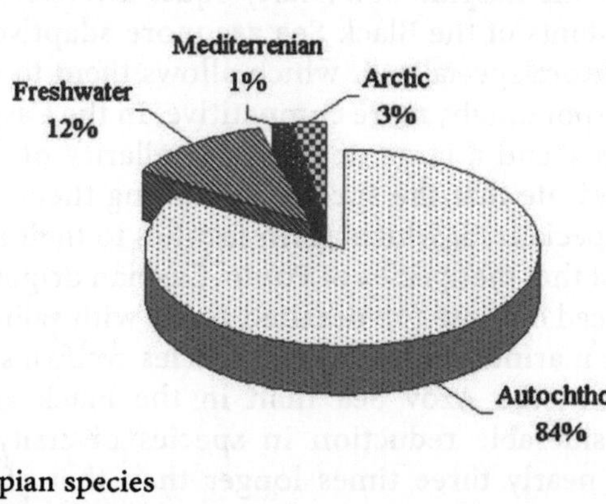

Fig. 3 Origin of Caspian species

groups as polychaetes (140 in the Black Sea and 6 in the Caspian), turbellarian worms (152 and 34), sponges (42 and 5), decapod crustaceans (35 and 5), coelenterates (29 and 5), nemerteans (18 and 1), isopods (15 and 2), and bryozoans (12 and 4), and there are no tunicates (in the Black Sea 16 species) or echinoderms (4) in the Caspian. Comparing only benthos species, the Caspian species make up 32% of the number of those in the Black Sea.

The biocenoses comparison is revealing and very demonstrative for the benthos fauna. Nineteen shelf biocenoses are identified in the Black Sea [12]—there are only ten species in the Caspian Sea—and autochthonous species dominate only in six of them: eight sestonophagous biocenoses and two of Mediterranean-introduced deposit feeders. Besides, the number of species in the Black Sea biocenoses usually makes up 10–20% and seldom up to 25% of the total, whereas in the Caspian it is 40–75%; it might be said that practically all species are present in one part of the biocenoses, while in the other part some groups are missing [13, 14].

Similarity in geomorphologic structures of depressions, current patterns, production of organic matter and sedimentation, insufficient salinity, common climatic zone, long-term common history and even the fact that depths more than 100–150 m are inhabited poorly because of the absence of deep-water fauna in the Caspian and are lifeless owing to hydrogen sulfide contamination in the Black Sea; all this means that the environmental factors of both seas are similar both in qualitative characteristics and in their diversity. Unfortunately, there is still no accurate definition of biotope diversity because of the impossibility of singling out a unit of a biotope; this is a hierarchical notion: both a sea and a water drop can be aquatic biotopes, which is why one has to be content with the subjective expert estimation of the comparison. Nevertheless, the mentioned similarity of the diverse environmental factors gives many reasons to suppose that the diversity of biotopes in these seas is almost the same in spite of the fact that the dimensions of the analogous biotopes might be different.

In that case, the considerable superiority of the species diversity of the Black Sea fauna over the Caspian at a nearly equal diversity of biotopes indicates that the inhabitants of the Black Sea are more adaptive to the environmental conditions: more specialized, which allows them to use potentials of biotopes, and, correspondingly, more competitive. In the Caspian, the smaller number of biocenoses and a large degree of similarity of the species composition in them indicate that the species inhabiting them are more plastic, universal, and less specialized, which again testifies to their low level of competitiveness. The fact that the species of Ponto-Caspian origin in the Black Sea turned out to be forced out into the desalted zones with salinity of 1–3% that are avoided by most marine and freshwater species confirms this situation.

Salinity is lower in the Azov Sea than in the Black and the Caspian, and there is a considerable reduction in species diversity; the Black Sea species checklist is nearly three times longer than that of the Azov. However, the greater part (about 60%)of the Azov checklist, as usual, is made up of Mediterranean species, 20% are of the Caspian and freshwater, brackish-water and lacustrine halophilous species make up the rest. The Azov fauna is inferior to the Caspian in number of species, other than in such "marine" groups as polychaetes (38 species), isopods (7) and decapod crustaceans (9); it is inferior in the traditionally "Caspian" fauna— amphipods (49), cumaceans (12), turbellarians (5)—while outnumbering the Caspian fauna in benthos organisms and being inferior in plankton. Such a comparison reveals that the checklist of Azov fauna is 92% of that of the Caspian [15, 16]. In the Azov Sea, 13 benthos biocenoses were identified [17]; biocenoses of boreal *Mya arenaria*, the Far East *Cunearca cornea* [18] and other invaders have been added to them recently.

In the Azov shallow water in the absence of permanent currents and independent water masses, biotope diversity is undoubtedly inferior to that of the Caspian. Although, as mentioned above, there is no accurate definition, based on which different biotopes might be distinguished and estimated, a simple comparison of resources of the potential habitats testifies to a considerable, at least twofold or even threefold, superiority of the Caspian to the Azov in the number of biotopes.

Such a reduction in biotope diversity at nearly analogous species diversity, as in the comparison with the Black Sea, testifies again to the fact that the Azov species are more specialized and better adapted to the biotope conditions than those of the Caspian. The species composition of the Azov biocenoses is not great (10–40 species [17]), which is very inferior to that of the Caspian Sea, where the species number in samples is usually 10–15 and in the biocenoses, 40–90 [14]. Again this shows that the Caspian species adapt more easily to different environments, which is possible at a lower level of specialization. Though only a small part of the Mediterranean species penetrated into the Sea of Azov, their higher degree of specialization and, correspondingly, competitiveness made it possible to force out species of Caspian origin into the desalted zones and to form quite settled communities in the rather unstable conditions of the Azov.

The brackish-water species might exist at salinity from 0 to 17–20‰; however, it is possible if there is no competition with species of other ecological groups. At the area of contact between marine and freshwater species they usually live at 2–5‰ salinity. The Caspian brackish-water fauna was formed under conditions of long-term isolation from other seas and the oceans, and thus without any competition with marine species, but in constant contact and, correspondingly, competition with freshwater species. As a result, in the Caspian Sea, the brackish-water species occupied the bulk of the reservoir with salinity of 12–13‰; however, in the Northern Caspian, where salinity varies from 0 to 8–10‰, most of them were forced out by freshwater species. This means that when marine species penetrate the Caspian, they meet favorable conditions and no real competition with native species, which makes it possible for them to acclimatize successfully. The effect of such an introduction is considered in detail in the next chapter [Karpinsky, Shiganova, Katunin, this volume].

In their turn, the brackish-water species, having colonized the Northern Caspian, turned out to be able to compete with the freshwater species. Owing to strong adaptive abilities many of them managed to penetrate through the deltas of the Volga and the Urals into the rivers themselves and then into lakes and even into the brackish-water seas. Gammaridae and Cumacea settled most successfully. The same species as live in the Caspian also penetrated from the Azov–Black Sea basin as well, into such rivers as the Danube, the Dnepr, the Don, and the Bug. The Caspian species started to penetrate into rivers rather recently [19, 20], and this process has became more active with the growth of human activity. There are 44 species of crustaceans, 18 species of fishes, 8 species of mollusks and 2 species each of coelenterates and polychaetes [15] that have penetrated freshwaters. The Caspian species live and reproduce constantly in the lower rivers, but often come to the upper flow as well, inhabiting water storages and reservoirs with no direct connection with the Volga-Caspian basin. Gammaridae, Corophiidae, and Cumacea were the first to colonize the deserted biotopes after the breakdown of benthos biocenoses at dumping works in the Kujbyshev reservoir. Polychaetes *Hypania invalida* in the Ivan'kovo and Gorkij artificial reservoirs constitute much more of the biomass than in the Caspian. The 21 Ponto-Caspian species that have penetrated the Baltic Sea make up 40% of all introduced species appearing there in the last century. Zebra mussel *Dreissena polymorpha* has spread most widely; it was one of the first to begin invading the rivers and lakes of Europe and the Baltic Sea, and even crossed the Atlantic Ocean where it was found in the Great Lakes. This is not, however, the only representative of the autochthonous Caspian fauna in North America: *Cercopagis pengoi* and *Neogobius melanostomus* [21] have also penetrated there. Besides, a great number of Caspian species, initially *Paramysis lacustris*, *P. baeri*, *P. intermedia*, *P. ullskyi*, and *Limnomysis benedeni* acclimatized to the freshwater reservoirs and in some brackish-water seas, where they competed successfully and forced out the native species in meso- and oligotrophic lakes and reservoirs, but were inferior to them in eutrophics [22].

The species of the Azov–Black Sea estuarine complex descended from the Caspian autochthonous fauna, and some species of which do not distinguish from the Caspian, possess the same strongly adaptive abilities.

Thus, as a result of its development and conditions of the fauna formation, the Caspian Sea turned out to be poorly protected from penetration of marine species, and at the same time it serves as a source of species penetrating actively into freshwaters.

References

1. Kasymov AG (1987) Caspian Sea. Gidrometeoizdat, Leningrad (in Russian)
2. Levshakova VD (1985) In: Yablonskaya EA (ed) Caspian Sea. Fauna and biological productivity. Nauka, Moscow, p 25 (in Russian)
3. Makarova IV (1965) Bot Zh 50:498 (in Russian)
4. Borodin VE (1991) In: Fisheries-related studies of plankton, Pt II Caspian Sea. VNIRO, Moscow, p 102 (in Russian)
5. Sars GO (1914) In: Proceedings of the Caspian Expedition 1904, vol 4, Petrograd (Sankt-Petersburg), p 5
6. Kazantcheev EN (1981) Fishes of the Caspian Sea, Legkaya i pischevaya promyshlennost, Moscow (in Russian)
7. Bogatchev VV (1933) Materials on study of Tertiary ichtiofauna of Caucasus. Transactions of the Azerbaidjan Petroleum Investigation Institute, 15. Aznefteizdat, Baku (in Russian)
8. Kisselevitz KA (1923) The clupeids of the Caspi-Volga District, Proceedings of Astrakhan research-fisheries expedition, Astrakhan
9. Zenkevitch LA (1963) Biology of the USSR seas. AN SSSR, Moscow (in russian)
10. Rychagov GI (1997) Pleistocene history of the Caspian Sea. Moscow State University, Moscow (in Russian)
11. Grimm OA (1876) Caspian Sea and its fauna. Proceedings of Aral-Caspian expedition issue 2. Sankt-Petersburg Naturalists Society, St Petersburg (in russian)
12. Kiseleva MI (1981) Benthos of the Black Sea soft soils. Naukova Dumka, Kiev (in Russian)
13. Karpinsky MG (2002) Ecology of the benthos of the Middle and Southern Caspian. VNIRO, Moscow (in Russian)
14. Karpinsky MG (2003) Okeanologiya 43:400 (in Russian)
15. Mordukhai-Boltovskoy (1960) Caspian fauna in Azov-Black Sea basin. AN SSSR, Moscow (in Russian)
16. Mordukhai-Boltovskoy (1960) Zool Zh 39:1454 (in Russian)
17. Vorobev VP (1949) Benthos of Azov Sea, AzCherNIRO. Proceedings 13. Krymizdat, Simferopol, Ukraine (in Russian)
18. Frolenko LN, Dvinyaninova OV (1998) In: The main problems of fisheries and protection of waterbodies with fisheries in the Azov Sea basin. AzNIIRKH, Rostov-on-Don p 115 (in Russian)
19. Beklemishev VN (1923) Russ Hydrobiol Zh 2:213 (in Russian)
20. Birstein YaA (1935) Zool Zh 14:749 (in Russian)
21. Orlova MI (2000) In: Matishov GG (ed) Species introducers in the European seas in Russia. Apatity, p 58 (in Russian)
22. Karpevitch AF, Bokova EN (1970) In: VNIRO Proceedings, vol 76. VNIRO, Moscow, p 163 (in Russian)

Hdb Env Chem Vol. 5, Part P (2005): 175–190
DOI 10.1007/698_5_009

Published online: 6 October 2005

Introduced Species

Mikhail G. Karpinsky[1] (✉) · Tamara A. Shiganova[2] · Damir N. Katunin[3]

[1]Russian Federal Research Institute of Fisheries & Oceanography (VNIRO),
17 V. Krasnoselskaya, 107140 Moscow, Russia
karpinsky@vniro.ru

[2]P.P. Shirshov Institute of Oceanology, Russian Academy of Sciences,
36 Nachimovsky Pr., 117997 Moscow, Russia
shiganov@sio.rssi.ru

[3]Caspian Research Institute of Fisheries (CaspNIRKh), 1 Savushkina, 414056 Astrakhan, Russia
d.n.katunin@mail.ru

Abstract Five species of diatom algae, ten species of epiphyte-fouling algae, seven species of plankton invertebrates, 21 benthic organisms and three species of fish were introduced during last 100 years and established in the Caspian Sea. In addition, a minimum of 18 species penetrated into the Caspian but did not get established. A very fortunate introduction of two species of invertebrates, polychaetous worms *Nereis diversicolor* and the bivalve mollusk *Abra ovata*, into an empty ecological niche became a classic example of successful acclimatization and has obtained worldwide recognition. Just after the opening of the Volga–Don Canal 19 species from the fouling community were introduced in the Caspian Sea, but now planktonic organisms (15 species) predominate among nonindigenous species. On one hand, this is explained by the fact that most of the possible fouling has already penetrated into the Caspian; on the other hand, a new way of organism delivery—with ballast waters—has appeared. The introduced species represented by a small number of species occupy the dominating position by biomass in communities of phytoplankton, zooplankton, benthos, and fouling. All introductions were accompanied by a considerable rearrangement of communities, which was manifested in changes of indices. The most impressive example is the introduction of comb jelly *Mnemiopsis leidyi* causing the most serious changes in the whole ecosystem.

Keywords Acclimatization · Caspian Sea · Introduced species · Invaders · Nonindigenous species

1 Introduced Species

The problem of introduced or nonindigenous species, some of that name also as species-invaders, acclimatizants, is usually considered as one of the aspects of the biodiversity problem, but the species having penetrated in the Caspian Sea during the last 100 years play such an important role that it should considered in a separate chapter.

As mentioned in Karpinsky (this volume) the Caspian fauna was formed from a limited number of species under the conditions of brackish-water content and long-term isolation, as a result of which competitiveness of the Caspian brackish-water species became lower than that of the marine species. So, invasion of marine species and other conditions being equal, the Caspian species have to let the more adapted species have the major environmental areas and the main sources of supply and have to be satisfied with worse areas and food, to which they can adapt more easily owing to their plasticity, universality and grater versatility, or they will become extinct.

Actually, seven Mediterranean species penetrated along the Kumo–Manych depression into the Caspian Sea about 60 000 years ago and 14 representatives of the Arctic complex at the time of glaciation, about 13 000 years ago, were the first species introduced. However, these species have already become full members of the benthic, pelagic,and ichthyologic communities and have evolved considerably, having generated new species and subspecies. In the twentieth century a vast number of new species of algae, invertebrates and fishes penetrated as a result of human activity, either accidentally with ship fouling, in ballast waters, on introduction of other species, or were introduced intentionally. The accurate number of such organisms is hard to determine, as the process is still continuing. Some species manifest themselves immediately, others distribute widely only after several years of incubation, the others appear for a short time, sometimes even having an outburst in abundance and then becoming extinct.

The first introduced species of the new wave appeared in the Caspian, strictly speaking, as far back as the end of the nineteeth century, when in 1897 the merchant Makedonskij tried to introduce two species of the Black Sea oyster (*Ostrea sp.*) into the Caspian. This attempt as well as the next attempts to introduce mussel (*Mytilus galloprovicialis*) in 1899 and the Black sea flounder and gray mullet in 1902 failed [1], which, however, did not affect the desire of man to do nature over at his discretion.

The mussel (*Mytilaster lineatus*) was the first invader that having got into the Caspian managed to get established. It appeared in the Caspian as a part of the fouling of the boats transported by train from the Black Sea into the Caspian in the Baku bight in 1919 and turned out to be the only one out of all species that was able to survive that transportation. Having appeared in the Baku bight, it started to settle by the whole sea gradually except for the the desalted regions of the Northern Caspian. In the first 15 years *Mytilaster* settled gradually, but then in 1934–1938 the biomass rose sharply from 11 to 42% of the total biomass, which reached 7 kg m^{-2} [2] at some places. *Mytilaster* inhabits, mainly, rocky grounds and sediments, the zone of strong currents, at a depth of 30–50 m from the surface, in some cases down to 80 m. Two endemic Caspian species *Dreissena elata* and *D. caspica* inhabited the same ecological niche. These two species of *Dreissena* were still present in the surveys of 1938; however, they were not found in the data of the first postwar surveys of 1955–1957, having been replaced completely by *Mytilaster*. The mechanism of *Dreissena* replacement is interesting. The filtrating apparatus of *Mytilaster* is more specialized than that of *Dreissena*, but it is obviously not enough for such rapid replacement. Judging by the peculiarities of the species biology, *Mytilaster* has completely replaced both species of oxyphilous *Dreissena* owing to its better ability to endure local oxygen deficit that it caused itself forming aggregations [3]; this happened during a sharp increase of its biomass. *Mytilaster* introduction also had an unfavorable impact on the Caspian autochthonous bivalves of family Cardiidae as well as on the Mediterranean *Cerastoderma lamarcki* that are "suffocated" by *Mytilaster* settling on the shells near siphons [4]. After complete settlement of the Caspian in the 1970s, the biomass of *Mytilaster* decreased, though now it has stabilized and makes up about 40% of the biomass of the total benthos (Fig. 1). Its main ac-

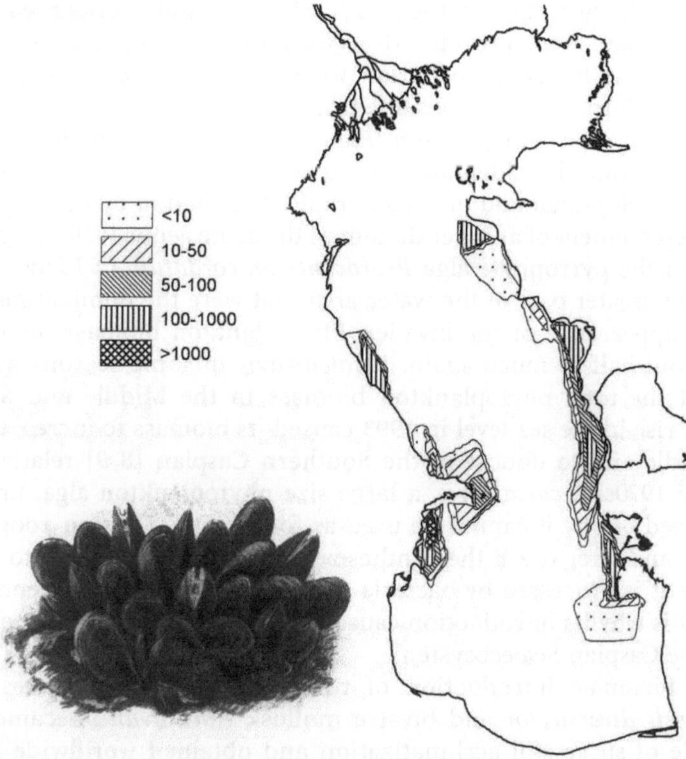

Fig. 1 Biomass distribution ($g\,m^{-2}$) of *Mytilaster lineatus* in the Middle and Southern Caspian Sea

cumulations are observed along the eastern shore of the Middle Caspian as well as in the area of Derbent and towards the north from the mouth of the Kura, at a depth of 10–25 m.

In order to increase fish productivity of the Caspian Sea, some fish and invertebrate species were acclimatized in the 1930s. The first acclimatization was carried out between 1930 and 1934, when two species of gray mullets, leaping gray mullet [*Liza* (=Mugil) *saliens*] and golden gray mullet [*L.* (= *M.*) *aurata*], were introduced from the Sea of Azov. These species got successfully acclimatized in the Middle and Southern Caspian and became the mass commercial species [5]. The future of the other species introduced to the Sea of Azov–Black Sea, the Black Sea flounder [*Platichthys* (=*Pleuronectes*) *flesus luscus*] was not so successful: individuals were found in catches, but did not have practical significance, and nothing is known for sure now whether that species exists in the Caspian Sea [1]. The attempt to introduce the Sea of Azov–Black Sea fishes—red mullet (*Mullus barbatus*), Black Sea turbot [*Psetta* (=*Rhombus*) *maeotica*], Black Sea anchovy, khamsa (*Engraulis encrasicolus*)—failed. Two species of shrimps, *Palaemon adspersus* and *P. elegans*, that are distributed all over the Sea but that do not form large populations, though their biomass reaches 150 g m^{-2} [6] in the southeastern part of the Southern Caspian, were introduced accidentally from the Sea of Azov together with these fishes. Turbellarian worm *Pentacoelum caspium* [7] was quite probably introduced at the same time.

One more species—the diatom *Pseudosolenia* (= *Rhizosolenia*) *calcar-avis*—was introduced accidentally during acclimatization of gray mullets. Its impetuous development and unusually rapid distribution by the Caspian Sea led to the replacement of another diatom of the same genus *P.* (*R.*) *fragilissima* and press of the pyrrophyte alga *Prorocentrum cordatum* (= *Exuviaella cordata*) on the greater part of the water area that were the dominating species before the appearance of the invader. Phytoplankton biomass increased as a result about half as much again. *P. calcar-avis* in some seasons makes up 80–92% of the total phytoplankton biomass in the Middle and Southern Caspian. A rise in the sea level in 1993 caused its biomass to increase 4 times in the Middle and to double in the Southern Caspian [8, 9] relative to the level of the 1970s. *P. calcar-avis*, a large size phytoplankton alga, unlike the species forced out by it cannot be used as food by the Caspian zooplankton organisms, and after death the synthesized organic matter sinks to the bottom, where it is processed by bacteria and is included into the benthic food chain. That is why its introduction caused a serious rearrangement of energy fluxes in the Caspian Sea ecosystem.

A very fortunate introduction of two species of invertebrates, bristle worms *Nereis diversicolor* and bivalve mollusk *Abra ovata*, became a classic example of successful acclimatization and obtained worldwide recognition [10]. The uncommonness of that acclimatization is that the introduced species were not commercial ones, but were feeding ones, and they were introduced to increase the sturgeon food reserve.

The idea of acclimatization of invertebrates in the Caspian Sea in order to increase the sturgeon food reserve was expressed for the first time at the beginning of the 1930s [11]. Then 3 years of study in the Black Sea and the Sea of Azov followed, after which two objects of acclimatization— polychaetous worms *Nereis succinea* and bivalve mollusks *Abra* (=*Syndesmya*) *ovata*—were chosen. That choice one was guided by the following considerations: both species were used to enduring low salinity and desalting, they were a preferred food of sturgeons, they were used to inhabiting slimy grounds poorly colonized in the Caspian and they were used to feeding on plant residues—practically unlimited food resources [12]. So, speaking the modern language, the new species that should not have competed with the native Caspian species were introduced into the empty ecological niche of deposit feeders living in the soil.

Nereis was introduced into the Northern Caspian in three stages, in 1939, 1940, and 1941, and as early as 1944 it was found in sturgeon guts [13] in big quantities. *Nereis* colonized very rapidly the Northern Caspian and started to penetrate into the Middle and Southern Caspian (Fig. 2a) getting into the food composition of all benthivorous fishes [14, 15]. However, the result of the introduction was unexpected: when in the 1958 samples of Caspian polychaetous worms were examined, no specimens of *N. succinea* were found, only *Nereis* (*Neanthes*) *diversicolor*. Further detailed study showed that there was no *N. succinea* in the samples from the Caspian Sea. That is why a supposition was expressed that on collecting of the materials for the introduction erroneously only *N. diversicolor* [16] was gathered.

The case was somewhat different. The actual situation might be restored now only on the basis of the available literature data. The two species live together in the Sea of Azov, from which they were taken. Outwardly they are poorly distinguished, the morphological differences concern only paragnaths structure, which is indistinguishable on sample collection, and most likely both species were gathered for introduction. But they have an essential difference in biology: *N. succinea* has a heteronereid epitokous stage, when the spawners change color, sometimes even the body form, at sexual product spawning, they come from the sediment up to the surface and further into the water column, while *N. diversicolor* does not have this stage. In the first years in the Caspian, both species were present and reproduced, at least, epitokous *Nereis* [14, 17], though one of them became extinct later and the reasons remained uncertain.

Originally, a supposition was expressed that epitokous *Nereis* as more accessible were completely eaten by sturgeons [18]. But evidence has been found that sturgeons were not to blame for the species extinction. There is a report by Birstein [15] on the role of *Nereis* in the feeding of Caspian fishes, besides sturgeons, that reveals the stomachs of Saposhnikovi's shad (*Alosa saposh-*

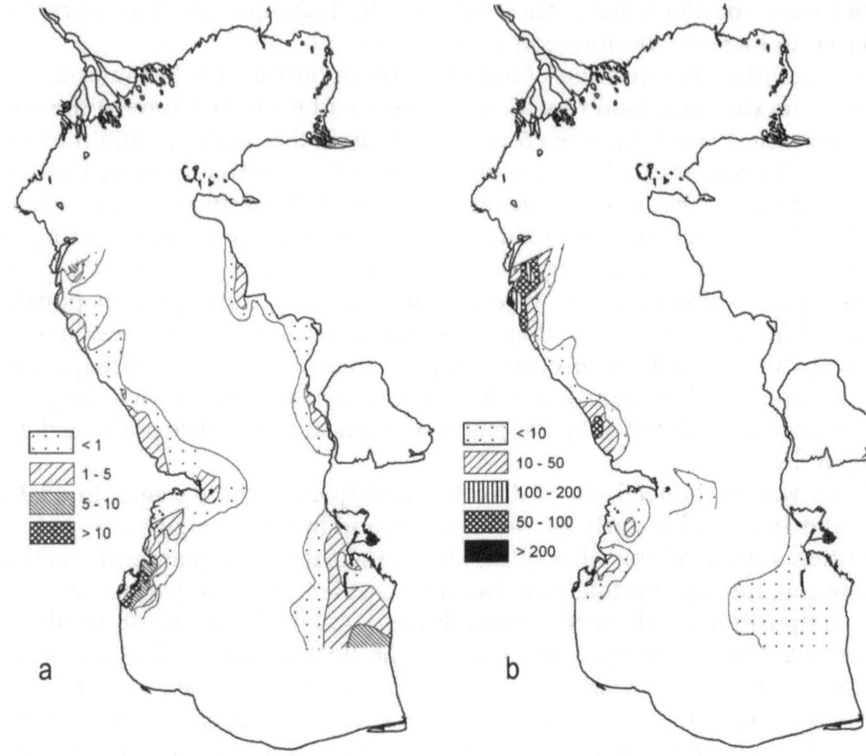

Fig. 2 Biomass distribution ($g\,m^{-2}$) of *Nerei diversicolors* (**a**) and *Abra ovata* (**b**), in the Middle and Southern Caspian Sea

nikovi) were completely filled by a vast amount of *Nereis* in April 1949. That species of shad as well as some others carnivorous shads (eight subspecies *Alosa brashnikovi* and two *Alosa kessleri* ones) are referred to the group of herrings that feed mainly on fish (75%, mainly kilka, sand smelt), mysids and shrimps and may feed on, though seldom, benthic organisms, mainly Corophiidae (Amphipoda). *Nereis* live in the soil and are accessible to herrings only when they come to the surface of the sediment. That is why they appear in the food composition of herrings, but very rarely. And the fact that the stomachs were filled in full by only *Nereis* means that herrings fed on large concentrations of swimming worms, which is possible only when spawning *Nereis* come to the water column. Herrings being universal in food selection fed only on *Nereis*, giving preference to them against more habitual food, kilkas. As a result the population of *N. succinea* was simply exterminated under the pressure of the intensive grazing by a large group of mass species. In further papers on the feeding of herring [19, 20], the mention of finding *Nereis* in herring's stomachs was very seldom, from which it might be concluded that the period of epitokous *Nereis* pelagic emergence was not long and ended rather quickly owing to extinction of the feeding object. And *N. diversicolor* reproduced well, became one of the mass species and have no competition with any of the native species at that time. Thus, nature itself has chosen the species most appropriate to the Caspian ecosystem.

At the same time as the *Nereis* bivalve *Abra ovata* was introduced, the mollusks were released into the Ural furrow that was soon destroyed by fresh water during the flood, and they most probably became extinct. That is why the introduction was repeated in 1947 and 1948 [1]. In 6–7 years, in 1955, the first settlements of that mollusk were recorded at Kulaly Island [21], and then it started to distribute widely along the coasts (Fig. 2b). Because they inhabit soft and slimy grounds, *Abra* forced out the other Mediterranean species *Cerastoderma lamarcki* that also inhabits slimy grounds when the *Abra's* area was reduced, as happened in 1971. These species compete not only for the habitats, but for the food as well; though one filtrates bottom waters and the other collects detritus from the sediment surface, both of them actually feed on plankton detritus, the reserves of which are nevertheless huge [22].

Both acclimatized species became numerous on the vast area of slimy grounds. They were included in the food composition and food supply of sturgeons and starred (stellate) sturgeons as one of the favorite foods and the benthos biomass and food reserve increased considerably in the Northern Caspian, at the western coast of the Middle Caspian and in the Southern Caspian. And the region of Agrakhan shoal, in the northwestern part of the Middle Caspian, where these species form the bulk of the biomass, turned into one of the most important feeding fields, where sturgeons graze food organisms almost to zero during spring and autumn migrations (Fig. 7 in "Biological features and resources").

A new stage of introduction into the Caspian Sea began after the construction of the Volga–Don Canal in 1952. An intensive expansion of a vast number of fouling species started along the direct short route on the vessel bottoms from the Sea of Azov–Black Sea basin into the Caspian Sea, and together with the ballast waters plankton species began to penetrate very actively. All these species either inhabited the Sea of Azov and the Black Sea constantly or had been introduced there earlier and were already well established.

Probably, two species of acorn barnacles *Balanus improvisus* and *B. eburneus* were the first invaders to penetrate the Volga–Don Canal; they were recorded in the Caspian in 1955 and 1956 [23]. Owing to the shell they are able to endure unfavorable conditions very well. In the Caspian Sea the first species live in areas with strong current and inhabited all possible rigid substrates, even Caspian seal hairs, and became the leading species creating major biomass in the fouling. The second species, first recorded in Turkmenbashy Bay and near Ogurchinskij island, was not found in further studies. Introduction of barnacles increased considerably the total fouling biomass and that of the native species. Bryozoans *Conopeum seurati* [24] and *Lophopodella carteri* [25], hydrozoans *Blackfordia virginica* [26], *Bougainvillia megas*, and

Moerisia maeotica [27], and kamptozoans *Barentsia benedeni* [28] followed the barnacles. Polychaete *Mercierella enigmatica* in 1958–1961 reproduced very rapidly in Turkmenbashy Bay, its biomass (measured together with the shell) reached 30 kg m^{-2} [29], but then it became extinct and had never been found, as is the case for *B. eburneus*. Fouling plants penetrated into the Caspian Sea together with fouling animals. An accurate number is impossible to determine, as the species composition was studied in detail only in the 1960s, but epiphyte fouling from green, brown, and red algae (*Arochaete parasitica, Ectochaete leptochaete, Enteromorpha tubulosa, E. salina, Ectocarpus confervoides* f. *fluviatilis, Entonema oligosporum, Acrochaetium daviesti, Ceramium diaphanum, Polysiphonia variegata, Monostroma latissimum*) appeared in the 1950s [30].

The only real predator, the crab *Rhithropanopeus harrisii tridentate*, got to the Caspian together with fouling. Its native area is the Atlantic coast of North and South America, from where it was taken to the North Sea, the Baltic Sea, the Black Sea, and the Sea of Azov by vessels. Having been first discovered in 1958, the crab colonized rather rapidly the whole water area of the Northern Caspian and then the whole of the Caspian Sea. It inhabits the the Northern Caspian and forms major accumulations in the Southern Caspian, but it is rare in the Middle Caspian. The crab feeds on *Nereis*, bivalve mollusks (*Mytilaster*, Cardiidae, *Dreissena, Abra*), oligochaetes, and dead fish, and the young eat corophiids, which means that the crab is a polyphage from its feeding spectrum which is close to that of benthivorous fishes and might be a competitor to the sturgeons. In turn it itself is a food object for sturgeon and starred sturgeon [31]. Before the crab was introduced there were no benthic predators usually supplying benthic communities with 10% of the biomass in the Caspian Sea, as their function was fulfilled by the sturgeons grazing benthos actively [18].

In the first years after the Volga–Don Canal was opened cladoceran *Pleopis (=Podon) polyphemoides* [32] appeared out of plankton species, and out of the benthos the bivalve *Hypanis colorata* [33] and, probably, the amphipod *Corophium volutator* penetrated into the Caspian Sea. The gastropod *Lithogliphus naticoides*, the intermediate host of several species of parasitic trematodes, that appeared with it [34] emerged and colonized the Volga delta rapidly in 1971.

Besides invaders, some species of fish were acclimatized in the Caspian in that period. The attempts to rear pink salmon (*Oncorynchus gorbuscha*), chum (*O. keta*), as well as coho salmon (*O. kisutch*) [35] at hatcheries on the Samur river are considered to be the most interesting, but these species failed to survive without the support of hatcheries. In the Volga delta the grass carp (*Ctenopharyngodon idella*) and the silver carp (*Hypophthalminchthys molitrix*) were acclimatized, but these fishes do not live in the Capian Sea. In 1965 at the Aleksandrovsk hatchery because of small numbers of spawners of Caspian inconnu (*Stenodus leucichthys leucichthys*) spawners of another subspecies were used, Siberian inconnu nelma (*S. l. nelma*) [1].

The process of introduction of new species continued in the last few decades, although it occurred mainly in the pelagic zone. The copepod *Acartia tonsa* determined previously as *A. clausi* inhabiting the coastal zone was discovered in the eastern part of the Southern Caspian in August 1981. In 2 years it became one of the mass species in the Middle and Southern Caspian, especially in the coastal zone because it is more competitive that autochthon copepods, making up a quarter in abundance and biomass, and it penetrated into the Northern Caspian, in the eastern part of which it plays an important role in kilka feeding. In the following years its share on the coast increased, and the copepod was found in the central part as well. The cladoceran *Podon intermedius* [36] was also found in the coastal area, but it is quite rare. A widely spread boreal-arctic neretic species of diatom *Pseudonitzchia (=Nitzchia) seriata* was found in the Mediterranean Sea as well, the abundance of which reached up to forty-eight million individuals in 1 m^3 at some stations, and was recorded near the eastern shore of the Middle Caspian in 1990. It is found everywhere in both parts of the Caspian Sea; the highest development was noted in the central part of the Middle Caspian [9]. In the Southern Caspian the nudibranchiate mollusk *Tenellia adspersa* [37] was recorded, and the mussel *Dreissena bugensis* was recorded in the in the Northern Caspian, singled out earlier as a subspecies *D. rostriformis* having reproduced before in the Volga [38].

In the Southern and Middle Caspian in 1998 fishermen observed some unknown jelly organisms, and in 1999 jellyfishes medusa *Aurelia aurita* and ctenophore *Mnemiopsis leidyi* were recorded by underwater video surveys at banks located on the border of the Middle and Southern Caspian. The native habitat *Mnemiopsis* is temperate-to-subtropical estuaries along the Atlantic coast of North and South America, where it is found in an extremely wide range of environmental conditions. At the beginning of the 1980s this species was introduced into the Black Sea, where it caused great changes in the ecosystem. Most probably it was transported with ballast water aboard oil tankers or other ships from the Black Sea or the Sea of Azov (where *M. leidyi* occurs in warm months), which had passed through the Volga–Don Canal and the shallow freshwater Northern Caspian Sea, and was released into the salty Middle or Southern Caspian. The fact that the first individuals of *Mnemiopsis* were found with large individuals of *Aurelia* support the hypothesis of the arrival of *M. leidyi* from the Black Sea [39].

Mnemiopsis is a zooplanktivorous ctenophore, the feeding intensity of which depends not on the degree of the organism saturation, but on the food availability. As a result, for large amounts of food *Mnemiopsis* swallow plankton organisms and belch them half-digested, which increases many times the amount of the destroyed zooplankton, so large concentrations of *Mnemiopsis* are able to cause great losses of plankton populations.

In 2000 *Mnemiopsis* spread across all areas of the Caspian Sea and in October it appeared in the western part of the Northern Caspian with salinity not less than 4.3‰. In 2001 it greatly increased in population size, especially in the Southern Caspian in August, where the abundance was 2 times higher than the highest values in the Black Sea, in 1989. And in 2002 this biomass doubled.

The *Mnemiopsis* population study in the 5 years since its appearance in the Caspian Sea has shown that its major area of distribution is the Southern Caspian, mainly the Iranian area, which it inhabits for around 1 year and where it spends winter, and where its population size strongly decreased in December–January, as temperatures drop to 8–10 °C, while surviving individuals shrink back to a small size. In spring (or earlier in an abnornally warm year as it was in 2004), *M. leidyi* begins to grow and reproduce here, soon spreading to the north. The main factors that determine the increase of the *Mnemiopsis* population are temperature and food, zooplankton availability. It appears in the Middle Caspian in July and in the Northern Caspian in late July to early August. In the Northern Caspian *M. leidyi* occurs in the

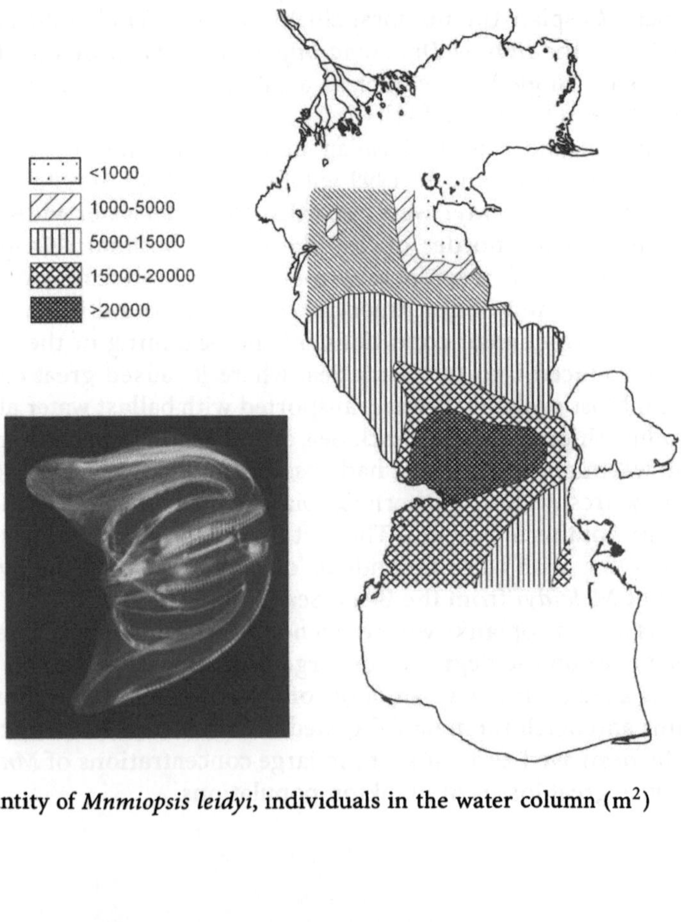

Fig. 3 Quantity of *Mnmiopsis leidyi*, individuals in the water column (m^2)

western part, where the salinity not less than 4.3‰ (Fig. 3). It is probable that as a result of winter cooling, *Mnemiopsis* completely vacates the Northern Caspian by late November. Next it disappears from the Middle Caspian, and only a small population survives winter in the Southern Caspian.

The consequences of the introduction of that species into the Caspian are great, if not disastrous: in 2000, when *Mnemiopsis* colonized actively the whole Caspian Sea, zooplankton biomass decreased 5–6 times in all groups and 6–20 times [39, 40] in some groups compared with the previous years, and the decrease only continued in later years. Zooplankton is a key link in the trophic chain and a decrease of its biomass affected all trophic levels. Three species of kilka feed on zooplankton, and such a sharp decrease of the food reserve resulted in the same sharp decrease of their population (Fig. 4). The intensive fishing of the kilka that was carried out in the Caspian Sea has practically ceased now [40]. Kilka itself serves as a food reserve for beluga sturgeon and Caspian seal, and its consumption by the seal, at least, was commensurable with the value of the commercial catch; other species of sturgeons might also feed on it. A decrease in the kilka abundance caused nutritional deficiency of seals and beluga and changed its food composition. Zooplankton decrease, in turn, brought changes in the phytoplankton community: the biomass of autochthon phytoplankton increased because it was not consumed so intensively as before, but the biomass of *P. calcar-avis* decreased, especially in the Southern Caspian [41]. After invasion of *Mnemiopsis* in 2001 first of all the Cladocera were affected; they decreased almost to zero and among them only invader *P. polyphemoides* predominated and remained in samples in small numbers. Copepods became represented almost completely by the nonindigenous *A. tonsa*. By August to September Copepoda species represented 95–99% of *A. tonsa*. The main resource for kilka-feeding copepods *E. grimmi* and *E. minor* completely disappeared from samples.

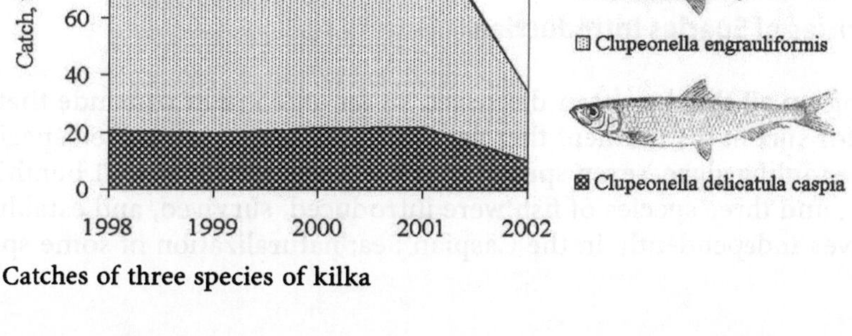

Fig. 4 Catches of three species of kilka

The decreased abundance, biomass, and species diversity of the holozooplankton also involved the pelagic eggs and larvae of all species of pelagic fish. But *M. leidyi* first and foremost consumes zooplankton, including meroplankton. Field and experimental data allowed us to estimate its predation impact. We found that the grazing rate of *M. leidyi* during intensive development population was higher than the seasonal zooplankton production, such that by the end of summer the zooplankton stock became extremely poor.

The introduction of *Mnemiopsis* affected benthic organisms, as mollusks, polychaetous and some crustaceous organisms have pelagic larvae that are grazed together with the other planktonic organisms by comb jellies. Most of benthic crustaceous floating to the water column from the bottom surface might become their prey.

In order to suppress the *Mnemiopsis* population a possibility is to acclimatize in the Caspian Sea its obligatory predator—comb jelly *Beroe ovata*, which, exceptionally, feeds only on zooplanktivorous comb jellies. The experiments carried out showed that *B. ovata* might live at a salinity not less than 7‰, it might feed actively on *Mnemiopsis*, and reproduce in the Caspian water with a salinity of not less than 10‰. In that case *B. ovata* would be able to inhabit the Middle and Southern Caspian and control the abundance of *M. leidyi*.

Aurelia aurita having been introduced together with *Mnemiopsis* in 1999 lived in the Caspian Sea for some time: its ephyrae were discovered in the spring plankton samples in the Middle Caspian in 2000, but since then it had not been found anymore.

Introduction of new species cannot be considered complete. New for the Caspian Sea, the diatom alga *Cerataulina pelagica* was found in 2002, and two another diatom species, *Chaetoceros pendulus* and *Tropidoneis lepidoptera*, were found some time later. Zooplankton species determined beforehand as the copepods *Oithona similis* and *Calanus euxinus* and arrow worm (Chaethognata) *Sagitta setosa* and not recorded before in the sea were recorded in the Middle Caspian in April 2004. It is difficult to say now whether these species will be established in the Caspian, but they play an important role as food objects for planktivorous fishes in the Black Sea [42].

2 Peculiarities of Species Introduction

Summing up all that has been discussed so far, one might conclude that it is known for sure at the moment that five species of diatom algae, ten species of epiphyte-fouling algae, seven species of plankton invertebrates, 21 benthic organisms, and three species of fish were introduced, survived, and established themselves independently in the Caspian Sea; naturalization of some species is in question. In addition, a minimum of 18 species penetrated into the Caspian but did not get established. For the present it is premature to speak about a great number of introduced species in the Caspian Sea. Compared with the Baltic Sea, 51 new species of only invertebrates [38] were recorded. Most of the nonindigenous species are of Mediterranean origin, but some species got into the Black Sea and the Sea of Azov from the western part of the Atlantic (*B. virginica, B. megas, B. improvisus, B. eburneus, A. tonsa, R.h. tridentate, M. leidyi*) or from the northern European seas (*M. enigmatica*), are were established successfully there and only then penetrated into the Caspian. Obviously, biological peculiarities of these species favor their distribution with vessels and acclimatization at new places. As there is a group of species in the Black Sea and the Sea of Azov that penetrated there from other seas, it is possible that they might appear as well in the Caspian soon. Another three introduced species (*H. colorata, C. volutator, D. bugensis*) are autochthonous Sea of Azov–Black Sea species close to the Caspian that formed after separation of the seas, and for which the conditions in the Caspian Sea correspond to natural ones.

Most of the accidentally introduced species, 19 species (two did get established), were included in the fouling community; another four (*Mytilaster, Rhithropanopeus, Tenellia, Corophium*) are related to it. That testifies to the fact that just after opening the Volga–Don Canal the main method of introduction was fouling from vessel bilge. Recently a principal change has occurred: now plankton organisms (15 species, only one out of them not established) predominate among introduced species. On one hand, this is explained by the fact that most of the possible fouling species have already penetrated into the Caspian; on the other hand, a new method of organism delivery—with ballast waters–has appeared. And if the role of fouling in the sea ecosystem is considerable, but not great, then the main organic matter is formed in the plankton community and its functioning determines the major processes occurring in the Caspian Sea. Few benthic nonindigenous species are either related to a considerable degree to the fouling community or are brought by man intentionally or accidentally. Man acclimatized all fishes that have got into the Caspian.

A small number of benthic introduced species attract one's attention. The main reason is the peculiarity of the Caspian community. It was formed under the constant effect of the strong grazing pressure by sturgeons, resulting in small sizes of benthic organisms, with juveniles being predominant in the population, and low total biomass at high productivity [18]. As a result the species capable of escaping from grazing by sturgeons (*Mytilaster*, barnacles that are difficult to separate from the substrate) or the species capable of an *r*-strategy and having been introduced into empty ecological niches (*Nereis, Abra, Rhithropanopeus*) managed to get acclimatized successfully. Now with a sharp decrease of the sturgeons' abundance and, correspondingly, of the pressure of their grazing, new benthic introduced species are able to reduce and even force out the native species. If the abundance of sturgeons is not restored in the near future, such a result is possible.

The introduced species presented by a small number of species occupy the dominating position in the Caspian Sea communities. *P. calcar-avis* forms the main biomass of phytoplankton. *A. tonsa* is one of the leading species in the zooplankton. The biomass of *Mytilaster, Abra, Nereis,* and *Balanus* totals more than 60% of the total benthos. Fouling is almost completely from introduced species; the native species predominate only in ichthyocenosis.

That happens, as described by Karpinsky (this volume), because the Caspian species were formed from a very limited number of species under the conditions of brackish water and long isolation and they acquired versatile features, but with a low degree of specialization, and that is why less successful. As a result, the brackish autochthonous Caspian species give in to the introduced species—marine species better adapted to both typical and new conditions. The introduced species have a considerable advantage over the native ones other things being equal. The brackish species inhabit, mainly, the Middle and the Southern Caspian at a salinity of 12–13‰ as do most marine species. However, in other seas they are pushed back by marine species into more desalted zones: in the Baltic Sea with a salinity of 3–5‰, in the Sea of Azov and the Black Sea with a salinity of 1–2‰. On their appearance in the Caspian, the marine species introduced found themselves under favorable biotopic conditions with no real competition from the native fauna.

The Caspian fauna was formed over 1.8 million years under isolation conditions from a very limited number of species. Together with the fauna and on its basis, communities—plankton, benthos, fouling and of the Caspian Sea as a whole—were formed. The limited number of species and their weak specialization caused high stability of communities and their tolerance to changes of the environment. At the same time the Caspian communities turned out to be poorly adapted to the introduction of new, more competitive, marine species. All introductions were accompanied by a considerable rearrangement of communities, which was manifested in changes of indices. The most impressive example is the introduction of comb jelly *Mnemiopsis* causing the most serious changes in the whole ecosystem. And only in cases of introduction into the empty ecological niches, the quantitative indices increased.

Is introduction of a new species into the Caspian Sea useful or harmful? From the standpoint of a productivity increase the appearance of some nonindigenous species is useful. Thus, *Nereis* and *Abra* became the favorite food of starred sturgeons and sturgeons. Crabs competing with the sturgeons for food are themselves a food resource for them. Even *Mytilaster*, unapproachable for the sturgeons in the Northern and Middle Caspian owing to the formation of sedentary druses, is included in the food composition in the Southern Caspian, where it is found on soft ground. Copepods *A. tonsa*, *O. similis*, and *C. euxinus* might serve as a food for planktivorous fishes. Gray mullet fishery is not great, but this is the result of acclimatization. At the same time *Mytilaster*, which forms the bulk of the benthos biomass, is used extremely rarely. *P. calcar-avis* having increased phytoplankton biomass decreased its nutritive value for pelagic phytophagous. The introduced copepods decreased the numbers of native species but are not less attractive in a nutritive respect than native species. And the appearance of comb jelly had disastrous consequences for the whole ecosystem. As a whole, rearrangement of relationships in the community sometimes accompanied by variation of abundance was observed if species were introduced into occupied ecological niches.

From the point of view of biodiversity, introduction of many species is unfavorable, as they force out less competitive aboriginal species that sometimes even become extinct. In perspective, if marine species start to penetrate actively, "mediterranization" of the Caspian Sea, that is replacement of the aboriginal species by Mediterranean species and forcing out of the brackish-water species into estuaries, as happened with the Sea of Azov and Black Sea after the discovery of the Bosporus, is possible.

Thus, in order to protect the unique Caspian brackish-water fauna and the original community it is necessary to carry out measures to protect the Caspian Sea its aboriginal flora and fauna from the introduction of more competitive marine species.

References

1. Karpevitch AF (1975) Theory and practice of aquatic organism acclimation. Pishchevaya promyshlennost, Moscow (in Russian)
2. Brotskaya VA, Netsengevch MR (1941) Zool Zh 20:79 (in Russian)
3. Logvinenko BM (1965) Nauchn Dokl Vyssh Shk Biol Nauki 4:14 (in Russian)
4. Romanova NN (1960) Zool Zh 39:811 (in Russian)
5. Babayan KE (1957) Zool Zh 36:1505 (in Russian)
6. Lavrov-Navozov NP (1939) Zool Zh 18:443 (in Russian)
7. Beklemishev VN (1954) Bull MOIP Dep Biol 59(6):41 (in Russian)
8. Levshakova VD, Sanina LV (1973) In: VNIRO proceedings, vol 80. VNIRO, Moscow, p 18 (in Russian)
9. Ardabieva AG, Tataritseva TA, Terletskaya OV (2000) In: Species introducers in the European seas in Russia, Abstracts, Murmansk, p 16 (in Russian)
10. Elton C (1958) The ecology of invasions by animals and plants. Methuen, London
11. Zenkevtch LA, Birstein YaA (1934) Rybn Khozyajst 3:38 (in Russian)
12. Shorygin AA, Karpevich AF (1948) New introducers of the Caspian Sea and their significance in the biology of this reservoir. Krymizdat, Simferopol (in Russian)
13. Spassky NN (1945) Zool Zh 24:23 (in Russian)
14. Belyaev GM (1953) In: Nikitin VN (ed) Acclimatization Nereis in the Caspian Sea. MOIP, Moscow, p 243 (in Russian)
15. Birstein YA (1953) In: Nikitin VN (ed) Acclimatization Nereis in the Caspian Sea. MOIP, Moscow, p 115 (in Russian)
16. Khlebovich VV (1963) Zool Zh 42:129 (in Russian)

17. Navozov-Lavrov NP (1948) Rybn Khozyajst 9:40 (in Russian)
18. Karpinsky MG (2002) Ecology of the benthos of the Middle and Southern Caspian. VNIRO, Moscow (in Russian)
19. Piskunov IA (1961) Vopr Ikhtiol 1(18):79 (in Russian)
20. Elizarenko MM, Andrianova CB (2002) Vopr Rybolovstva 1(9):53 (in Russian)
21. Saenkova AK (1956) Zool Zh 35:678 (in Russian)
22. Romanova NN (1979) In: 6th Meeting on the investigation of molluscs. Main results of their study, Leningrad. Nauka, Moscow, p 109 (in Russian)
23. Zevina GB, Starostin IV (1961) In: Transactions of Shirshov Institute oceanology, vol 49. Nauka, Moscow, p 97 (in Russian)
24. Abricosov GG (1959) Zool Zh 38:1745 (in Russian)
25. Abricosov GG, Kosova AA (1963) Zool Zh 42:1724 (in Russian)
26. Logvinenko BM (1959) Zool Zh 38:1257 (in Russian)
27. Naumov DV (1968) In: Atlas of Caspian Sea invertebrates. Pischevaya promyshlennost. Moscow, p 43 (in Russian)
28. Zevina GB (1968) In: Atlas of Caspian Sea invertebrates. Pischevaya promyshlennost. Moscow, p 65 (in Russian)
29. Bogoroditsky PV (1963) In: Transactions of Shirshov Institute oceanology, vol 70. Nauka, Moscow, p 26 (in Russian)
30. Zevina GB (1959) Priroda 7:79 (in Russian)
31. Reznchenko OG (1967) In: Transactions of Shirshov Institute oceanology, vol 85. Nauka, Moscow, p 136 (in Russian)
32. Mordukhai-Boltovskoy PD (1962) Zool Zh 41:289 (in Russian)
33. Saenkova AK (1960) Priroda 11:111 (in Russian)
34. Pirogov AI (1972) Zool Zh 51:912
35. Kasymov AG (1987) Caspian Sea. Gidrometeoizdat, Leningrad (in Russian)
36. Tataritseva TA, Ardabieva AG, Terletskaya OV, Tinenkova DK, Malinovskaya LV, Tarasova LI, Petrenko EL (2000) In: Matishov GG (ed) Species introducers in the European seas in Russia. Apatity, p 169 (in Russian)
37. Antsulevtch AE, Starobogatov YI (1990) Zool Zh 69(11):138 (in Russian)
38. Orlova MI (2000) In: Matishov GG (ed) Species introducers in the European Seas in Russia. Apatity, p 58 (in Russian)
39. Shiganova TA, Kamakin AM, Zhukova OP, Ushivtsev VB, Dulimov VB, Musaeva EI (2001) Okeanologiya 41:542 (in Russian)
40. Shiganova TA, Dumont HJ, Sokolsky AF, Kamakin AM, Tinenkova DK, Kurashova EK (2004) In: Dumont HJ, Shiganova TA, Niermann U (eds) Aquatic invasions in the Black, Caspian and Mediterranean Seas. Kluwer, Dordrecht, p 71
41. Tataritseva TA, Terletskaya OV (2004) In: Fisheries researches in the Caspian. CaspNIRKh, Astrakhan, p 123 (in Russian)
42. Shiganova TA, Musaeva EI, Pautova LA, Bulgakova YV (2005) Izv RAN Ser Biol 1:78 (in Russian)

Hdb Env Chem Vol. 5, Part P (2005): 191–210
DOI 10.1007/698_5_010

Published online: 28 September 2005

Biological Features and Resources

Mikhail G. Karpinsky[1] (✉) · Damir N. Katunin[2] · Vera B. Goryunova[1] · Tamara A. Shiganova[3]

[1]Russian Federal Research Institute of Fisheries & Oceanography (VNIRO), 17 V. Krasnoselskaya, 107140 Moscow, Russia
Karpinsky@vniro.ru, veragor@vniro.ru

[2]Caspian Research Institute of Fisheries (CaspNIRKh), 1 Savushkina, 414056 Astrakhan, Russia
D.N.Katunin@mail.ru

[3]P.P. Shirshov Institute of Oceanology, Russian Academy of Sciences, 36 Nachimovsky Pr., 117997 Moscow, Russia
Shiganov@sio.rssi.ru

Abstract The Caspian Sea is of enormous importance, because of its extremely valuable fish resources. Thousands of tons of kilka, roach, bream, herring, common carp, pike-perch, perch and others, were fished there, but the sturgeons were and remain the most valuable asset of the Caspian Sea. The fish productivity is based on the productivity of phyto- and zooplankton and the benthos, and on the trophic structure of the Sea. However, for various reasons, commercial reserves and catches of some fish have been reduced in the last 10–15 years. After the breakup of the USSR and subsequent formation of several independent states, unregulated, illegal fishing of sturgeons in the Caspian Sea began, causing a sharp reduction in commercial stock and catches. As sturgeons are reproduced mainly at hatcheries, the fry now predominate in the population. The introduction of the comb jelly *Mnemiopsis leidyi* caused serious changes in the Caspian Sea ecosystem that have affected fish productivity. It grazes zooplankton, food reserve, eggs and larvae of two species of kilka, as a result of which their abundance and catches have also reduced sharply.

Keywords Benthos · Biological resources · Caspian Sea · Fish · Plankton

1 Biological Features and Resources

The Caspian Sea, inhabited by a unique fauna and possessed of extremely valuable fish resources is of enormous importance. Hundreds of thousands of tons of kilka, dozens of thousands of tons of roach and bream, and thousands of tons of herring have been caught; a large number of valuable freshwater fish, such as common carp, pike-perch, perch and others, have been fished out annually and seals fisheries have also been run in the Northern Caspian. Nevertheless, the sturgeons were and remain the greatest asset of the Caspian Sea. In USSR times, about 90% of the world catch of these valuable fish were obtained here, and almost all black caviar on the world market was of the Caspian. However, in the last 10–15 years commercial reserves and catches of some fish have sharply reduced for various reasons. In order to identify and estimate the consequences of these changes, it is necessary to analyze the biological features of, and relationships in the Sea. The feeding and trophic relationships have always been the most important factors, and the food reserve affects the population development in the first place; that is why the analysis has been carried out by classifying the organisms by their trophic levels: the lowest (plankton, benthos, some planktivorous fish) represents the food reserve and the upper (carnivores and some planktivores), its consumers, commercial fish.

An analysis of ecosystem functioning must include the production, transformation and processing of organic matter or transport of organic matter and the energy contained within it to the different trophic levels (Fig. 1). The bulk of organic matter, more than 90%, is produced in the pelagic community by phytoplankton (producer) that serves as a food for the organisms of the following trophic level, the zooplanktonal phytophages (consumers of the first level). The most numerous Caspian fishes are zooplanktivorous: three species of kilka, the young of all herrings, one of which feeds on zooplankton even when it is adult, sand smelt and some zooplankton organisms, of which the comb jelly *Mnemiopsis* is notable. Finally, kilka serve as a food for the organisms of the higher trophic level: sturgeons (mainly, beluga), seals, herrings, and sea birds. Besides this, man also consumes kilka by fishing. Not all organic matter synthesized at this trophic level moves to a higher level. In the absence of consumers the organisms die and their bodies and also the products of alive organism metabolism are included in the detrital food chain, where they are utilized by the vital activity of bacteria (reducers). At the same time, some bacterial organic matter returns to the pelagic food chain as a food for phytophages. However, the bulk of organic matter, with the bacteria on it, sinks to the bottom and is incorporated into the benthic food chain in the form of detritus. Bottom-dwelling organisms collect detritus from the deposit surface, from its depths, or as filtrate from the benthic water layers. Sturgeons, gobies, roach and bream feed on the benthos. A small part of the organic matter is synthesized by benthic macrophytes and is also incorporated into the benthic food chain with the detritus. The organic matter transported with the runoff—the remains of macrophytes and of the freshwater organisms that died as a result of salinity increase—plays an important role in the pre-mouth space of rivers. Non-utilized organic matter is buried in the deposits.

This is a brief description of the trophic relationships in the Caspian Sea, and is quite typical for all seas. In reality it is much more complicated and

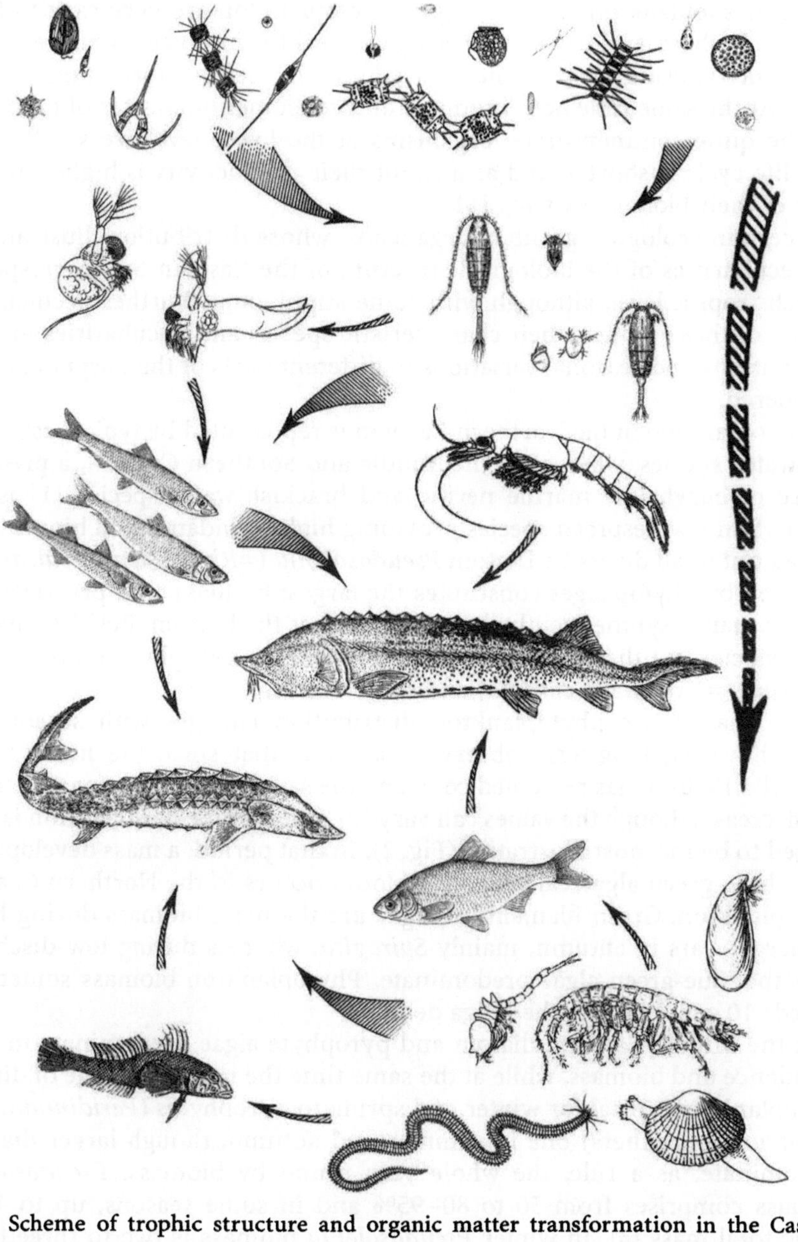

Fig. 1 Scheme of trophic structure and organic matter transformation in the Caspian ecosystem

ramified; a larger number of species are actually involved, but the main routes taken by organic matter are as described. Most biomass (90%) is lost when passing from one trophic level to another, which is to say that 10 kg of kilka,

100 kg of zooplankton and 1000 kg of phytoplankton are necessary to form 1 kg of a highest level consumer (beluga or seal for instance), and even more taking into consideration the necessary losses in the transformation into detritus. At the same time both producer and consumer biomasses of each level may be quite commensurate: organisms at the lower level are smaller and their life cycle is shorter, and as a result their productivity is higher but the value of their biomass existing is lower.

A certain ecological group of organisms, whose distribution illustrates all the peculiarities of the biological structure of the Caspian Sea, corresponds to each trophic level, although with some stipulations. Further peculiarities of each of these groups, their characteristic species and peculiarities in their distributions and seasonal variations in different parts of the Caspian Sea are considered.

Phytoplankton in the Northern Caspian is represented by typical estuarine freshwater species, whereas in the Middle and Southern Caspian, a predominance of euryhaline marine neritic and brackish-water species [1] is observed. Some widespread species providing high abundance and biomass are singled out of all diversity. Diatom *Pseudosolenia* (=*Rhizosolenia*) *calcar-avis* not eaten by phytophages constitutes the largest biomass, and practically all organic matter synthesized by it accumulates at the bottom. Besides this, the mass species are diatom *Actinocyclus ehrenbergii*, pyrophyte *Prorocentrum cordatum* (=*Exuviaella cordata*) and some others.

The quantitative phytoplankton distribution changes with season and space; however, long-term observations show that since the first surveys (from the 1930s, it has remained constant: the same zones of biomass increase and decrease, though the values can vary [2]. The summer distribution is considered to be the most illustrative (Fig. 2). In that period, a mass development of the blue-green algae causing algal bloom occurs in the Northern Caspian phytoplankton. Green filamentous algae are the main biomass during high-discharge years in autumn, mainly *Spirogira*, whereas during low-discharge years the blue-green algae predominate. Phytoplankton biomass sometimes exceeds 10 g m^{-3} near to the Volga delta.

In the Middle Caspian, diatom and pyrophyte algae predominate in both abundance and biomass, while at the same time the natural change of diatom phytoplankton typical for winter and spring to pyrophytes (*Peridinium*, *Prorocentrum* and others) one in summer and autumn, though larger diatoms predominate, as a rule, the whole year round by biomass. *Pseudosolenia* biomass comprises from 50 to 80–95% and in some seasons, up to 100% of the total mass [3]. In winter *Pseudosolenia* biomass is two to three times higher than in summer at lower abundance, which occurs due to a considerable increase in cell volume [4]. In the coastal waters of the eastern part of the sea, phytoplankton develops most intensively in winter under the effect of the warm current from the Southern Caspian. Vegetation intensity dampens by August–September, when water stratification is established, and a new

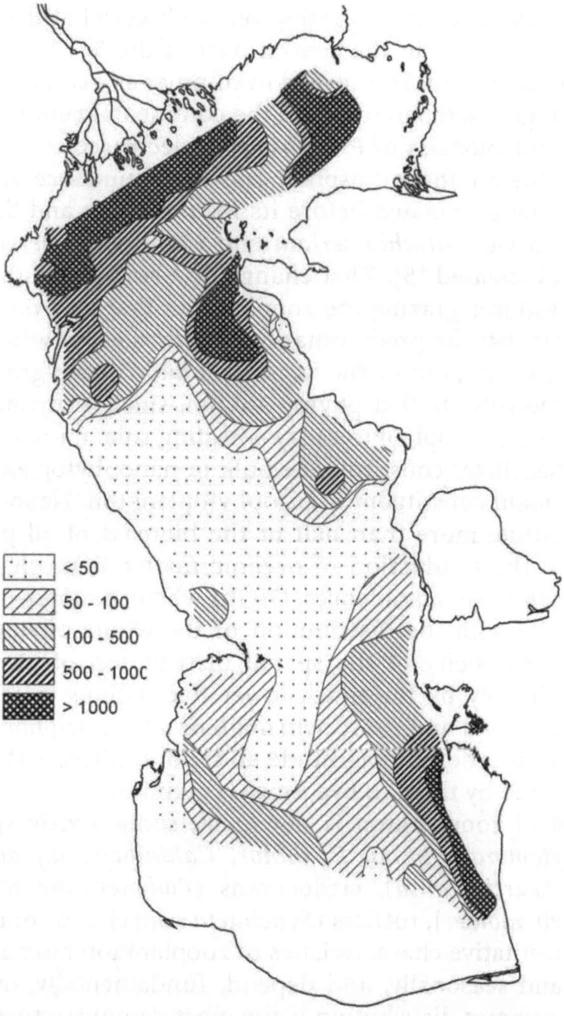

Fig. 2 Summer phytoplankton biomass distribution in surface water in milligrams per meter cubed

flare occurs in October–November. In the central part of the Sea phytoplankton develops intensively in the periods of vertical circulation strengthening (in early spring and autumn) and its development eases slightly in summer, in the period of stratification.

The Southern Caspian phytoplankton is characterized by the constant numerical predominance of pyrophytes (*Prorocentrum*) and in some years by the predominance of the blue-green algae over the diatoms. Phytoplankton biomass is three to five times lower than in the Middle Caspian, but *Pseudosolenia* remains the dominant species with almost the same share. A vast abundance of species peculiar to freshwater is observed at the western coast in spring. Here, as well as in the central part of the Sea the most intensive phytoplankton development is recorded in summer as a result of the favorable combination of higher temperature and the continental runoff [3].

In 2001–2003 the biomass of *Pseudosolenia* decreased to 77% in the Middle and 21% in the Southern Caspian, and the abundance and biomass of small-cell algae that dominated before its introduction, and the Azov–Black Sea introduced species *Nitzchia seriata* and *Cerataulina bergonii* also used by zooplankton, increased [5]. That change occurred under the effect of the *Mnemyopsis* introducer grazing the zooplankton that feed on these algae. It thus became clear that the predominance of *Psedosolenia* was caused not by its high competitiveness, but by the fact that it hadn't been grazed.

Besides traditionally studied phytoplankton, smaller organisms referring to as nano- and picophytoplankton play a leading role in organic matter production. Small flagellates constitute the bulk of nanophytoplankton, whereas bacteria are the main constituent of picophytoplankton. Nano- and picophytoplankton constitute more than half of the biomass of all phytoplankton, and their role in the production of organic matter is much higher due to their small sizes and high abundances. On the whole, nanophytoplankton distribution coincides with the distribution of the whole phytoplankton. The considerable heterogeneity of the spatial distribution of phytoplankton is smoothed considerably by the much lower fluctuations in the nanophytoplankton: *Pseudosolenia* and *Prorocentrum* biomass at neighbouring stations often differed by an order of magnitude and more, whereas the small flagellate biomass differed by three to five times (maximum) [4].

The checklist of zooplankton is not great; some widely spread species, copepods (*Eurytemora grimmi, E. minor, Calanipeda aquae-dulcis, Heterocope caspia, Acartia tonsa*), cladocerans [*Podonevadne trigona, Pleopis* (=*Podon*) *-polyphemoides*], rotifers (*Synchaeta vorax*) and some others make up the bulk. Quantitative characteristics of zooplankton biomass vary widely both annually and seasonally, and depend, fundamentally, on the climatic conditions; the summer distribution is the most demonstrative (Fig. 3).

Zooplankton distribution in the Northern Caspian is determined by variations in the temperature and river flow value. In winter, the species composition is poor and the biomass small; in summer, owing to freshwater species development such as different rotifers and water fleas, both species number and the biomass of 150–500 mg m^{-3} and in the pre-delta areas up to 1 g m^{-3} and more, increase sharply. *C. aquae-dulcis, H. caspia,* and *P. trigona* dominate, as do benthic animal larvae in some periods.

Copepods, such as *E. grimmi, E. minor, C. aquae-dulcis,* predominate in the zooplankton of the Middle and Southern Caspian in all seasons, though in summer, especially in the coastal zone, the share of cladocerans (*P. polyphemoides*) as well as of benthic invertebrates is great. Seasonal variations of the species composition are not observed, whereas biomass, both total and of separate

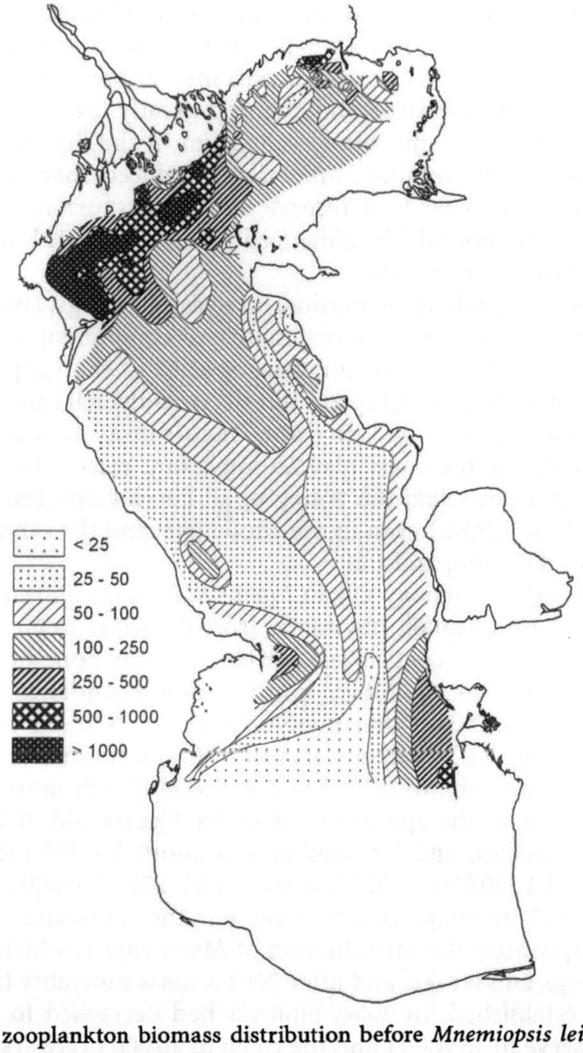

Fig. 3 Summer zooplankton biomass distribution before *Mnemiopsis leidyi* introduction in milligrams under meter squared

species, varies considerably. That occurs as a result of the fact that cladocerans go through the wintertime at the resting egg stage, but there is a succession of generations in copepods. The amplitude of seasonal biomass variations is determined here by winter meteorological conditions. In the central parts, biomass averages to 50–100 mg m^{-3}, which gives 5–10 g m^{-2} in terms of the total volume of water column. In spring, by April the zooplankton biomass increases to 25–50 g m^{-2}; however, it decreases slightly by August and more considerably by autumn. Near the coast, biomass is much higher at 1 m^3; however, it remains the same as in the central part, under 1 m^2, due to shallow depths [6].

Living in coastal waters, carnivorous cladocerans of the family Polyphemidae feed on zooplankton; however it is relatively scarce, and is grazed by zooplanktivorous fish. In 1998, a new consumer of zooplankton, the comb jelly *Mnemiopsis leidyi* appeared and changed greatly the relationships both in the pelagic zone and in the whole ecosystem of the Sea (for more detail see Karpinsky et al., this volume). *Mnemiopsis* reduced sharply zooplankton having undermined the kilka food reserve, which in its turn resulted in the reduction of their numbers, which influenced the next trophic level and almost completely destroyed the fishery.

The planktonic–nektobenthic mysids, on which herrings, sturgeon fry and other carnivore fish feed, are the most important component of zooplankton.

Three species of kilka, the major mass species of the Caspian Sea, have been the main consumers of zooplankton till recently. All kilkas feed chiefly on copepods, mainly *E. grimmi* (after *Mnemiopsis* invasion mainly *A. tonsa*); cladocerans inhabiting the coast play an important role in the ration of the common kilka. Besides kilka, the young of all herring species, of which the Caspian shad (*Alosa caspia*) even in the adult state, and the sand smelt (*Atherina mochon*) feed on zooplankton.

The more abundant anchovy kilka *Clupeonella engrauliformis* inhabits the Middle and Southern Caspian, staying far from the shore in places more than 20 m deep. Its highest concentrations are found in the zone of the main circulation above depths of 50–200 m, and in lesser abundance it is found at lower depths (Fig. 4). It is a thermophilic species; its main concentrations are confined to areas with temperatures of 8 °C and higher. The life span of this species as well as of two others is 7 years, although more than 90% of the population falls in the age group of up to 3 years old. It is an important commercial species, and its number was about 2×10^8 individuals and biomass averaged 1 360 000 t and its annual catch 250–260 000 t in the period 1960–1995. In 1996, an insignificant decrease in the abundance occurred that intensified sharply after the introduction of *Mnemyopsis*, which grazed their food reserve, eggs and larvae, and after 2001 a mass mortality began for reasons not fully established. In 2004, biomass had decreased to 430 000 t, the commercial reserve to 165 000 t and the catch to 20 000 t [7, 8] (see also Fig. 4: Karpinsky et al., this volume).

The common kilka (*C. delicatula caspia*), an euryhaline subspecies of the Black Sea species, having penetrated the Caspian via the Kumo-Manych depression is found throughout the whole of the sea, but mainly, in the coastal zone at depths of 10–60 m. This kilka has migrated into the lower Volga, Ural and Terek, and penetrated into the Volga reservoirs and formed stable populations there. In spite of high abundance and biomass its share in the kilka fishery did not exceed 1%, as the specialized catch in the coastal zone was banned because of a considerable by-catch of sturgeon young. The introduction of *Mnemyopsis* influenced it to a considerably lesser degree, as the spawning of the kilka does not concur with the period of its mass develop-

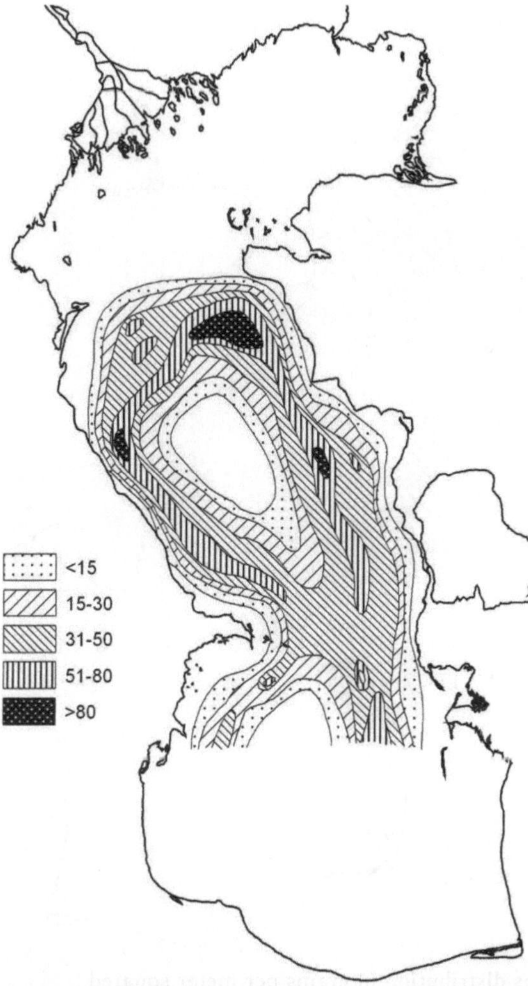

Fig. 4 Anchovy kilka distribution before *Mnemiopsis leidyi* introduction, in kilograms per one catch of conic net

ment. That is why its commercial reserve is estimated now as 232 000 t and the forecasted catch at 58 000 t, which led to improvement in the methods of specialized fishing and greatly enhanced its role in the fishery [8, 9].

The area of big-eye kilka (*C. grimmi*) includes the open parts of the Middle and Southern Caspian with depths of more than 40–50 m, where it concentrates in winter and spring in the northwest and east of the Southern Caspian, and in autumn in the northwest of the Middle and in the Southern Caspian. It avoids the near-surface water layers, dwelling at a depth of 20–200 m and deeper, down to 500 m. In winter concentrations of big-eye kilka are observed at depths of 80–150 m, while in spring and summer it dwells at 40–60 m and

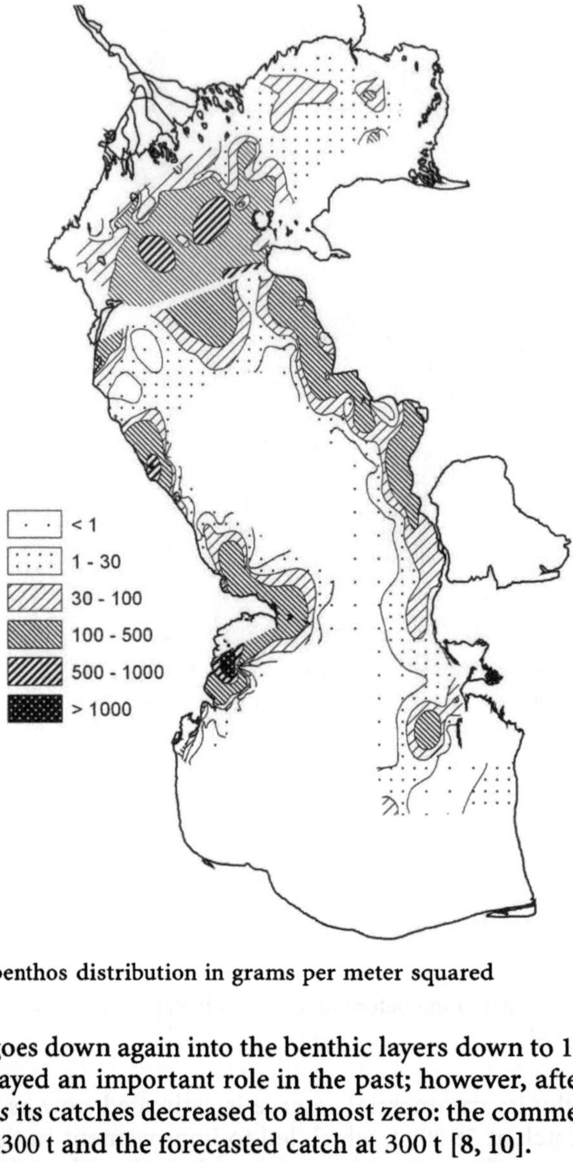

Fig. 5 Autumn benthos distribution in grams per meter squared

in autumn it goes down again into the benthic layers down to 100–150 m. The species has played an important role in the past; however, after introduction of *Mnemyopsis* its catches decreased to almost zero: the commercial reserve is estimated at 4300 t and the forecasted catch at 300 t [8, 10].

Benthos distribution (Fig. 5) is quite stable, especially in the Middle and Southern Caspian, though seasonal variations are more expressed than annual. Benthic organisms in the Northern Caspian are represented chiefly by freshwater species, as well as plankton organisms, though in zones of salinity of 7–8‰ marine-introduced species (*Mytilaster*, *Nereis*, *Abra*) form most of the biomass. Interannual variations are mainly due to the introduction of new species (*Mytilaster*, *Nereis*, *Abra* [11]) and by further fluctuations in their abundance. The bulk of the biomass is concentrated within the limits of the shelf at depths of 70–100 m, whereas deeper than 200 m it is low and stable, which is why mainly these depths were studied. The distribution of the benthos total biomass is determined, chiefly, by bivalve mollusks. The highest biomass is formed by *Mytilaster* inhabiting depths down to 100 m, whose vast settlements, not necessarily very dense, are concentrated at a depth of 10–25 m off the eastern coast of the Middle Caspian, in the region of Derbent and northward of the Kura mouth near the western coast (Karpinsky et al., this volume Fig. 1). Biomass per square unit of *Dreissena*, the next most important mollusk, is about five to eight times smaller than that of *Mytilaster*, though it occupies a greater territory, and predominates at depths of 25–50 m in the Middle and Southern Caspian and at 3–20 m in the Northern Caspian. Large, but comparatively few, autochthonous Cardiidae (*Didacna* and *Hypanis*), whose biomass value may vary greatly, form a rather large biomass near the eastern coast of the Middle Caspian and in the Northern Caspian. *Abra* distribution is even, and its seasonal changes are stable , the major changes being due to sturgeon grazing. The biomass of crustaceans and worms, which chiefly inhabit the soft soils, is much less than that of mollusks; however, these groups play an important role in sturgeon feeding.

It remained unclear for a long time, why there is a clear predominance of sestonophages off the eastern coast of the Middle Caspian, where they constitute the maximal benthos biomasses in the almost complete absence of deposit feeders. Only after heavy cyclic alongshore currents at a velocity of 0–70 cm/sec [12] were discovered did the reason for both this and the fact that there is no accumulation of modern sediments [13] become clear. The almost complete absence of the carnivores that are found in all benthic systems constituting 10% of the overall biomass [14] should be considered the most amazing anomaly in the trophic grouping distribution. The introduction of the crab (*Rhithropanopeus harrisii*) was the first appearance of a real carnivore in the bottom fauna of the Caspian, and even so, its share in the overall biomass of the Northern Caspian comprises 0, 5–1% and of the Southern 1–2%, which is obviously low. The absence of carnivores is explained by the powerful, intensive grazing pressure of the sturgeons, to which the usual functions of the carnivore passed in the bottom-dwelling communities.

Roach, bream and all species of gobies feed on the benthos in the Caspian, although its main consumers are the sturgeon and star sturgeon, as well as the fry of all sturgeons. Only roach of the attached sestonophages is able to separate *Dreissena* from the substrate, and the sturgeon in the Southern Caspian feeds on single *Mytilaster*, as they settle on firm silts. Large bivalves, *Didacna* and *Hypanis*, after they reach a weight of 1.5 g, become inaccessible to the sturgeon, although their fry is eaten actively. Of the rest of groups, only oligochaetes are poorly utilized.

Roach (*Rutilus rutilus caspicus*) forms some isolated stocks within the limits of the Caspian Sea. The abundance of the Azerbaijan roach is not great; it inhabits the southwestern part of the Southern Caspian and spawns in the

Kura and its branches. The area of the Turkmenian stock is the southeastern part of the Southern Caspian. The most abundant is the Northern Caspian stock: its area is the entire Northern Caspian and it spawns in the deltas of the Volga, the Ural and the Terek. It is an important species, its commercial reserves varying from 33 000 to 170 000 t and the catch from 5000 to 25 000 t; the major fishing region is the Volga delta. Females dominate in the commercial catches making up 83% of the catch, sometimes rising to 95–97%, although female dominance in the young age groups is expressed at a considerably lesser degree. Such a sex ratio ensures great population fecundity and high abundance of the stock in conditions of limited food supply [15, 16].

The eastern bream (*Abramis brama orientalis*) inhabiting the Caspian Sea also forms some stocks: the Kura stock in the Southern Caspian, the Ural, Terek and the more abundant Volga stock in the Northern. The bream spends most of its life cycle in the sea and avandeltas of the rivers, where the fry fats until maturation and the adult fish live after spawning. By the end of summer and in autumn the bream migrates to the shallow area of the Sea and the river avandelta for hibernation; in spring the fish start spawning migrations into the lower rivers. It is a commercial species, although, depending on productivity, its reserves vary strongly from 15 000 to 75 000 t and the catch from 5000 to 30 000 t . Roach, bream and sturgeon have a similar food spectrum, but they are distinguished by feeding fields, dates and salinity [16, 17].

A vast abundance of gobies of different species (family Gobiidae) inhabits the entire Sea area, mainly the coastal shallow water zone. These bottom-dwelling fish that are not mobile live sparsely, not forming aggregations, although their feeding and reproduction places are close to each other. Although bullhead has no commercial value, some species may be the object of sport fishing. In various years the gross gobies reserve in the Northern Caspian was within 7000–16 000 t, and as their food spectrum is similar to that of other benthivorous fish, bullhead species are rather strong competitors to them, and at the same time the very food for sturgeons, carnivorous herrings and seals [18].

The main benthos consumers are sturgeons; these are of great importance not only as the most valuable commercial asset, but as unique fish—some of the oldest on Earth having been preserved without changes for millions of years. The fry of all sturgeons feed on crustaceans and worms, but their food spectrum changes with time. The adult sturgeon (*Acipenser guldenstadti*) prefers to feed on mollusks and the star sturgeon (*A. stellatus*) on worms and crustaceans [19], while the beluga (*Huso huso*) and the spine sturgeon (*A. nudiventris*) switch to a chiefly fish diet. However, not all species give up other food forms, but move them into the minor or forced category, which occurs very often at a time of dietary deficiency. The sturgeons are benthic fish. Beluga and spiny sturgeon, active predators, come up above the bottom chasing their prey, kilka, sand smelt, gobies, herring fry and carps. The Caspian sturgeons spend most of their lifespan in the sea, conducting migrations into

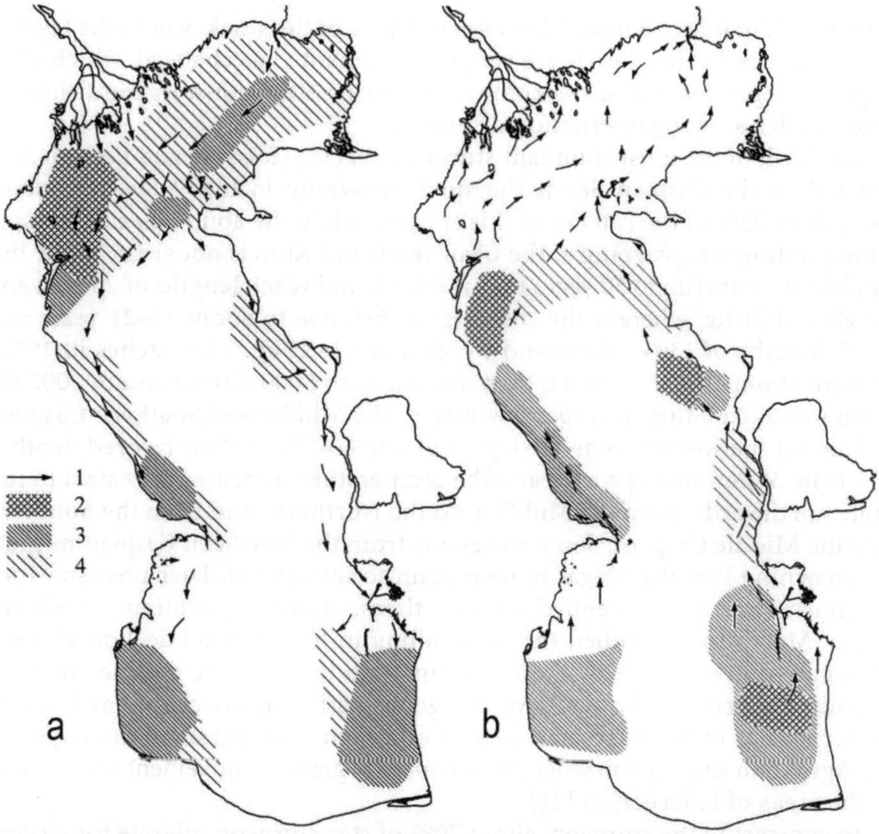

Fig. 6 Distribution and migrations of sturgeons in summer and autumn (**a**)and in winter and spring (**b**). *1* Migrations, *2* high density, *3* average density, *4* low density

rivers: seasonal variations are caused by temperature variations and spawning (Fig. 6).

The sturgeons are the most valuable fishin the Caspian; maximal catches of 26 000–27 000 t were recorded in 1976–77, after fishing had been banned in the Sea in the 1960s and a large part of the fry was produced in hatcheries, but a drop in abundance followed. The first sick fish appeared at the beginning of the 1980s; later a disease of unclear origin, a myopathy, struck the bulk of the sturgeon stock, reducing it significantly. After the breakup of the USSR, and the subsequent formation of several independent states, unregulated, illegal fishing of sturgeons in the Sea began. Now the official catch does not exceed 500 t, though in reality it is higher. As a result the size of the commercial stock has very sharply reduced by 20 times compared to 1976 (by the experts' estimations); sturgeon calling at the Volga are only parts of a percent of the number of calling at the same place in 1976 [20]. As chiefly large, mature individuals are removed, the age structure of the stock, where the fry that are reproduced at hatcheries are very predominant , has changed, which gives hope for reproduction of the commercial stock after the sturgeon fishery is taken under a strong international control.

The bulk of the most abundant sturgeon species (Russian and Persian subspecies) in the Caspian Sea is the stock spawning in the Volga, producing more than 95% of the catches of this species, while the abundance of the specimen calling for spawning at the Ural, Terek and Kura is not great. These fish can live 40 years (up to 50 years in some fish), and reach lengths of 200 cm and weights of 70 kg, whereas the commercial fish live to about 13–21 years and reach lengths of 130–150 cm and weights of 13–25 kg. The catches in 1975–88 were about 10 000–13 000 t, then decreased to 3000–7000 t and in 2002 the catch was 220 t. Sturgeons spend winter in the Middle and Southern Caspian, mainly off the western coast at depths of 2 to 60–70 m; their favored depth is 10–30 m. When spring comes and the temperature increases they start to migrate northwards: from the Middle into the Northern, and from the Southern into the Middle Caspian. Some sturgeons from the Northern Caspian migrate for spawning into the Volga, some remain to forage, and later post-spawned specimens and the descended fry join them. Spawning continues from the end of March to November, the peak falling in July at flood decline and water temperatures of 21–27 °C. In autumn, with temperature decline and food organism grazing at the shallows, sturgeons gather at the central and southwestern parts of the Northern Caspian and then concentrate in the region of the Agrakhan shoal, from where they continue gradual movement southwards to the areas of hibernation [21].

In contrast to the sturgeon, about 70% of star sturgeon migrate for spawning to the Ural and only 28% to the Volga; the rest go to the Kura, Terek, Sulak, and Sefid-Rud. The maximal age is 31 years, with length 195 cm and weight 25 kg. At the average age (16 years for females and 13 years for males) the average length is 145–150 cm and the average weight is 10–12 kg. Catches in 1976–77 were about 12 000–13 000 t, and by 1980–92 had decreased to 3000–4500 t; in 2002 the catch was 140 t. The star sturgeon spends winter in the Middle and Southern Caspian, keeping to depths of 20–50 m, but forming large aggregations also off the eastern coast in the Kazakh Bay, where it feeds on Corophiidae, pulling them from the coquina. In spring the star sturgeon migrates to the Northern Caspian, where it aggregates in the region of the Ural Furrow, and to the Middle Caspian, forming aggregations in the Agrakhan shoal. The spawning migrations are timed to the spring flood in April–May, though they continue throughout the summer to a small extent. In summer the star sturgeon is abundant along the entire Northern Caspian, concentrating in the western area, though by July it is already migrating away from the northern part of the Sea. The fry descend from July to November. In August they are already migrating away towards the south, forming aggregations at the Agrakhan shoal. In autumn the star sturgeon migrates to the Middle and then to the Southern Caspian, where it inhabits the shallow located southwards from Ogurchinskij Island [22].

During spring and, especially, autumn migrations, the sturgeons continuing to feed form dense aggregations over small territories at the bottom, and so the major factor affecting the velocity of their migration becomes the state of the food fields. At that period sturgeon and star sturgeon feed on minor and even substituted foods, since their preferred food has been grazed to practically zero. The Agrakhan shoal is very illustrative of this aspect, since it became the most important fatting area after colonization by *Abra* and *Nereis*. Comparison of the benthos survey data in August, October and February shows the almost complete disappearance of *Nereis* from August to October and gradual restoration by February (Fig. 7). In other parts of the Sea benthos grazing by sturgeons might not be so intensive over a short period, but may continue for longer as, for instance, in the Northern Caspian.

Sturgeons are a very old fish, which have existed for a very long time without change. In the Caspian Sea the only real enemy capable of affecting their abundance was man. That is why during their existence in the Caspian the only factor restraining their development was the food reserve, primarily the benthos. In turn, the whole period of formation of the modern bottom community (1.8 million years) (Karpinsky, this volume) was influenced by a severe, intensive grazing by sturgeons that has no analogy in any other sea. As a result some specific features of particular feeding categories developed.

The constant elimination of a considerable part of the community because of sturgeons' grazing pressure resulted in morphophysiological adaptations permitting populations to restore rapidly in organisms making up that community: small size, high abundance and fecundity, and relatively short life cycle.

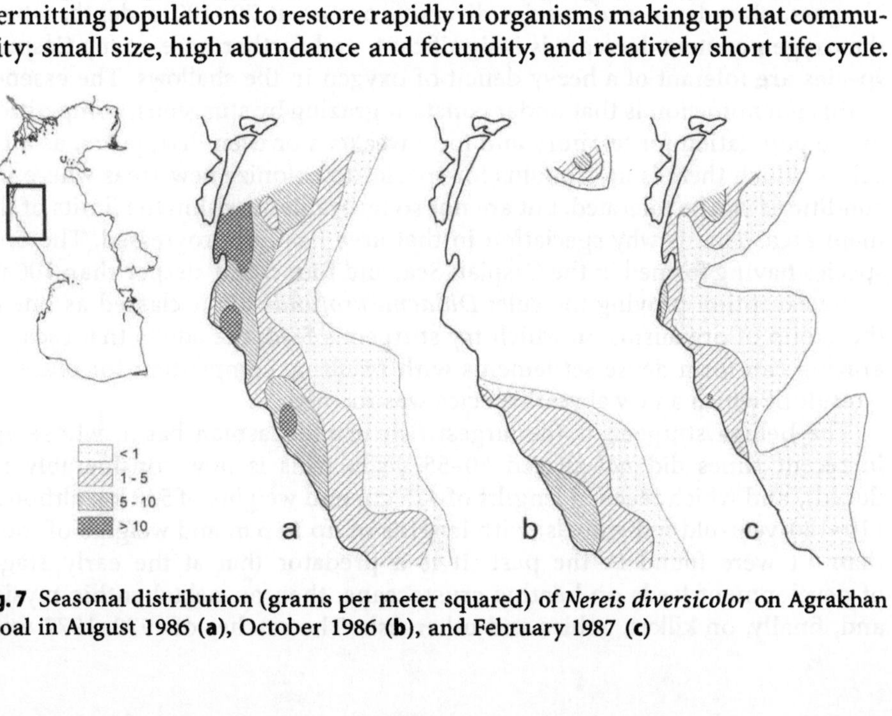

Fig. 7 Seasonal distribution (grams per meter squared) of *Nereis diversicolor* on Agrakhan shoal in August 1986 (**a**), October 1986 (**b**), and February 1987 (**c**)

Usually, species with such features predominate at the early stages of succession and are named *r*-strategists; however other features of such Caspian species characterize them as species of the later stages of succession, or *K*-strategists. In a bottom community formed under intensive grazing pressure, besides the above-mentioned features, a constant predominance of the fry and low total biomass at high productivity, which constitute a stable food reserve for a large number of benthivorous fish, are observed. The comparison shows that in the Northern Caspian benthivorous fish production is 1.5 times higher and benthos biomass is 8 times lover per the square unit than in the Azov Sea.

But in such a case, weakening of the grazing pressure produces significant changes. Most of the organisms are then able to reach adult size, resulting in intensified intraspecies competition for territory and food. However, the Caspian autochthonous organisms are poorly adapted to such competition, giving way to marine introducers. And here the danger arises: a relatively small number of marine introducers are able to force out the native Caspian fauna without the restraining factor of grazing.

The other peculiarity is the absence of other benthos predators and carnivores, whose functions are performed by the sturgeons.

Finally, one more specific feature of the Caspian benthos is its vertical distribution. Early researchers [23, 24] recorded a considerable decrease in quantitative indices of organisms with depth and related it to oxygen distribution, or more precisely with the decrease going into the Sea basins, which corresponds quite well to plankton and fish distribution (Fig. 8). However, it turned out to be more complicated with the benthos: a sharp reduction in biomass, abundance and species diversity occurs at 70–100 m depth, where the oxygen content decrease is insignificant, and furthermore, many Caspian species are tolerant of a heavy deficit of oxygen in the shallows. The essence of this phenomenon is that under constant grazing by sturgeons, competition in the population for territory and food weakens or even disappears, as a result of which there is no stimulus for species to colonize new areas where the conditions can be tolerated, but are not so favorable as within the limits of the main area. That is why speciation in that area has not progressed. The only species having formed in the Caspian Sea, and found only deeper than 100 m, is the exception proving the rule: *Didacna profundicula* is classed as one of the group of organisms on which fry sturgeons feed; the adults that escaped grazing can form dense settlements with resulting competition for place, as a result of which a new abyssal species was formed.

The beluga sturgeon is the largest fish in the Caspian basin, whose age in recent times did not exceed 50–55 years (this is now considerably reduced), and which reached lengths of 425 cm and weights of 520 kg, although 110–120-year-old individuals, with lengths of up to 5 m and weights of more than 1 t were found in the past. It is a predator that at the early stages of development feeds on benthic crustaceans, then on nektobenthic mysids and, finally, on kilkas, gobies and other fish. The catches of 1970–1977 were

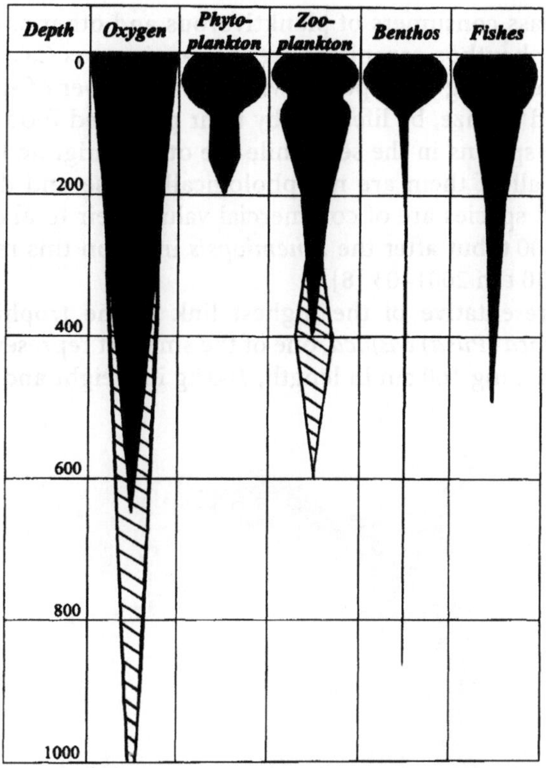

Fig. 8 Scheme of vertical distribution of oxygen (*black* - period of hight sea level, 1900–1930, 1990–present time; *hatching* - low sea level, 1940–1980), phytoplankton, zooplankton (with dumal vertical migration), benthos, and fishes

about 1700–2500 t, but decreased gradually to 500–900 t in 1980–1990 and to 130–170 t in 2000–2002 . In comparison with the sturgeon and star sturgeon, beluga as a predator is dispersed across a wider forage area. Like all sturgeon, it hibernates, in this case off the western coast of the Middle and Southern Caspian, chiefly at depths of 20–50 m and sometimes down to 180 m. In spring it migrates closer to the coast to depths of 2–30 m, and to the Northern Caspian, keeping to the southwestern part, where it spends summer. Spawning occurs over almost the whole year, although there are spring and autumn peaks. In autumn, as the coastal waters cool, beluga migrates to the southern and deeper parts of the sea. Owing to such migrations the bulk of the beluga live in winter at temperatures of 8–16 °C and in summer at 20–26 °C [25].

The fourth species of Caspian sturgeon is quite rare: the spiny sturgeon. It mainly inhabits the southern part of the Sea, migrates seasonally, and spawns at the Kura, Ural and Sefid-Rut, and seldom at the Volga; the adult fish feeds on kilka and gobies. The spiny sturgeon is close to the beluga by biological features, but is inferior to it in size and lifespan. Its commercial value is very small.

One of the mass consumers of planktivorous and other small fish, as well as of some other benthos organisms are the carnivorous Caspian herrings of genus *Alosa*, comprising four species with a vast number of subspecies. They are differentiated by size, by lifespan, by their preferred food organisms and migrations (one spawns in the sea, while the others migrate for spawning to the rivers), but all of them are morphologically close and descended from one ancestor. All species are of commercial value, their total catches varying from 1500 to 4500 t, but after the *Mnemiopsis* invasion this recently reduced sharply to 70–230 t in 2001–03 [8].

The last representative of the highest link of the trophic chain is the Caspian seal, *Phoca (Pusa) caspica*, one of the smallest representatives of family Phocidae, reaching 160 cm in length, 100 kg in weight and a maximal age

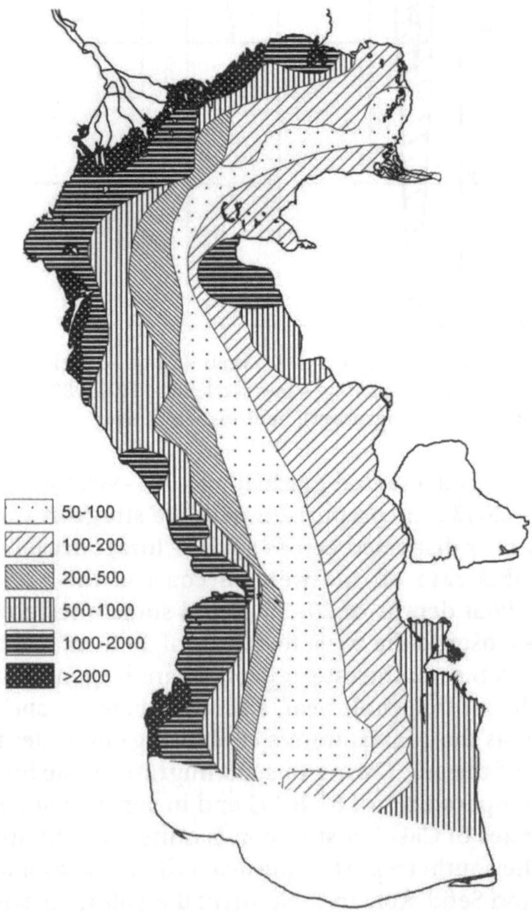

Fig. 9 Plankton bacteria distribution in suface water in thousands of cells per milliliter cubed

of 37 years. It is distributed throughout the Caspian, but its highest abundance is in the Northern Caspian. Mating and reproduction occur on the ice of the Northern Caspian. Kilka (25–95%) comprises the bulk of its food during the summer forage migrations to the Middle and Southern Caspian, while roach (50%), gobies (30%) and bream fry (13%) do so for the Northern. Fishing of adult seals was banned in 1967, and in 1970 a catch quota on the newly born calves was introduced, which made it possible to increase the school up to 500 000–600 000 individuals. In 2000, a mass mortality of seals caused by a viral epizooty, carnivorous distemper, occurred and heavy pollution provoked higher female barrenness. Now they number about 375 000 individuals and the possible commercial catch is 18 000 individuals [26, 27].

Bacteria complete the cycle of organic matter transformations. They process a vast mass of unutilized phytoplankton, returning nutrients into the water column. The highest development of bacteria in water and soil is observed in the regions of contact between river and sea water in the Northern Caspian, where their abundance reaches 2 million—4 million cells per 1 ml of water. Microorganism content decreases to 100 000–400 000 cells per 1 ml away from the Volga delta. A relationship between bacterial distribution in the surface waters and phytoplankton distribution consisting of a considerable abundance increase off the shores and decrease in the central parts of the Sea is observed. However, there is a significant difference in abundance off different shores: it is five to ten times higher off the western shore than off the eastern (Fig. 9) [28]. This is difficult to explain: it may be due to more developed terrigenous runoff near the western shore, intensive currents off the eastern shore or other factors.

References

1. Proshkina-Lavrenko AI, Makarova IV (1968) Plankton algae of the Caspian Sea. Nauka, Leningrad (in Russian)
2. Sanina Lv, Levshakova Vd, Tataritseva Ta (2000) In: Neyman Aa, Tarverdieva Mi (Eds) Marine Hydrobiological Researches. Vniro, Moscow, p 38 (In Russian)
3. Ardabieva AG, Voloshko LN, Klimova AN, Levshakova VD, Sanina LV, Tataritseva TA (1985) In: Yablonskaya EA (ed) Caspian Sea. Fauna and biological productivity. Nauka, Moscow, p 31 (in Russian)
4. Borodin VE (1991) In: Fisheries-related studies of plankton, part II Caspian Sea. VNIRO, Moscow p 102 (in Russian)
5. Tataritseva TA, Terletskaya OV (2004) In: Fisheries researches in the Caspian. CaspNIRKh, Astrakhan, p 123 (in Russian)
6. Kuzmicheva VI, Kurashova EK, Kortunova TA, Tinenkova DKh, Epshtejn BM, Abdullaeva NM, Vladimirskaya EV, Badalov FG, Mamaev MM (1985) In: Yablonskaya EA (ed) Caspian Sea. Fauna and biological productivity. Nauka, Moscow, p 86 (in Russian)
7. Paritski YuA (1989) In: Belyaeva VN, Vlasenko AD, Ivanov VP (eds) Caspian Sea. Ichthyofauna and commercial fish resources. Nauka, Moscow, p 83 (in Russian)

8. Sedov SI, Paritski YuA, Zykov LA, Kolosyuk GG, Aseinova AA, Andrianova SB, Kanatiev SV, Gazizov IZ (2004) In: Fisheries researches in the Caspian. CaspNIRKh, Astrakhan, p 360 (in Russian)
9. Derevyagin VA, Rychagova TL (1989) In: Belyaeva VN, Vlasenko AD, Ivanov VP (eds) Caspian Sea. Ichthyofauna and commercial fish resources. Nauka, Moscow, p 81 (in Russian)
10. Aseinova AA (1989) In: Belyaeva VN, Vlasenko AD, Ivanov VP (eds) Caspian Sea. Ichthyofauna and commercial fish resources. Nauka, Moscow, p 77 (in Russian)
11. Karpinsky MG (2002) Ecology of the benthos of the Middle and Southern Caspian. VNIRO, Moscow (in Russian)
12. Bondarenko AL (1993) The Caspian Sea currents and forming the salinity field in the Northern Caspian. Nauka, Moscow (in Russian)
13. Kholodov BN, Khrustalev YuP, Lubenchenko IYu, Kovalev VV, Turovskij DS (1989) Caspian Sea. Problems of sedimentogenes. Nauka, Moscow (in Russian)
14. Kuznetsov AP (1980) Ecology of bottom communities of World Ocean. Nauka, Moscow (in Russian)
15. Strubalina NK (1989) In: Belyaeva VN, Vlasenko AD, Ivanov VP (eds) Caspian Sea. Ichthyofauna and commercial fish resources. Nauka, Moscow, p 124 (in Russian)
16. Kushnarenko AI, Fomichev OA,Sidorova MA, Chernyavski VI, Rodionova OV, Kuznetsov YuA, Mankova IYu (2004) In: Fisheries researches in the Caspian. CaspNIRKh, Astrakhan, p 293 (in Russian)
17. Sidorova MA (1989) In: Belyaeva VN, Vlasenko AD, Ivanov VP (eds) Caspian Sea. Ichthyofauna and commercial fish resources. Nauka, Moscow, p 153 (in Russian)
18. Ragimov DB, Stepanova TG (1989) In: Belyaeva VN, Vlasenko AD, Ivanov VP (eds) Caspian Sea. Ichthyofauna and commercial fish resources. Nauka, Moscow, p 190 (in Russian)
19. Tarverdieva MI (1965) In: Zenkewits LA (ed) The changes of the Caspian Sea biological complexes. Nauka, Moscow, p 234 (in Russian)
20. Zhuravleva OL (2000) In: Ivanov VP (ed) The international conference "Sturgeons on the threshold of the 21st century", abstracts. CaspNIRKh, Astrakhan, p 52 (in Russian)
21. Pavlov AV (1989) In: Belyaeva VN, Vlasenko AD, Ivanov VP (eds) Caspian Sea. Ichthyofauna and commercial fish resources. Nauka, Moscow, p 51 (in Russian)
22. Dovgopol GF (1989) In: Belyaeva VN, Vlasenko AD, Ivanov VP (eds) Caspian Sea. Ichthyofauna and commercial fish resources. Nauka, Moscow, p 66 (in Russian)
23. Grimm OA (1877) Caspian Sea and its fauna, Proceedings of Aral-Caspian Expedition, issue 2. Sankt-Petersburg naturalists society, Sankt-Petersburg (in Russian)
24. Knipovich NM (1921) Hydrological investigations in the Caspian Sea in 1914–1915, Proseedigs of the Caspian expedition 1914–1915, vol 1. St. Petersburg (in Russian)
25. Raspopov VM (1989) In: Belyaeva VN, Vlasenko AD, Ivanov VP (eds) Caspian Sea. Ichthyofauna and commercial fish resources. Nauka, Moscow, p 27 (in Russian)
26. Khuraskin LS (1989) In: Belyaeva VN, Vlasenko AD, Ivanov VP (eds) Caspian Sea. Ichthyofauna and commercial fish resources. Nauka, Moscow, p 198 (in Russian)
27. Khuraskin LS, Zakharova NA,Kuznetsov VV, Khroshko VI, Valedskaya OM (2004) In: Fisheries researches in the Caspian. CaspNIRKh, Astrakhan, p 400 (in Russian)
28. Salmanov MA (1999) Ecology and biological productivity of the Caspian Sea. Ismail, Baku (in Russian)

Hdb Env Chem Vol. 5, Part P (2005): 211–221
DOI 10.1007/698_5_011

Published online: 28 September 2005

Kara-Bogaz-Gol Bay

Aleksey N. Kosarev[1] · Andrey G. Kostianoy[2] (✉)

[1]Geographic Department, Lomonosov Moscow State University, Vorobievy Gory, 119899 Moscow, Russia

[2]P.P. Shirshov Institute of Oceanology, Russian Academy of Sciences, 36 Nakhimovsky Pr., 117997 Moscow, Russia
kostianoy@online.ru

Abstract Kara-Bogaz-Gol Bay is a large lagoon east of the Caspian Sea and normally covers an area around 18 000 km^2 and is only a few meters deep. It is several meters lower than the Caspian Sea, so water flows from the Caspian Sea through a narrow strait into the bay, where it evaporates. Kara-Bogaz-Gol Bay is one of the saltiest bodies of water in the world: salt concentrations up to 350 $g L^{-1}$ have been documented. The salt in this natural evaporative basin has been used commercially since at least the 1920s. In March 1980, in order to restrict the losses of the Caspian waters and to decelerate the fall of the Caspian water level, which in 1977 was the lowest at any time over the last 400 years (– 29 m), the Kara-Bogaz-Gol Strait was dammed. In response to this anthropogenic intervention, the bay had already dried up completely by November 1983. In 1992, the dam was destroyed. Kara-Bogaz-Gol Bay has been filling up with Caspian water at a rate of 1.7 $m yr^{-1}$ up to 1996 as observed by the TOPEX/Poseidon altimetry mission. Since then, Kara-Bogaz-Gol Bay level evolution with characteristic seasonal oscillations and a decreasing tendency with a rate of 6 $cm yr^{-1}$ has been similar to that of the Caspian Sea.

Keywords Caspian Sea · Ecology · Kara-Bogaz-Gol Bay · Mineral salts · Sea level

1 Nature and Riches of the Bay

Kara-Bogaz-Gol Bay deeply penetrates into land on the eastern coast of the Caspian Sea (Fig. 1, Fig. 1 in [11]). In Turkmen, its name means "the bay of the black mouth." According to ancient legend, a deep was located in the bay, which absorbed the Caspian waters and even the ships that dared to visit this vast lagoon. The local people were sure that in this way the water passed either to the Aral Sea or even to the ocean. Maps of the Caspian

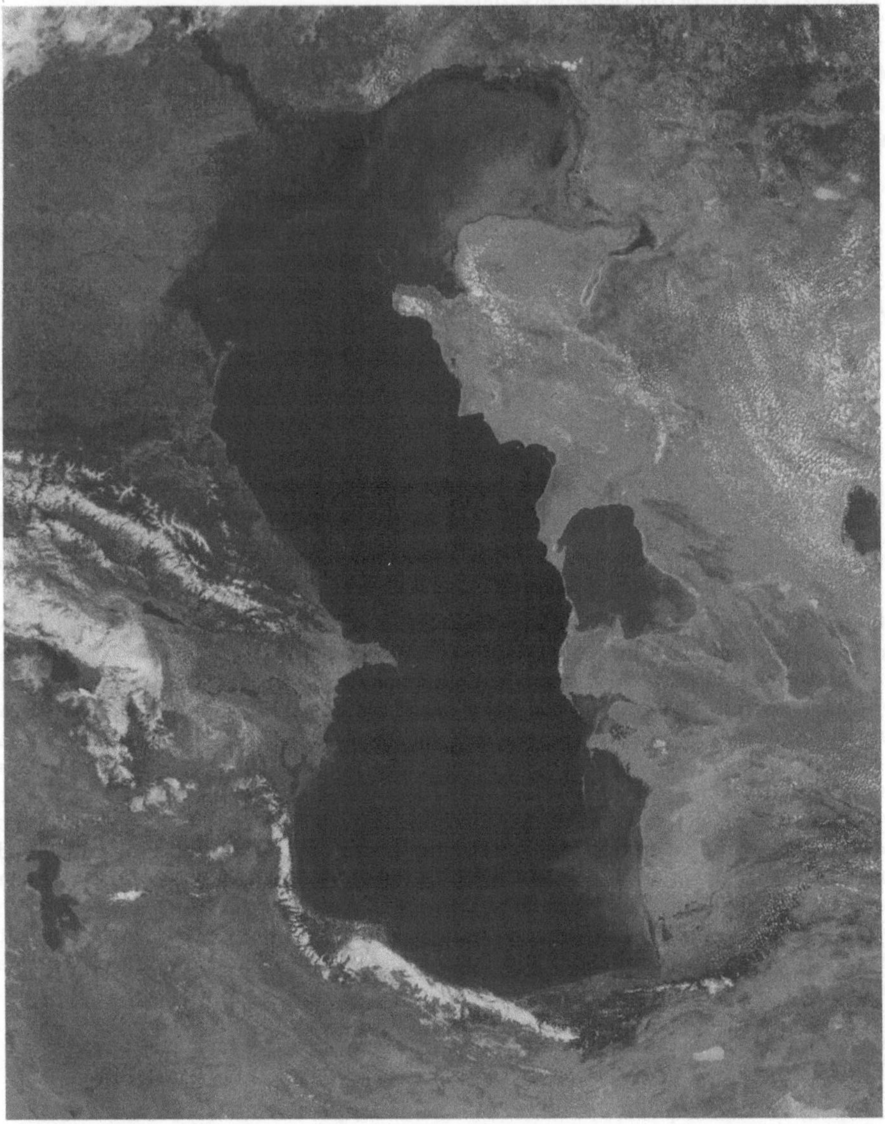

Fig. 1 High-resolution satellite image of the Caspian Sea and Kara-Bogaz-Gol Bay obtained from *MODIS-Terra* on June 11, 2003 (image courtesy Jeff Schmaltz, MODIS Land Rapid Response Team, NASA GSFC, http://eol.jsc.nasa.gov). The Volga River and its vast delta are recognizable in the northwestern part of the image

coast in this region were compiled for the first time in 1715 according to the orders of Peter the Great, by the expedition headed by Prince Alexander Bekovich-Cherkassky; this mission required great personal courage from its participants. The enigma of Kara-Bogaz-Gol Bay was explained 120 years later in 1836 by the well-known explorer G.S. Karelin, who dared to enter the bay and revealed that the Caspian water strongly evaporates under the hot and dry climate of the surrounding desert rather than disappears. In 1847 I.M. Zherebtsov, a lieutenant of the Russian Navy on board the steam corvette *Volga*, was the first to follow the shores of the bay, to describe them, and to plot a geographic map. His voyage is brilliantly presented in the story by K.G. Paustovsky "Kara Bogaz" (1932). In 1897, the Ministry of Trade and Industries of Russia directed an expedition to Kara-Bogaz-Gol Bay on board the steamship *Krasnovodsk*, it was headed by the hydrologist I.B. Shpindler and the expeditions purpose was a study of the salt resources of the bay and to determine the reason for the mass death of fish in the bay [1].

So, what does Kara-Bogaz-Gol Bay represent in itself? It is a shallow-water topographic depression with a flat floor and a variable coastline. It is the largest lagoon of the Caspian Sea separated from the sea by two sand spits (Fig. 2). Between them, a strait 7–9 km long, 120 to 800 m wide, and 3–6 m deep is located (Fig. 3). The morphometric characteristics of Kara-Bogaz-Gol Bay strongly vary depending on the position of the levels of the sea and the bay. Due to the difference in these levels and depending on its magnitude,

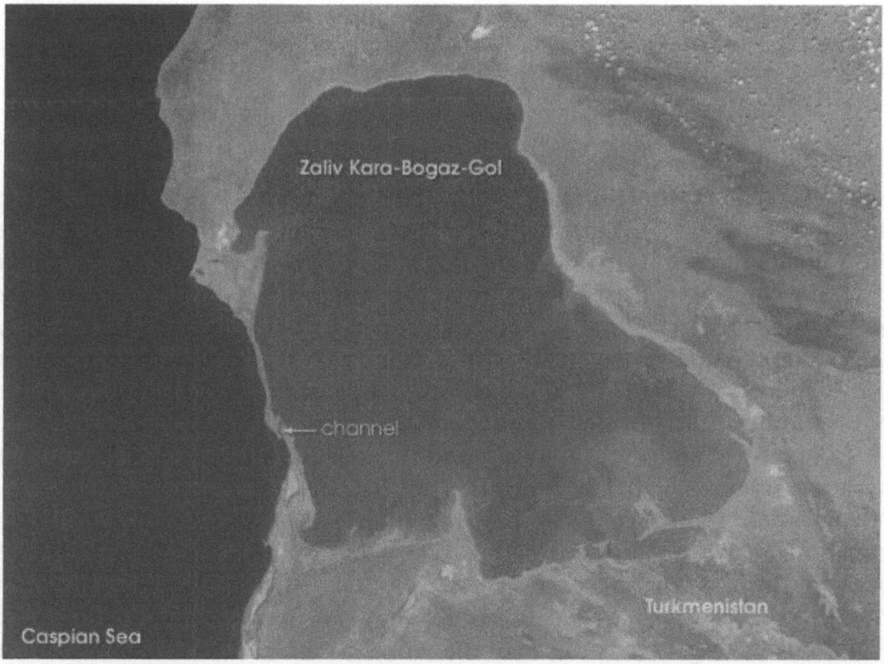

Fig. 2 Magnified high-resolution satellite image of Kara-Bogaz-Gol Bay obtained from *MODIS-Terra* on June 11, 2003 (image courtesy Jeff Schmaltz, MODIS Land Rapid Response Team, NASA GSFC, http://eol.jsc.nasa.gov)

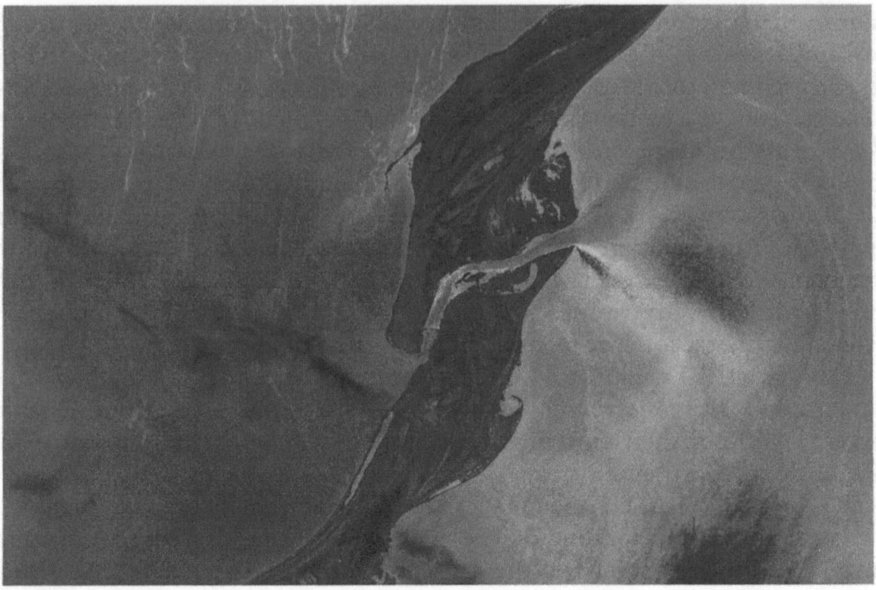

Fig. 3 Photo of the strait connecting the Caspian Sea and Kara-Bogaz-Gol Bay shot by American astronauts from the *Space Shuttle* spacecraft on June 17, 2002. The Caspian Sea is on the left, Kara-Bogaz-Gol Bay is on the right. In the mouth of the strait, one can see the site of the former dam (image courtesy of the Earth Sciences and Image Analysis Laboratory, NASA Johnson Space Center, Mission-Roll-Frame STS111-E-5485, http://eol.jsc.nasa.gov)

the seawater flows via the strait to the bay at a velocity of 50–100 cm s^{-1}; once there, it is subsequently completely evaporated (at a mean rate of 800–1000 mm $year^{-1}$). Thus, with the mean annual precipitation in this region not exceeding 110 mm, Kara-Bogaz-Gol Bay represents an enormous natural evaporator of seawater. Due to the evaporation of the Caspian water and to the "secular" stocks of the bay proper, its basin is filled with a brine (leach) represented by seawater with a salinity of 270–300‰. It is a concentrated solution of chlorides of potassium, magnesium, and sodium; of sulfate of magnesium; and of small amounts of rare-earth elements [1, 2].

The physical and geographical conditions of Kara-Bogaz-Gol Bay and its natural regime, which are controlled by the level of the Caspian Sea and related water transport to the bay, are subjected to significant interannual variations. During the first decades of the 20th century, the level increment between the sea and the bay was small (about 0.5 m) and they were hydraulically linked. The level of Kara-Bogaz-Gol Bay was positioned at an altitude of −26.5 m with respect to the level of the World Ocean and its annual variations differed from those of the sea level due to the more intensive summertime evaporation and to the wind surges, which reached an amplitude of 2 m. The area of the bay equaled 18 000 km^2, the water volume was 130 km^3, and the dominating water depths ranged from 8 to 10 m. The water supply to the bay comprised 18–25 km^3 $year^{-1}$. Annually, about 330–380 Mt of salts were delivered to the bay; meanwhile, its role in the desalination of the Caspian Sea was insignificant - 0.2–0.3‰ over the past 100 years [2, 3].

The availability of enormous stocks of minerals and the steady water and salt balance between the sea and the bay favored the initiation of industrial mining of the salt resources of Kara-Bogaz-Gol Bay. The overall salt stocks in the bay are estimated at billions of tons. This site is the largest field of sodium sulfate in Eurasia and is the only location in the world where industrial mining of mineral salts was performed along with their crystallization under natural conditions. The bay contains the richest stocks of mirabilite (Glauber's salt)—decahydrate of sodium sulfate $Na_2SO_4 \cdot 10H_2O$. Sodium sulfate is a valuable raw material used in the chemical, pulp-and-paper, textile, and glass industries as well as in agriculture and medicine. About 500 enterprises of the Soviet Union used the production of Kara-Bogaz-Gol [1, 4].

In the first decades of the 20th century, the mining of the resources was based on natural processes. The upper layers of the waters of the bay contain sodium chloride NaCl, magnesium chloride $MgCl_2$, and sodium sulfate Na_2SO_4. Every year, at the end of November, as the water temperature decreased to 5.5–6 °C, the water was saturated with sodium sulfate; mirabilite was released in the form of colorless crystals precipitating over the floor of the bay and delivered onto the coasts by the wintertime storms. By the middle of March, with the bay heating up to 6 °C and greater, its waters started to take back their riches. By July–August, all the mirabilite released was dissolved in the waters of the lagoon. During the period from November to March, the mirabilite precipitated was collected from the shores of the bay. In the summertime, under the conditions of the dry climate, it easily lost water and transformed into a water-free mineral thenardite Na_2SO_4 used in soda production and glass-work; it was collected at the end of the summer.

In the autumn of 1926, the "*Turkmensol*" trust was set-up; it performed sodium sulfate mining on the southern coast of the bay near Cape Umchal. In 1929, the united "*Karabogazsul'fat*" trust was created for industrial salt mining. The production was based on a newly developed basin technique. It consisted of pumping the surface brine to adjacent dry depressions; mirabilite precipitated on their bottom due to evaporation. The sulfate-free brine was removed to another basin and mirabilite was collected using special mechanisms and sent to consumers [1, 2].

2
Changes in the Regime

Starting from the 1930s, the nature of Kara-Bogaz-Gol Bay has experienced significant changes related to the sharp fall in the level of the Caspian Sea.

In 1939, the supply of seawater to the bay decreased down to 6 km^3 $year^{-1}$ and the water balance was distorted. In August 1939, crystallization of sodium chloride in the bay started because of the oversaturation of its brine. The decrease in the water volume of Kara-Bogaz-Gol Bay, accompanied by the long-term sea level fall and related growth in the salt concentration in the brine and changes in its chemical composition negatively affected the conditions of sulfate mining. The industrial mining of salts from the surface brine became unprofitable; it was replaced by a method of mining from buried interstitial brine occurring between the salt layers beneath the bottom of the bay and characterized by a constancy of their salt composition. Beginning from 1954, the "*Karabogazsul'fat*" trust operated exclusively with buried brines. In addition to sodium sulfate, sodium chloride (bischofite), magnesium sulfate (epsomite), and Glauber's salt were produced. In the future, it was planned to perform an integrated processing of the brine in order to obtain a wide range of chemical production (including brome, boron, magnesium oxide, selected rare metals, and others).

With further sea level fall, the area of the bay continued to decrease, the depths reduced, and the length of the strait increased, while it's bottom was eroded. Meanwhile, in the 1960s, when rigid rocks were exposed at the surface, the erosion of the floor of the strait almost stopped. The level increment between the sea and the bay increased and, in the mouth of the bay, a marine waterfall unique in the world with a height of 3.7 m (1970) appeared. The maximum flow velocity in the waterfall was greater than 2.5 m s^{-1} [5].

Because of the combination of the strait with its bluish seawater flowing over yellow sands, together with the waterfall and the adjacent part of the bay, a particular fauna of birds and animals developed creating a unique natural environment on the deserted eastern seaside of the Caspian Sea. The delta of the strait and the waterfall were the most populated sites on the dead coasts of Kara-Bogaz-Gol Bay. The fish stocks (pike perch, mullet, and sturgeon) passing through the strait accumulated near its mouth. Dead fish, mollusks, algae, and abundant live fish in the strait served as a food base for various bird species: sea gulls, eagles, sandpipers, Dalmatian and Roseate pelicans, geese, swans, and flamingos. When the crustacean *Artemia salina (L.)*, which formed the feeding base for flamingos, disappeared from the brine of the bay the flamingos stopped nesting in Kara-Bogaz-Gol Bay, and instead merely stopped during migrations. The flamingo, similar to many other birds migrating via Kara-Bogaz-Gol Bay, has been placed on the list of the Red Book of endangered species and requires protection. It is interesting to note that recently the crustacean *Artemia* has reappeared in the bay and one can expect that flamingos will return here. In the autumn of 1957, a family of seals penetrated into the bay via the strait and settled near the waterfall; there is information that the seals had dwelled there up to the spring of 1965. Numerous animals—saigas, Persian gazelles, hares, foxes, and even wolves—gathered on the shores near the strait and the waterfall.

3
Human Intervention

By the end of the 1970s, the supply of Caspian water to the bay decreased down to 5–7 km^3 $year^{-1}$, and the level fell down to –32 m, its area and water volume reduced to 10 000 km^2 and 20–22 km^3, respectively, while the salinity of the brine increased up to 270–300‰. In March 1980, in order to restrict the losses of the Caspian waters and to decelerate the fall of the level of the Caspian Sea, which in 1977 was the lowest it had ever been over the last 400 years (–29 m), the strait was closed by a nonoverflow sand dam; in this way, the delivery of the seawater to the bay was stopped. However, by this time, the Caspian Sea level had already risen by 0.5 m and continued to rapidly rise up to 1995. Meanwhile, Zherebtsov was the first to suggest the closure of the bay by a dam with a lock to maintain the level of the Caspian Sea. Over 125 years of instrumental measurements—from 1847 to 1971 —the strait proper has suffered significant morphological changes: its length has increased almost threefold (from 2.9 to 8.7 km), the width of its top part has decreased from 170 to 120 m, while the mouth became wider—from 400 to 800 m [2, 5, 6].

After the separation of the bay from the Caspian Sea, it rapidly dried up. By December 1, 1982, the absolute mark of the lagoon level comprised –33.5 m. At the end of 1983, the area of the bay was as small as 1000 km^2 at a volume of 0.2 km^3 and a depth of 0.1–0.3 m. The brine's salinity reached a value of 330–380‰. By the middle of 1984, the bay was almost completely dried up and transformed into a "dry" salt lake (Fig. 4b). The surface layer consisted of the salt precipitated from the brine and represented a solid crystalline rock composed mostly of five minerals: halite, astrakhanite, epsomite, glauberite, and gypsum. The dry bottom of the bay demonstrated special topographic features such as various hummocks, columnar stocks, and bedding caves. The separation of the bay from the Caspian sea resulted in chemical changes in the salt composition and deterioration of the chemical raw materials, which caused additional difficulties in the technology of their processing [2, 6].

4
Revival of the Bay

In order to protect and develop the unique salt field on the Caspian Sea under the conditions of the rapid sea level rise, it was decided to rehabilitate the water supply to Kara-Bogaz-Gol. In September 1984, the feed of the Caspian water to the bay was resumed at a rate of 1.5–1.6 km^3 $year^{-1}$. This restricted seawater supply did not result in an active restoration of the hydrological and hydrochemical conditions of the bay. In April 1992, the area of the bay reached 4600 km^2, the absolute level mark was –33.71 m, and the depth varied from 0.2 to 1.4 m. In the deepest part of the bay, the absolute mark of the

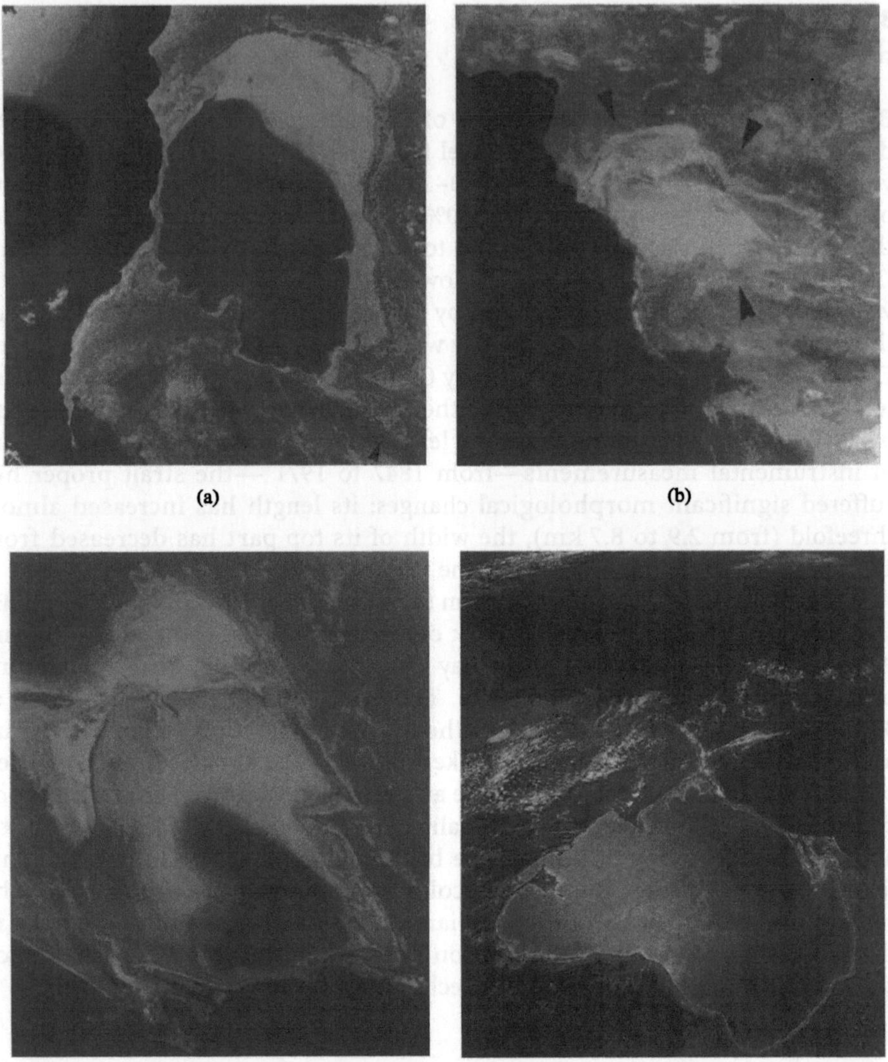

(a) (b) (c) (d)

Fig. 4 Photos of Kara-Bogaz-Gol Bay taken by the Soviet satellites **a** *Soyuz-9*, 1970 and **b** *Meteor-30*, August 20, 1984, and from the American *Space Shuttle* **c** March 25, 1992 and **d** April 14, 1995 (image courtesy of the Earth Sciences and Image Analysis Laboratory, NASA Johnson Space Center, Mission-Roll-Frame STS045-81-57, STS059-L17-73, http://eol.jsc.nasa.gov)

bottom was −35.1 m. In June 1992, the dam was destroyed and the natural seawater runoff to the bay resumed (Fig. 4c). The topography of the lagoon gradually acquired its former outlines. In 1993–1995, when the Caspian Sea level was high, the annual water discharge reached 37–52 $km^3\,year^{-1}$, which significantly exceeded its former rates. After the middle of 1996, when the pool of the bay had been completely filled (Fig. 4d), water discharge via the strait began to be controlled only by the evaporation from its surface and, by 1999, the discharge had reduced down to 17 $km^3\,year^{-1}$. The level increment between the sea and the bay decreased from 6.9 m in June 1992 to 0.2–0.6 m in 1996 (depending on the season). A direct correspondence between the water levels in the Caspian Sea and the lagoon was established—the water level in the bay rose when the water level in the Caspian Sea rose and vice versa [7, 8].

The process of the filling of the bay and its acquisition of the climatic regime is well traced in the satellite altimetry data (*TOPEX/Poseidon* and *Jason-1*) with high spatial (5–6 km) and temporal (5–10 days) resolutions (Fig. 5) [9]. The height of the satellites above the sea's surface are measured within an accuracy of 1.7 cm. This method of measuring sea surface heights with respect to the center of the earth's mass allows one to remove vertical movements of the earth's crust from the interannual level variations. The high level of efficiency of this method for the purposes of sea-level monitoring

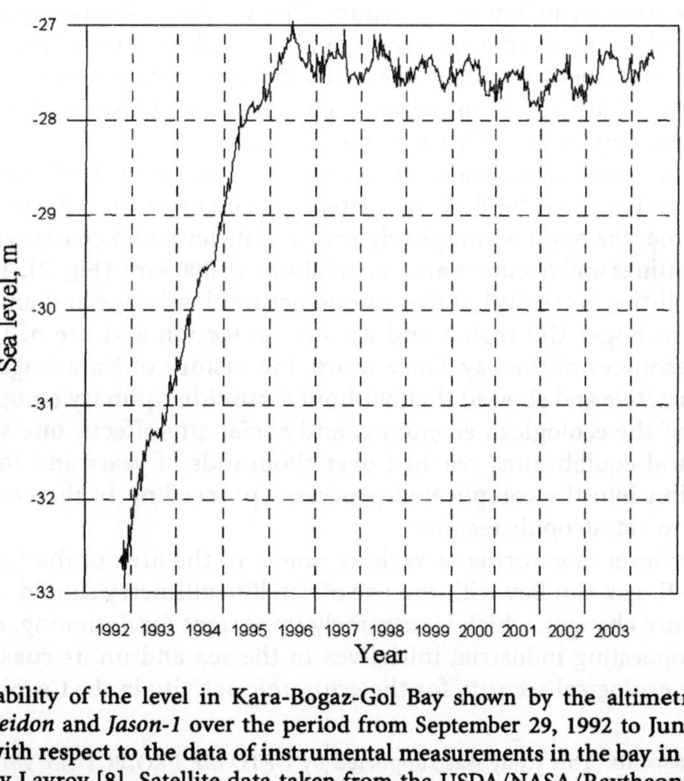

Fig. 5 Variability of the level in Kara-Bogaz-Gol Bay shown by the altimetric data of *TOPEX/Poseidon* and *Jason-1* over the period from September 29, 1992 to June 12, 2004, calibrated with respect to the data of instrumental measurements in the bay in 1992–1999 presented by Lavrov [8]. Satellite data taken from the USDA/NASA/Raytheon/UMD site (http://www.pecad.fas.usda.gov)

(seasonal and long-term variabilities) and for the studies of the water dynamics of enclosed seas and lakes, in particular, of the Caspian Sea is proved by the results of recent investigations [10]. The data from the *TOPEX/Poseidon* and, later, *Jason-1* satellites represent the longest continuous measurement time series (from September 1992), which successfully coincided with the beginning of the filling of the bay.

The area of the bay is crossed by two tracks of the above-mentioned satellites; the level altitude is measured precisely at the point of their intersection. Up to the middle of 1996, the bay was rapidly filled with water from the Caspian Sea (Fig. 5) causing a level rise rate of about 168 cm year^{-1} [9]. Then, the level rise stopped and its variations started to reflect seasonal changes (Fig. 5) well correlated with the seasonal level changes in the Caspian Sea. Because of this, the rate of the level fall in both of the basins comprised approximately 6 cm year^{-1}. At present, the level of the bay oscillates near an absolute mark of −27.5 m.

It is extremely interesting to note that, starting from 1996, a wedging out of mineralized water and its supply from the bay to the sea was observed and the thickness of the near-bottom current varied depending on the season from 1 m in the summer to 3 m in the winter. The waters of Kara-Bogaz-Gol were encountered in minor marine depressions located in front of the strait. According to the data of Turkmen scientists, this phenomenon may result in the mass deaths of fish stocks in the near-shore zone of the sea and in the strait proper, which was repeatedly noted by natives.

Significant volumes of seawater discharge to the bay in 1992–1995 favored the deceleration of the level of the Caspian Sea by more than 30 cm. Up to the present time, the bay has completely restored its level and coasts within their former outlines and occupies an area of about 18 000 km^2 (Fig. 2). This shows the possibilities for revival of the unique natural landscape and environment of the Kara-Bogaz-Gol region and for the protection and use of the richest natural resources of the bay. Once more, the history of Kara-Bogaz-Gol has been instructive and showed that, without a multidisciplinary comprehensive research of the ecological, economic, and social aftereffects, one should not alter natural equilibriums reached over thousands of years and that human intervention into the complicated processes proceeding in the environment may lead to catastrophic results.

Further level monitoring at various points of the area of the Caspian Sea and Kara-Bogaz-Gol Bay with the use of satellite altimetry should allow us to follow future changes, which is extremely important for designing, constructing, and operating industrial initiatives in the sea and on its coasts and for providing ecological security for the economic activity in the Caspian region.

Acknowledgements This study was supported by NATO SfP PROJECT No. 981063 “Multidisciplinary Analysis of the Caspian Sea Ecosystem” (MACE) and the Russian Foundation for Basic Research Grant No. 04-05-64239.

References

1. Dzens-Litovskii AI (1967) Kara-Bogaz-Gol Nedra, Leningrad (in Russian)
2. Zonn IS (2004) Kosarev AN (ed) The Caspian Encyclopedia. Mezhdunarodnye otnosheniya, Moscow (in Russian)
3. Kosarev AN (1975) Hydrology of the Caspian and Aral seas. Mosk Gos Univ, Moscow (in Russian)
4. Fedin VP (1983) In: Earth Sciences. Kara-Bogaz-Gol 10:6 (in Russian)
5. Varushchenko AN, Lukyanova SA, Solovieva DG, Kosarev AN, Kurayev AV (2000) In: Lulla KP, Dessinov LV (eds) Dynamics Earth Environments. Remote Sensing Observations from Shuttle-Mir Missions. Wiley, New York, p 201
6. Terziev FS, Goptarev NP, Bortnik VN (1986) Vodnye Resursy 2:64 (in Russian)
7. Bortnik VN, Luchkov VP (1988) Meteorologiya i gidrologiya 9:113 (in Russian)
8. Lavrov DA (2000) In: Ecological Problems of the Caspian Sea. International Scientific Seminar, Moscow, Russia, p 17 (in Russian)
9. Lebedev SA, Kostianoy AG (2004) Vestnik Kaspiya 3:82 (in Russian)
10. Kostianoy AG, Zavialov PO, Lebedev SA (2004) In: Nihoul JCJ, Zavialov PO, Micklin PhP (eds) Dying and Dead Seas. Climatic versus Anthropic Causes. Kluwer, Dordrecht p 1
11. Kosarev AN (2005) Physical and geographical features (in this volume). Springer, Berlin Heidelberg New York

Hdb Env Chem Vol. 5, Part P (2005): 223–242
DOI 10.1007/698_5_012

Published online: 20 October 2005

Environmental Issues of the Caspian

Igor S. Zonn

Engineering Research Production Center on Water Management,
Land Reclamation and Ecology, 43/1 Baumanskaya ul., 105005 Moscow, Russia
igorzonn@mtu-net.ru

Abstract Environmental problems in the Caspian and its coastal zone shaped, and reflect the consequences of, the historical and economic development of the Circum-Caspian countries. The Caspian Sea is a meeting place of the different economic and ethnopolitical interests of littoral states. National and transnational oil and gas corporations are involved in the extension and intensification of utilization of the commercially attractive Caspian natural resources, which are fraught with the increased stress loads on the natural environment and biota. The ecological situation on the Caspian is being shaped in conditions of long-term natural change (century-wise fluctuations of the sea level, climatic changes) and also acute recent socioeconomic issues (a transitional period, economic crisis, conflicts, etc.). Among key environmental issues are pollution of the sea and coastal ecosystem, threats of its degradation, qualitative and quantitative depletion of natural resources (including renewable bioresources) involved in economic activities, ecological risks in on-land natural and anthropogenic ecosystems, deterioration of living conditions and health of the population, depletion of landscape and biological diversity, desertification, and intrusion of alien species that distort the normal functioning of ecosystems in the Caspian Sea.

Keywords Caspian Sea · Coastal zone · Desertification · Intruders · Oil and oil products · Pollution

1 Introduction

The Caspian Sea can boast of unique world reserves of endemic sturgeons giving 85% of world black caviar supplies, many other valuable commercial fish species, and other renewable bioresources. At the same time the

Caspian Sea is a place of prospecting and utilization of non-renewable resources, such as hydrocarbons. New projects of hydrocarbon transportation are developed and realized here. The situation with simultaneous utilization of these key natural resources became most acute after disintegration of the Soviet Union. Considerable poorly forecasted fluctuations of the Caspian Sea level essentially affect the socioeconomic activities and make more acute the environmental problems associated with the sea and coastal zone.

It can be seen that many problems of the Caspian may be settled after study of two closely interrelated natural-anthropogenic complexes, the coastal territory of the Caspian watershed and the sea proper, as a single geodynamic system.

The ecology of the Caspian Sea, which depends on the input and diversity of pollutants, is determined by economic development of littoral states and the condition of the Caspian watershed basin. The landlocked nature of the sea and volume of the river flow are the main sources of pollution here. The landlocked nature of the Caspian Sea increases the role of transborder processes, both natural and the varied consequences of man's activities. The unique nature of the biosphere complex and its vulnerability turns the problem of the maximum admissible level of a man-made load on the natural environment and biosphere of the sea into a key issue of the environmental strategy of the regional countries. In the Caspian region natural resources have been managed so far in conditions of unsettled political and legal norms of activities in the coastal zone, and in the open sea of five independent Circum-Caspian countries.

2 Pollution Sources in the Caspian Region

Environmental problems of the Caspian have been shaped in the course of economic development of the sea, coastal territories, and watershed basins of the rivers flowing into the sea. Each of these three components of the Caspian basin has its own environmental problems that, in many instances, have similar consequences, but they "amass" in its terminal part – a drainless water body. The key environmental issue of the Caspian region is pollution (Table 1).

At the same time, an anthropogenic load on the Caspian ecosystem occurs against a background of varied natural (endogenous and exogenous) processes. First of these are the sea level fluctuations, morpholithodynamic processes in the coastal zone, periodical occurrence of seismic events, storm surge phenomena, mud volcanism, and others. At the end of the previous century the general environmental situation in the Caspian Sea basin, attended with sharp deterioration of the sanitary-toxicological and fishery situations, was appraised as precrisis. In 1992 the Volga River basin and coastal terri-

Table 1 Pollution sources in the Caspian region

Anthropogenic (technogenic) sources in the watershed basin of the Caspian Sea	Objects of activities	Results of activities	Environmental consequences
Watersheds of rivers flowing into the Caspian	1. Industrial enterprises 2. Agricultural lands and processing plants 3. Cities and settlements 4. Shipping	Disposal of wastewaters, domestic-utility discharges and wastes, disposal from ships, washoff of fertilizers, pesticides, etc.	Pollution, degradation of river ecosystems
Sea coastal territory	1. Petrochemical plants and other industrial objects 2. Thermal and nuclear power plants 3. Exploration and extraction of oil from wells 4. Port terminals 5. Agricultural lands and processing plants 6. Cities and settlements 7. Oil pipelines and tank areas	Disposal of wastewaters, domestic-utility discharges and wastes, emergency disposals, discharge of thermal waters, washoff of dumps, stacked wastes	Pollution, salinization, degradation of coastal ecosystems, loss of biodiversity
Sea	1. Offshore oil prospecting and extraction 2. Construction of sea platforms, jetties 3. Construction and operation of underwater oil pipelines 4. Shipping and transportation of oil by barges, tankers, etc. 5. Bottom dredging works 6. Naval training, maneuvers	Disposal of technological solutions, washoff of heavy metals from bottom sediments, oil spills during accidents, disposal of polluted waters from ships, discharges from wells	Pollution, degradation of marine ecosystems

tories of the Caspian were officially recognized as "zones of environmental catastrophe" [1].

In the 1990s the Circum-Caspian region was referred to as a region with high environmental stress, i.e. characterized by availability of nearly equal territories with acute and fairly acute environmental situations [2]. Other sources refer to it as a region with very high stress [3], due to the joint action of negative natural and anthropogenic factors and a growing threat of degra-

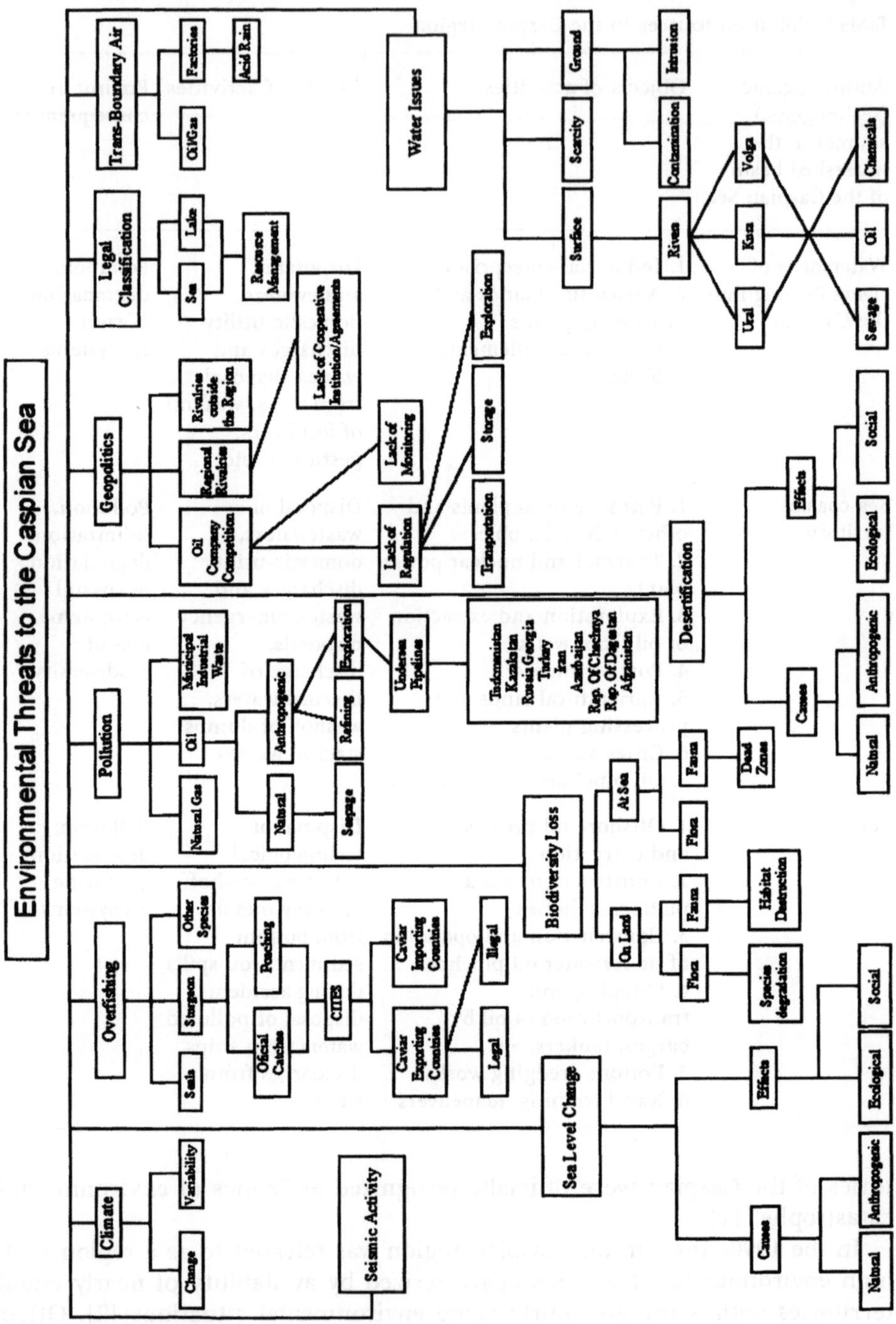

Fig. 1 Environmental Threats to the Caspian Sea

dation of natural complexes in the coastal zone and ecosystems of the whole Caspian Sea (Fig. 1). Some authors consider it the initial stage of environmental catastrophe [4].

In the recent years the environmental gravity center has shifted from industrial pollution (due to a drop in industrial production) to pollution caused by washoff of hazardous substances due to the sea level rise.

This environmental stress may grow with the increase in the economic potential of the Caspian countries due to hydrocarbon extraction, construction of new sea ports and rehabilitation of those existing, revival of the merchant tanker fleet, enhancement of the navy component, and construction of oil and gas pipelines. A risk of negative effects of hydrocarbon deposit development in the bottom and coastal regions of the Caspian is especially vivid in the shallow-water Northern Caspian, which is exclusively important for development of the unique commercial biological resources of the whole Caspian Sea and at the same time as a nature-reserve zone.

3 Pollution of the Caspian Sea Area

In general, the Caspian is one of the most polluted seas [5]. The main sources of pollution of the Caspian natural environment are transborder atmospheric and water transfer of pollutants from other regions, washoff with river flow, discharge of untreated industrial and agricultural wastewaters, municipal-domestic wastewaters from cities and settlements in the coastal zone due to an insufficient number of treatment facilities, operation of oil and gas wells on land and offshore, oil transportation via sea, navigation over both river and sea, secondary pollution during bottom dredging works, and transborder atmospheric and water transfer of pollutants from other regions.

While making general assessments of the Caspian Sea pollution, one should keep in mind the following circumstances. On the one hand, there is an uneven distribution of pollution sources over the sea perimeter, which leads to uneven pollution of its separate parts. On the other hand, there are specific features of the hydrological condition of the sea (mostly cyclonic currents along the shore and intensive vertical water exchange) due to which the pollution of one of the sea parts invariably leads to pollution of other water areas. It should be remembered also that pollutants accumulate in the surface layer, localize in the transitional zones water–atmosphere and water–bottom sediments, and tend to marginal areas of the sea. Getting into bottom sediments they may later become a source of secondary pollution of water. Extensive shallow water areas, being the most biologically significant parts of the sea, are very prone to secondary pollution.

The Caspian has always been distinguished by a high rate of physical, chemical and biochemical processes of self-purification that facilitated year-round quick decomposition of many pollutants. However, in the recent decades external hydrometeorological changes have affected all chains of the sea ecosystem. Thermohaline fields, the whole vertical hydrological structure, have been transformed [6].

Waters of the Caspian Sea contain all basic pollutants: oil and petroleum products, phenols, synthetic surface-active substances (SSASs), heavy metals, and pesticides (especially DDT). The amount of pollutants getting into the sea varies from year to year and from season to season.

River flow, the main supplier of pollutants, presents an integral basin index of a number of natural and technogenic factors that include: a volume and duration of high-water periods; flow regulation; the effect of industrial, domestic and agricultural wastewaters containing more than 1000 chemical compounds; chemical, thermal and acid pollution; forest; irrational fishing; intrusion into water protection zones and violation of the regimes of economic activity in them; and increasing uncontrolled surface runoff of pollutants. Affected by these impacts, practically all rivers in the Volga–Caspian basin have been transformed. The greatest amounts of pollutants are brought to the sea with the flow of rivers of the Caspian basin, such as the Volga, Ural, Terek, and Kura. These bring approximately 85% of oil and phenols, approximately 80% of SSAS, the greatest portion of very hazardous heavy metals (lead, mercury, zinc, copper, cadmium, chromium, and others) and of chloride organic compounds (pesticides and DDT).

Every year the Caspian, more precisely its northern part, receives 40–45 km^3 of wastewaters: 23–25 km^3 from the Volga basin, the flow of which makes more than 80% of the total river inflow into the sea, and 17–20 km^3 from the rivers of Azerbaijan and Iran. During a year $300–350 \times 10^3$ tons of organic substances, $13–18 \times 10^3$ tons of petroleum products, $100–110 \times 10^3$ tons of ammonium nitrate, $90–92 \times 10^3$ tons of phenols, 1000–1200 tons of zinc, 350–375 tons of copper, and 750–850 tons of chromium are discharged into the water bodies of the Volga basin [7]. The Republic of Daghestan puts into the Caspian basin over 5×10^6 tons of geothermal and 1.5×10^6 tons of formation oil waters in which the phenol content more than 1000 times exceeds the admissible concentration [8]. As concerns oil pollution, special emphasis should be paid to the Terek River. Due to military events in the Chechen Republic the treatment facilities in the areas of oil extraction and processing were destroyed. Settling basins accumulate hundreds of tons of petroleum products that are periodically discharged into the Sunzha River, from which they get into Terek.

Azerbaijan discharges over 0.5 km^3 of heavily polluted wastewaters and over 0.3 km^3 of normally treated wastewaters into the sea [9]. As a result the sea receives over 3×10^3 tons of petroleum products, 28×10^3 tons of suspended matter, 74×10^3 tons of sulfates, 315×10^3 tons of chlorides, 520 tons of SSAS, 25 tons of phenols, and other hazardous highly toxic substances [10].

A serious source of the Caspian pollution is the Kura River. Draining the mining and industrial areas of Georgia and Armenia, this river mostly carries heavy metals, such as copper and molybdenum, and residual petroleum products in small quantities.

The most important function of a river flow, affecting conditions of a marine ecosystem, is removal to the sea of biogenous substances. The Caspian receives with river flow, as a result of agricultural runoff and drainage of irrigated lands, an average of 41×10^3 tons of phosphorus and 399×10^3 tons of nitrogen [11]. These compounds are one of the key factors in production–destruction processes and are a source of food for phytoplankton, which stirs development of eutrophication of marine waters. Over 80% of the area of the Northern Caspian is affected by anthropogenic eutrophication, which leads in summer to intensive oxygen deficiency, i.e. hypoxia in the near-bottom water layer. In the southwestern Caspian the zones of hypoxia reach $6–10 \times 10^3$ km^2 [12].

In the eastern part of the sea eutrophication processes are witnessed in the Kazakh, Turkmenbashi, and Turkmen bays, which are connected with the inflow of wastewaters from the ports of Aktau, Turkmenbashi, and others.

Recently a growing pollution of the Caspian with heavy metals has been observed, the greatest portion of which also comes with river flow. The content of heavy metals in the Northern Caspian is 2–7 times higher than in the Volga River, which may be attributed to the increased levels of dissolved forms of metals in comparison to the suspended form, as a result of mixing of river and sea waters [7]. The highest levels of nickel and cadmium are registered in the oil development zones. In addition, large amounts of heavy metals are found in places of disposal of industrial wastewaters, as they are widely applied in technological processes. Estimations of an average level of heavy metals in the Caspian (1996–2000) reveal their high concentrations, especially of zinc (3–6 times the admissible concentrations), titanium (9 times), nickel, lead, and copper (13 times) [7]. The tendency of their concentration growth from the surface to the bottom may be traced.

The most widespread and most hazardous pollution for the Caspian Sea is from petrochemicals. It revealed itself most vividly after the 1870s. Till 1930–1940 the leading role was played by the oil input originating in oil extraction and processing in Azerbaijan, and oil transportation over the Caspian and Volga. From the 1930s the volume of industrial wastewaters disposed into the sea has increased drastically and, beginning in the 1960s, the pollution of the basin with the products of organic synthesis started.

At present an average concentration of oil hydrocarbons in the Caspian exceeds 1.5- to twofold the norm for fishery water bodies. Fortunately, concentrations of other hazardous pollutants do not exceed the maximum admissible levels for fishery in the Caspian, if the incidents connected with local pollution and also with volley emissions and different technogenic accidents (the number of which is steadily growing) are not taken into account. As concerns volley emissions, it should be stressed that every year approximately 20–30 such cases are recorded.

The effects of oil and gas processing productions on land and marine ecosystems and their bioresources have a complex nature (Table 2). They are already detected at the stage of seismic surveys when a sharp growth of the sea water pressure creates the conditions for lethal damage of hydrobionts in water. Oil platforms in the shelf became a source of varied pollution with drilling mud, formation waters, drilling slag with a high content of petroleum products, heavy metals, radionuclides, and others. The effect of disposal and spills under a normal regime of drilling works may be felt within a radius of 3–12 km from a drilling place. Prospecting of oil and gas fields is invariably accompanied with accidents.

Below is a list of emergency events in the Caspian Sea related to oil accidents and spills; it is far from complete:

1983 Near Kazakh shores: the self-propelled mobile drilling rig "60 let Azerbaijanu" sank. Today its tanks still contain 187 tons of diesel fuel, 29 tons of machine oil and possibly 300 tons of chemical reagents.
1985–1986 Tengiz oilfield in the northeast of the sea: there was a catastrophic accident on well 37. Liquidation of consequences lasted 398 days, 3.5×10^6 tons of oil burnt out, and 900 tons of soot fell out
March–April 2000 Northern Caspian: oil spills occurred at drilling site "Sunkar" as well as during tests of the Eastern Kashagan well.
April 2001 Northern Caspian: at drilling site "Sunkar", oil spills and emission of hydrogen sulfide occurred during tests of Western Kashagan well.
December 2001 Port Aktau, Kazakhstan: during loading of the tanker "Islam Saparli" there was an oil spill in a port area.
July 2002 An explosion on the oil tanker "General Shikhlinsky".
October 2002 Catastrophe with the Azerbaijan cargo-passenger ferry "Merkuriy" that carried 18 tanks each with 60 tons Kazakh oil .

The hazardous effect of petroleum products on water organisms is well studied. However, cumulative changes (due to admixture inputs) in natural production–destruction processes are possible. In addition, the construction of underwater communication lines affects aqueous ecosystems, which is revealed in the repelling effects of noise and changing of migration routes of fish. One more group of impacts on the natural environment related to fisheries includes alienation of water areas and restriction of fishing (within a radius of 1–1.5 km from a platform), installation of tubes on the seabed without deepening into ground, contouring of mouths of suspended wells with special accessories, and keeping of out-of-use drilling platforms and other industrial objects in a shelf area.

The losses of oil and products of its processing during extraction, transportation, and utilization may reach 2% of the total amount. Low technological culture and wear-out of the equipment and pipelines lead to emergency spills causing irreparable damage to nature. As a result, great areas of land and water bodies fall out of use, and considerable damage is incurred

Table 2 Sources of Caspian Sea oil pollution

Oil production and transportation	Pollution sources
On land	1. Leaks from wells with long-time conservation 2. Leaks from earth pits for storage of drilling muds and oil 3. Planned spills from drilled wells 4. Drilling and emergency flowing of prospecting wells 5. Leaks from oil storage tanks and oil bases 6. Wastes of petroleum product production 7. Drifts of rivers in the basins of which oilfields have been developed 8. Breach of pipe sealing and greater permeability of near-well space 9. Oil spills into the sea from overfilled accumulation terminals
In the sea	1. Planned disposal of waste drilling mud and untreated wastewaters 2. Accidents on drilling platforms 3. Leaks from offshore drilled wells as a result of accidents or inadequate drilling and operation technologies 4. Breaking of trestles and foundation of drill wells 5. Construction of artificial islands as a base for drilling 6. Dumping
Technological processes in working of oilfields	1. Water flooding of producing formations (including chemical) 2. Gas injection into formation 3. Mecello-polymeric process 4. Thermal and thermal-chemical impacts on productive formation
Transportation	1. Damage of main and on-field pipelines 2. Leaks during loading of oil and petroleum products into tankers 3. Leaks during bunkering of ships 4. Disposals from transportation means 5. Accidents during collisions and running aground of transportation means

to the animal and plant world. One gram of oil products in water makes 20×10^3 L of water unsuitable for use. An increase of a film thickness on the water surface to 0.1 mm interferes with the processes of gas exchange and threatens the death of hydrobionts. This is already observed on reaching 1 g/m^2 of oil on the water surface. Oil produces a hazardous effect on marine organisms – from bacteria and phytoplankton to fish. A concentration of oil products of 0.01 mg/L becomes hazardous for fish, 100 mg/L for macroalgae, and 0.1 mg/L for phytoplankton. In these cases larvae and fries are the most vulnerable. It should be remembered that in combination with other pollutants the toxic effect of oil becomes stronger.

On the basis of on-land and offshore oil extraction major petrochemical productions were formed with their own infrastructure going far beyond the confines of the Caspian region. Also formed were the port–industrial complexes of Baku-Sumgait, Astrakhan, Tangiz-Aktau, and Turkmenbashi. Pollutants disposed by them into the sea are responsible, to a great extent, for heavy pollution of the sea.

We mean here, first of all, offshore oil production and underwater oil pipelines near to the Apsheron Peninsula and Mangyshlak Peninsula. From the late 19th century Azerbaijan started energetic development of coastal and offshore oilfields, and this republic was the first to face pollution issues in its water area. When the drilling rigs of the first large offshore oilfield "Neftyanye kamni" (now "Neft Dashlary") appeared in the sea, this was an undoubted achievement of the Soviet science and technology. Only a few scientists spoke about negative environmental consequences. Critical comments about development of offshore oil production in the Caspian were not published and the opinions of biologists were simply hushed up. Artificial springs were formed in 37 wells (uncontrolled oil spills to the sea surface) just during the period from 1941 to 1958 during drilling works in this oilfield. These springs were active from several days to 2 years, while the amount of spilled oil varied from 100 to 500 tons per day. Unfortunately today, 50 years later, we know that these fears have come true. According to Brandon [13], in the area of "Neftyanye kamni" the oil film on the sea surface covers an area more than 800 km^2. Here considerable areas of a migrating oil film should be added. The film prevents dissolution of oxygen in water and affects the whole marine flora and fauna. The only solace here is that Azeri oil occurs over salt domes and does not contain different modifications of sulfuric compounds.

Baku Bay is one of the most polluted areas in the Caspian and it is biologically dead. A thickness of bottom sediments permeated with oil wastes is 10–12 m. According to estimates, they accumulate approximately 200×10^6 ton of toxic substances, the concentrations of which exceed the maximum admissible by tenfold [10]. According to the Azerbaijan State Committee on Hydrometeorology, the oil content in the waters of the Baku Bay is as high as 1.88 mg/L (38 times the maximum admissible concentration) and in the ground is 31.96 mg/kg of dry weight [9]. The coastal area near Sumgait is also heavily polluted by its more than 115 industrial enterprises, mostly connected with petrochemistry, ferrous and non-ferrous metallurgy. Even today, when industrial production is on a decline, every year 120 to 130×10^6 m^3 of wastewaters (in the past up to 400×10^6 m^3) are discharged into the Caspian [9].

The Turkmenbashi and Turkmen Bays are also highly polluted areas of the Caspian Sea. The new oil and gas production centers – Tengiz (Kazakhstan) and Cheleken (Turkmenistan) – follow in the stead of Apsheron. The environmental situation in the Atyrau region, where the Tengiz oil production center is located on the shore and the major oilfield Kashagan is in the sea coastal area, repeat at large the situation established on the western coast of the Caspian. The situation is aggravated by the fact that oil here, by its chemical composition, refers to a nathphano-paraffin type – it is sour oil with a high level of mercaptan. Such oil requires special removal of mercaptans before its transit over pipelines, otherwise leakage of such oil into the sea may cause serious environmental hazards. Difficult conditions of oil extraction (high temperatures and pressure) demand additional expenditures to ensure trouble-free operation of wells.

In this context, even the standards of the US Environmental Protection Agency (EPA), applied by Western companies, are insufficient for Caspian oil projects because they are intended for open water bodies. The level attained by oilmen on land is 4–5 kg of disposed pollutants per ton of produced oil. By 2015 in the Kazakh area of the Caspian it is planned to construct 56 oil platforms and artificial islands and to drill more than 1100 wells from them.

One more likely source of the Caspian pollution is bottom-dredging works and dumping of ground, as conducted in the mouth area of the Volga and in the Volga–Caspian channel. As a result of these works some portion of metals buried in bottom sediments gets into the water and become involved in food chains. A concentration of many heavy metals in the zone of these works increases 1.5–4 times in comparison with the background values, while in benthos organisms it increases 100 times [14].

Notwithstanding a critical pollution of some water areas in the Caspian, the general level of Caspian water pollution may be assessed, so far, as average or not high.

Waters of the Northern Caspian are assessed as moderately polluted. Here catastrophes are most dangerous because one oil spill may lead to perishing of spawning grounds of unique sturgeons and, in general, to unforeseeable damage of the ecosystem.

Local heavy pollution and also 1.5- to twofold exceeding of the maximum admissible norms of hydrocarbons in general for the Caspian have started producing negative effects on the living organisms of the water body.

Here we mean changes of size–weight and age indices as well as morphological, physiological, and biochemical anomalies of Caspian hydrobionts. Affected by cumulative toxicosis due to many years of chronic pollution, the degradation of muscle tissue (myopathy) and total desorption of caviar have developed in the sturgeons. By the early 1990s such deformations were found in 60% of the sturgeons. Passing along food chains, the toxic substances may get into humans and cause distortions in the genetic apparatus, thus, causing hereditary and carcinogenic diseases. As a result of this, and despite imposition of fishing quotas by the Circum-Caspian countries, the problems of rational management of the sturgeons in the Caspian basin has turned into a problem of conservation and reproduction of sturgeon resources, including composition of sturgeon species.

In recent years some elementary intruders have threatened the biodiversity of the Caspian ecosystem. This is a case of biological pollution, i.e., bringing of foreign organisms in the ballast waters of tankers navigating from the Azov–Black Sea basin to the Caspian along the Volga–Don channel. In the mid-1990s near the Turkmen and Azerbaijan coasts were found jellyfish *Aurelia aurita*, combfish *Mnemiopsis leidyi*, and plankton crayfish *Penilia avirostris* (for more details see Karpinsky and Shiganova, this volume). More active shipping and transit between the Caspian and other seas due to widening of oil and gas production may lead indirectly to biological pollution.

Because of insufficient capacities of treatment facilities in many coastal cities, microbiological pollution is occurring. Thus, in recent years Makhachkala discharged about 100×10^3 m^3 of untreated wastewaters, as a result of which the bacteriological pollution of the sea in summer reached several dozen times the maximum admissible concentration of pollutants. Similar situations were observed near some other coastal cities. The cause of bacterial pollution is emergency disposal of domestic fecal wastewaters from cities. All the above invariably affect the sanitary-epidemiological and medical-demographic situations.

4 Pollution of the Caspian Coastal Zone

The environmental issues of the coastal zone of the Caspian become more acute due to fluctuations of the sea level, high natural seismicity (in the recent 100 years several destructive earthquakes with a magnitude of 9–10 points have been registered in this region), and mud volcanism. Here we can add abnormal modern activity of fractures in the Earth's crust, the displacement along which reaches 3–5 cm a year.

Beginning from 1978 (sea level −29 m) to 1996 (sea level −26.7 m) the sea level has risen by more than 2.5 m, accompanied by numerous negative environmental processes. These include flooding and submergence of the coastal zone, scouring of beaches and deltas of rivers, intensive pollution of sea waters, shore abrasion (at a rate up to 10 m a year, in Daghestan 25–200 m a year), overall groundwater rise, and deterioration of the living habitat for many biological components of the sea ecosystem. Dozens of kilometers of oil pipelines, power transmission lines, and settlements contributed additionally to pollution of the Caspian Sea.

As a result, in the coastal zone many industrial enterprises and buildings were flooded and the sewerage and water main systems were damaged. This situation is periodically aggravated by autumn–spring surge events with spreading of a surge wave 2–3 m high to 20 km inside a coast. The Caspian water level rise also adds to pollution when toxic substances contained in newly flooded areas get into the sea water. A serious headache is also caused by "nobody's" oil wells that number about 1500 in Kazakhstan, especially in the flooded zone (more than 140 wells are flooded). They are located in oilfields Pustynnoye, Tadjigali, and Pribrezshnoye. Quite often oil spills have been registered there. Table 3 presents likely environmental consequences of the rise and drop of the Caspian Sea level.

Although in the recent years the water level in the Caspian has remained relatively stable, further rises of the sea level in oil production areas may lead to emergency situations, flooding of drilling sites located in low-lying areas, breaking of protection embankments and levees around drilling sites, breach of on-field pipelines, and pollution of ground waters that will further on pollute the sea.

Oilfields in the northern part of the Caspian shelf, where intensive oil production is planned, are located in a seismic hazardous zone. In this water area the fold Paleozoic complex of the Epihercynian platform thrusts over the edge

Table 3 Environmental consequences of the water level rise and drop in the Caspian Sea

Environmental consequences	
Water level drop	Water level rise
• Changes of the halo-geochemical situation • Improvement of water quality in the coastal zone • Improved water exchange between separate parts of the sea • Extension of the desalted buffer zone into the sea • Transformation of a hydrographic network in river mouth areas • Hydromorphism and salinization • Halophytization of vegetation • Transformation of habitat of water fowl and spawning grounds in the coastal shallow-water areas • Transformation of biodiversity of on-land ecosystems; • Pollution of sea waters as a result of oilfield flooding • Loss of nature reserves, zoological and biological preserves • Changes in the landscape structure in coastal territories • Intensification of eutrophication • Changes in bioproductivity and biodiversity of river and marine organisms • Improvement of the potential productivity of the Northern Caspian	• Xerophytization of vegetation • Loss of habitat by water fowl • Intensification of dust and salt drift • Desertification of coastal areas • Intensification of wind erosion • Growth of land salinization • Changes in the landscape structure and borders of coastal territories • Intensification of river erosion • Formation of new morphostructures in the coastal-marine zone • Changes in microclimate • Changes of an ice regime (ice drift to the regions of oil extraction facilities) • Changes (worsening) of a hydrochemical regime • Weakening of water exchange between separate parts of the sea • Lowering of potential productivity of the Northern Caspian

of the Precambrian platform along a system of deep faults. Here is also found the deep Agrakhano–Embenian fault dividing the Northern Caspian into two parts – northwestern and southeastern – and playing an important role in the tectonics of this area [4].

The Southern and a greater part of the Middle Caspian are subject to considerable geodynamic hazard also related to seismicity. That is why installation of oil pipelines over the Caspian seabed is fraught with accidents and wide oil spills as a result of underwater earthquakes, which, finally, create the prerequisites for negative ecological and socioeconomic consequences [15]. In addition, mud volcanoes erupt on the Caspian seabed. Most dangerous by their environmental consequences are fields of oil and gas deposits with a high content of hydrogen sulfur (like Tengiz oilfield). During strong earthquakes in these regions millions of tons of hydrocarbons containing hydrogen sulfur may be released to the surface and into the atmosphere under a pressure of about 1000 atm, which may bring about a global catastrophe [16]. The grave environmental consequences of the long-term local impacts of hydrogen sulfur on the natural environment are well visible in the example of the Aksaraisky gas-condensate plant in Astrakhan that is engaged in extraction and processing of high-sulfur gas, condensate, and small quantities of oil [17].

In the Caspian region there are about 200 large cities with more than 220 sources of industrial pollution. The total population living in the coastal zone of the Caspian numbers approximately 100×10^6 people, out of which 50% is the urban population. All coastal cities are responsible for pollution of the sea area, but mainly large cities such as Astrakhan, Baku, Makhachkala, and Turkmenbashi, which dispose considerable quantities of wastewaters into the sea.

A high degree of oil pollution is also observed in the environment of the Apsheron Peninsula, particularly its coastal part, where oil production has long been practiced. The average content of oil products in a 0–5 cm soil layer exceeds 50–100 times the background level [18]. In some regions the ground of the peninsular reminds a cemetery of old oil derricks. There are leaks of oil products from oil pipelines and inappropriately plugged used wells. A similar situation became established in the Sumgait area.

In some old oilfields in Daghestan such phenomenon as spring formation, waterlogging, and oil shows in abandoned wells are witnessed. A sharp increase of groundwater salinity from 20 g/L to 40 g/L is observed. Ground waters polluted with toxic drilling mud along drainage systems are discharged into the Caspian Sea. In some areas pollution with radionuclides is detected, being a result of salt sedimentation from formation waters coming from self-flowing wells. There are some areas where natural radiation level reach 500–700 μR/h, which exceeds the admissible values by 15–20 times [19].

According to Kazakh ecologists, in the northwestern part of Kazakhstan the on-land industrial complexes (Tengiz, Emba, and others) are responsible for up to 48% of releases of hazardous substances, 36% of wastewaters, and 30% of solid wastes, etc. Oil and chemical pollution of soils is observed over all territories of operated oil and gas fields. The main sources of such pollution are breaks in oil pipelines, emergency flowing of exploratory wells, violations of technologies of storage, accumulation, separation, transportation via pipelines, and inadequacy of constructions and equipment used in production and transport. Oil production in the region goes on under very complicated conditions of paraffin deposition on the overground and underground equipment, salt deposition in the hole-bottom zone and in communication lines, and watering of wells and corrosion of the equipment. All the above lead to frequent breaches of pipelines, flowing of crude oil and saline waters onto the soil surface, and pollution of the natural environment. Studies have shown that a depth of oil and chemical pollution of soils in oilfields varies from 22 to 82 cm reaching 5–10 m in the oldest oilfields. This includes a bitumen crust of 25–50 cm thick formed as a result of a oil-high content of resins (silicogel hydrocarbons and paraffins) [20]. In addition, during spills while drilling cuttings, the initial soils get buried and specific technogenic cover forms.

As a result, in the areas around oilfields and oil pipelines there are formed zones of overall chemical technogenic desertification where up to 80% of the vegetation cover is destroyed. A negative impact of such zones on the natural environment is always of a long-term nature and leads to depletion and disappearance of various kinds of fauna.

In the early 1990s the coastal zone in the northeast of the Caspian Sea was classified as very polluted. The content of oil products in soil was 20 times more than the admissible one. Quite recently it was quoted that 6×10^6 ton of oil has been spilled in Western Kazakhstan from the beginning of oil extraction in this region. In the territories of oilfields in the Atyrau and Mangistau regions the soil is permeated with oil to a depth of 10 m, approximately 800 hectares are soaked with residual oil, and pits of old oilfields store approximately 200×10^3 tons of oil [21].

Both hypothetically and in view of a potential conflict situation in the Caspian region one should not neglect "technological terrorism" aimed at breaking or destroying offshore oil platforms or oil pipelines to cause damage to the natural environment.

As oil production is already underway, the intensity of its transportation by tankers over the Caspian grows considerably, which increases the risk of oil pollution, and still more so in view of the fact that the tankers are mostly single-walled. At present tankers move along the routes Aktau (Kazakhstan)–Dyubendy (Azerbaijan) and Aktau–Makhachkala. Specialists say that the Tengiz oil, due to its aggressiveness, cannot be prepared for transportation without causing damage to the boards of any tanker. That is why nobody is insured against likely accidents and oil spills. For the Caspian, being a land-locked ecosystem, even a small accident involving oil spill will be sufficient to cause sea "death".

In recent years thousands of units of the so-called short-sea fleet (boats, boats with hang motors) appeared in the Caspian and they became a serious source of seawater pollution with oil products. At present in the Northern Caspian more than 150 various ships used in prospecting drilling are dislocated. In the near future their number will increase 2–2.5 times. A rather significant source of pollution is wastewaters discharged from ships in some major Caspian ports such as Baku, Makhachkala, Turkmenbashi, and Aktau. The planned extension of these ports and also of former low-active ports, such as Okarem, Aladja, Ufra, and Bautino, will result in a greater number of large-capacity ships, tankers, and barges for transportation of crude oil and oil products. These will become potential sources of pollution and will increase the potential risk of accidents.

The navy forces and coastal bases created by each Caspian state should be added to the above [17]. The Caspian has great prospects to become an important segment of the complex transcontinental transport systems along the lines north–south and east–west, the development of which may earn good money for the Caspian states. We mean here, first of all, a transport corridor: port Olya (Astrakhan Region)–Caspian Sea–port Enzeli (Iran).

Most noticeable environmental changes are observed in the Caspian coastal zone. Natural conditions in the coastal zone are very vulnerable. That is why this zone with its dense population and intensive development of natural resources is affected most by desertification processes. Studies of coastal landscapes of the Caspian conducted in the last three decades (1970–2000) have demonstrated that the consequences of the sea level fluctuations mostly affect the soil and vegetation cover. Desertification is a dry echo of transgression. During a sea level drop vast areas become free of water and we can trace quite clearly two directions of successions of the soil-vegetation cover: meadow and desert. In the first stages of formation of natural complexes on these lands the determining factors are lithology of deposits, relief and coastal processes. Clay deposits are more saline than sands, which is why halophyte vegetation prevails on them. For relatively less saline sands the psammophyte vegetation is more typical. Desertification is largely a product of irrational nature management. The basic manifestations of desertification processes are degradation of the vegetation cover, soil water erosion, ground deflation, flooding and salinization of soils, and man-made desertification. Within a 100-km coastal belt the total area of land affected by varying extents of desertification is approximately 130×10^3 km^2. During the sea level rise vast lagoons, marsh, and coastal solonchaks and their partial inundation are observed. In general, the long-term sea level fluctuations are attended with shifting of natural complexes in the coastal zone and their subsequent transformation.

5 Perspectives of Ecology Improvement

The likely outcome of the oil and gas scenario of the development of Caspian riches will affect, first of all, the fishery and fish industry in general because construction of oil wells started in the Northern Caspian in sturgeon spawning grounds and on the paths of their migration. By Brandon's estimates [13], the Caspian countries will lose about US$6 billion every year just due to reduction of sturgeon catches. In addition, the caviar business, the annual turnover of which is several billion dollars, was damaged by over 90%.

A threat of deterioration of the environmental situation in the Caspian region and depletion of its natural resources is in direct dependence on the condition of economics and awareness of the society about a global nature and importance of these issues. As the National Security Concept of Russia stresses, this threat is especially great because of excessive development of the fuel-power industries, drawbacks of legal foundations of nature conservation activities, restricted application of nature-saving technologies, and a low ecological culture, which increases the risk of technogenic catastrophes. In general, the shaping situation in the Caspian region urges taking concrete actions for saving the Caspian ecosystem. It is impossible to stop economic development of the region, but it is necessary to make sure that all economic activities comply with the criteria of environmental safety.

Aggravation of the environmental situation in the Caspian region demands effective management of the natural environment of the Caspian Sea via regulation of anthropongenic impacts on the marine and coastal ecosystems. Multipurpose management of the Caspian environment is a prerequisite for sustainable development of the whole Caspian region, ensuring balanced solution of socioeconomic issues, addressing vital needs of present and future generations, and simultaneously preserving the favorable natural environment and natural resource potential. Such an approach should be based on key specific features of the Caspian marine environment as an object of management. They are as follows:

- The natural environmental of the Caspian Sea is a balanced system that functions and responds to basic external impacts as a single natural unit
- Effective understanding and utilization of a complex ecological system of the Caspian Sea may be attained only by taking integrated decisions on the basis of regional cooperation.

That is why among the priority actions of a state spearheaded to elimination of environmental threats in the Caspian region there are the following:

- Rational management of natural resources
- Prevention of environmental pollution
- Creation and introduction of environmentally safe technologies, first of all, in extraction and transportation of hydrocarbons (at present the Rus-

sian oil companies operating on the Caspian apply a "zero-disposal" technology prohibiting any disposals during drilling and operation)
- Improvement of a system of warning and liquidation of emergency situations
- Development of cooperation among the Caspian states for nature conservation

Nature conservation activities in the Caspian region should include the following:

- Environmental expertise and assessment of environmental impacts of projects related to utilization of natural resources, including oil extraction and hydrocarbon transportation
- Integrated environmental monitoring
- Response to emergency situations connected with negative changes in the natural environment
- Optimal management of commercial biological resources (firstly sturgeons)
- Preservation of biological and landscape diversity
- Integrated management of the natural environment in a coastal zone (most urgent in view of the Caspian level fluctuations)

Further expansion of prospecting and production of oil with poor control of operation in the coastal zone of the sea, and the poaching and pollution of the Caspian waters may lead to irreversible processes related to reduction of a resource potential of water and land ecosystems, disappearance of habitats of valuable commercial fish species (including sturgeon), and reduction of species diversity.

An unsettled delineation of the Caspian and uncertainty of its legal status are the main obstacles for successfully coping with many issues, including environment protection and preservation of biological resources. At the same time, the absence of coordinated regulatory and legal norms in relation to activities in the coastal zone and offshore involves the risk of attracting companies from non-regional countries to develop the Caspian Sea resources. For these companies the efficiency of energy resource recovery will be a priority issue in comparison to maintenance of an ecological balance in this very vulnerable landlocked water body. No less risky are projects for the construction of gas and oil pipelines on the seabed, and also for the intensification of tanker traffic carrying Kazakh oil for filling of the Baku–Tbilisi–Ceyhan pipeline.

In the absence of generally recognized delimitation lines on the Caspian, its militarization is one more serious threat to the situation here. In addition to a growing conflict potential it should be remembered that build-up by littoral states of the navy forces and construction of coastal bases for them will invariably produce negative effects on the natural environment.

Here the key issue should be national and international environmental safety, i.e., a system of coordinated state and interstate mechanisms, actions, and guarantees based on compliance by one and all states, with the common humanitarian principles and norms of international law that are called to guarantee effective solutions or to prevent emergence of environmental problems of interstate and world community dimensions.

Rapid settlement of the legal status of the Caspian is necessary for a transition to sustainable development capable of ensuring a balanced solution of the socioeconomic and nature-conservation issues in the interests of the Caspian countries and the whole world community. The first step in ecological diplomacy of the Caspian has been made. In November 2003 five Caspian countries met in Tehran where they signed the *Framework Convention on Protection of the Marine Environment of the Caspian Sea*. This convention states that its member-countries undertake "to take jointly all necessary actions to prevent pollution of the Caspian Sea, lowering of its present-day level and further control of its condition". This document outlines the requirements on prevention, diminishment, and control of pollution and also on protection, preservation ,and restoration of the marine environment the compliance with which becomes obligatory for the states that signed this document. The next step should be ratification of the Convention by all five Caspian states, who will have to take practical actions on its realization. One of basic mechanisms for settlement of these problems may be, and this was already proposed more than once, establishment of an international Caspian center for environmental monitoring. It is quite clear that on the Caspian Sea, where we see rather close interdependence of littoral states, a level of environmental cooperation should be very high and trustful.

References

1. State report (1993) On environment condition in the Russian Federation in 1992. Minekologia, Moscow
2. Zalikhanov MCh, Matrosov VM, Shelekhov AM (eds) (2002) Scientific basis of a sustainable development strategy in the Russian Federation. State Duma, Moscow
3. Kochurov BI (2003) Ecodiagnostics and balanced development. Madjenta, Smolensk, p 181
4. Diarov MD (2002) Influence of oil-gas activities on environment of Northern Caspian. Caspian Sea Bulletin 1:114
5. Kosarev AN, Gyul AK (1966) Pollution of the Caspian Sea. In: Assessment of environmental stress in European Russia: factors, zoning, consequences. MGU, Moscow, p 141
6. Kosarev AN, Kostianoy AG (2003) Problems of crisis seas and lakes. Earth and Universe 6:67
7. Kosarev AN, Tutilkin VS, Daniyalova ZH, Arkhipkin VS (2004) Hydrology and ecology of the Black and Caspian Seas. In: Geography, society, environment, vol VI. Atmosphere-hydrosphere dynamics and interaction. Gorizont, Moscow, p 218

8. Aliev ZM, Rybnikova VI, Tetakayeva EA (1997) Environmental and biological aspects of phenol and oil product recovery in the Caspian sea waters. In: Problems of environmental safety of the Caspian region, Makhachkala
9. Mamedov MM, Gadjieva SE, Ismail-zade TA, Mustafa-zade BV (2000) Caspian Sea ecology and basic factors of its stabilization. In: Forecast and control of geodynamic and ecological situations in the Caspian Sea region in connection with the oil and gas complex development. Nauchnyi mir, Moscow, p 38
10. Efendieva IM, Dzhafarova FM (1993) Ecological problems of Caspian Sea. Hydraulic Construction 1:22
11. Leonov AB, Nazarova NA (2001) Intake of biogenious matter into the Caspian with river flow. Water Resources 28, 6:718
12. Kosarev AN, Zalogin BS (1998) Seas pollution. Earth and Universe 6:19
13. Brendon S (1995) Oil on troubled water. Focus Central Asia 22:12
14. Gyul AK (2005) Hydrochemical parameters of water quality in the damping of the Caspian Sea. Caspian Bulletin 1:70
15. Zonn IS (1998) Seismic echo on the Baku-Ceyhan pipeline. Caspian Bulletin 5:50
16. Vostokov EN (1997) Natural environment destabilization in the Caspian region in connection with fuel-power resources development. RF Ministry of Natural Resources, p 76
17. Zonn IS, Zhiltsov SS (2004) Caspian region: geography, politics, cooperation. Moscow, p 637
18. Akhmedov AI, Aliyev MI (1992) Oil extraction is the main pollutant of the Caspian Sea. Izvestia VUZov Oil and Gas 5-6:89
19. Gazaliev IM (1997) Problems of environment protection in oilfield development in the Caspian coastal zone. In: Environmental safety issues of the Caspian region, Makhachkala. p 25
20. Auezova ON, Asanbaev IK (2002) Pertochemical pollutant in the Caspian Sea and possibility of its utilization by microorganisms. Problems of Desert Development 2:3
21. Kutsaiy IA (2003) In the Atyrau area 1.5 thousand ownerless wells are located. Caspian Bulletin 5:41

Hdb Env Chem Vol. 5, Part P (2005): 243–256
DOI 10.1007/698_5_013

Published online: 20 October 2005

Economic and International Legal Dimensions

Igor S. Zonn

Engineering Research Production Center on Water Management,
Land Reclamation and Ecology, 43/1 Baumanskaya ul., 105005 Moscow, Russia
igorzonn@mtu-net.ru

Abstract The Caspian Region has abundant hydrocarbon resources (oil and gas), as well as biological resources (fish), particularly the unique sturgeon, which gives black caviar. Seeking to attain a higher pace of economic development the new independent Caspian states are focusing on improvement of the fuel and power complex. Better utilization of the Caspian Sea resources depends, to a great extent, on settlement of the legal status of the sea. While before the breakup of the Soviet Union the legal status of the sea was regulated by the Treaties of 1921 and 1940 between the Soviet Union and Iran, each of the several new independent states on the Caspian claims its own "share" of the Caspians riches. Negotiations over more than a decade resulted in the adoption of the principle "shared seabed– common water" and signing of agreements between Russia and Azerbaijan, Russia and Kazakhstan, and Azerbaijan and Kazakhstan. However, complete consensus has not yet been reached, preventing the signing of the Convention on the Legal Status of the Caspian Sea.

Keywords Caspian Sea · Demography · Economics · Legal status · Oil and gas resources

1 Introduction

The breakup of the USSR and subsequent formation of new and sovereign independent states have radically changed the political and economic situation in the region. In addition to Russia and Iran, who had determined the situation on the Caspian for a long period, Azerbaijan, Turkmenistan and Kazakhstan are now interested parties, beginning a new stage in the historical development of the Caspian region. This increase in the number of the Caspian legal entities from two to five has given rise to a whole tangle of geopolitical, economic, international legal, ethnic and environmental problems, each of which demands its own approach and settlement mechanism.

After shaking off the guardianship of the Union center, the Caspian states, quite unexpectedly, attracted very close interest from the world's leading states, primarily because of their strategic position and (potentially) considerable hydrocarbon reserves.

One of the most disputed and acute issues was determination of the international legal status of the Caspian; negotiations have, with time, produced an effective instrument for the realization of the interests both of the Caspian states and extra-regional countries. However, it hasn't become a factor in regional integration so far.

2
Economic Potential of the Regional States

The present-day map of the territories surrounding the Caspian was established in 1992 after the formation of three new independent states on Caspian shores. Of the Caspian states, Russia is the largest, followed by Kazakhstan (the world's 9th largest country by territory), and the Islamic Republic of Iran (16th in the world). The Azerbaijan Republic is a Transcaucasian state, while Turkmenistan and Kazakhstan are Central Asian states. All of the states have a presidential republic form of government. The territory of the regional states is 21 996 600 km^2.

If we consider the Caspian region as that including territories directly bordering on the sea and connected with it, then it comprises twenty-two administrative entities: two republics, three districts, one velajat, three provinces and thirty regions. Development of economic and political relationships among all the countries of the region is very important, which is why after the breakup of the Soviet Union they have had to build a new system of relationships governing movement and employment among them.

Four of the countries are members of the Commonwealth of Independent States (CIS), which influences their interaction and cooperation in relation to many issues. These are countries differing in levels of economic development, types of established political regime, the nature of their foreign relations, religious aspects (Christians, Moslem, Buddhists), and the cultural and psychological mentality of their peoples.

Five countries of the Caspian region have 4% of the world territory and 4% of the world population. Nearly 3% of the world's gross national product (GNP) is concentrated here. The Caspian Sea region plays a significant role in the world economic system; the world's economics, in general, depends to a great extent on the economic situation and development in this region, and on the active involvement of the region in the worlds economic relations. General data on the regional countries are presented in Table 1.

According to the World Banks rating of countries by level of per capita GNP, in 2000/2001 this index for the Caspian states was (in US dollars): Russia, $2584

Table 1 General information on the states of the Caspian region [1]

Country	Population 1000s	Territory 1000 (km^2)	Population density (man/km^2)	GNP bill $	Per capita GNP $
Azerbaijan	8233[a]	86.6	88.3	4.45	576.2
Iran	67441	1633.2	41.3	101.073	499.3
Kazakhstan	14865	2712.3	5.5	15594	1053.6
Russia	145231	17075.4	77	375345	2584.4
Turkmenistan	5239	488.1	10.7	2708	516.89

[a] Trend News Agency, Baku, 17.07.2003

(70th place in the world); Kazakhstan, $1053 (100th place); Azerbaijan, $576 (119th place); Turkmenistan, $517 (122nd place); and Iran, $1499 (84th place). By the Human Development Index (HDI) of the United Nations Development Programme (UNDP) the Caspian region was classed as part of the group of "medium-developed countries". The rating of these countries in 2000 was as follows: Russia, 0.771 (61st place in the world); Kazakhstan, 0.754 (72nd); Azerbaijan, 0.722 (89th place); Iran, 0.709 (96th place); and Turkmenistan, 0.704 (99th place) [1]. From the data on population and economic potential, we can distinguish the two most economically powerful countries in the region: Russia and Iran. This is confirmed by the rating of these countries in the list of per capita production. Russia and Iran are in the first hundred countries, at 70th and 84th place, respectively. Kazakhstan is in an intermediate position; in recent years its economic potential has been growing, primarily due to the prospecting and extraction of hydrocarbons in the Kazakh sector of the Caspian Sea. Kazakhstan was the first of the CIS states to be given the status of a market-economy country. It maintains its leading position among the CIS countries by the volume of investments attracted . Of the$40 billion invested by the world community in the economies of the post-Soviet states in the period from 1991 to 2001, $13 billions were directed to Kazakhstan [2].

Azerbaijan and Turkmenistan are of the group of countries with a modest economic potential. They managed to avoid a "shock therapy" by improved state planning, thus, maintaining their administrative–command economics. However, they have an advantageous economic–geographical position and play a major role in servicing Russias transit links with the countries of Central Asia and the Near and Middle East.

The total population in the five Caspian countries is 240.4 million (2001). Despite a lack of data on many regions in Azerbaijan, on the Caspian shores 220 settlements are estimated to be populated with more than 10 million people.

Settlements on the Caspian shores vary greatly. Among the cities the largest are the Azerbaijan capital, Baku, with a population of 1.8 million; it is followed by such Russian cities as Astrakhan, the center of the Astrakhan Region, and Makhachkala, the capital of the Republic of Daghestan, with populations of 484 000 and 391 000 respectively; and then by cities located on the Kazakh Coast, the regional centers Aktau and Atyrau having approximately the same population of about 150 000; and, finally, by small cities on the Iranian coast, the largest of which are Enzeli with a population of 554 000, Babol with 55 000, and Bender-Torkemen with 173 000.

The greatest share of the population is concentrated in the 5-km-wide coastal zone in Azerbaijan and Turkmenistan, which is due to widespread extraction of hydrocarbons here, primarily, on the Apsheron Peninsula (Azerbaijan) and the Cheleken Peninsula (Turkmenistan). It should be noted that in Azerbaijan almost all of the coastal population (78%) live within a coastal zone up to 10 km wide and the majority of them are engaged in agriculture. Since the 1960s, after opening of the major Tengiz oilfield in Kazakhstan, a new settlement center began shaping.

The significance of the Caspian region is related, to a great extent, to its multiple mineral resources, the reserves of which differ greatly. Most essential are the hydrocarbon resources, represented by major fields with commercial reserves ensuring their profitable extraction. Other mineral deposits are of local significance only and are used mostly for meeting local needs in fuel, building materials, etc.

Prospective oil- and gas-bearing provinces here were determined in the early 1960s. But consideration of the global significance of these resources can be controversial. It was noted that "hydrocarbon resources of the Caspian are great, but the Caspian is not the Persian Gulf and it cannot claim the role of a complete alternative source for the world oil market" [3].

The U.S. Department of Energy reports that the total resources of the Caspian region are estimated at 100–200 billion barrels of oil (which exceeds North American oil resources as a whole) and 7.9 trillion m^3 of gas. This makes the region the worlds third largest for natural gas reserves [4].

The report, prepared by the Organization on Economic Cooperation and Development Organization (OECD), says that the proven oil resources of the Caspian constitute approximately 3% of the world reserves.

By most optimistic estimates using the most ideal combination of political and economic circumstances, in 10–15 years this region will have a share of no more than 6% of present world production.

For the countries of the Caspian region (excluding Russia and Iran) hydrocarbons are: part of a strategy for economic development [it should be noted that the share of the fuel–power complex (FPC) in the overall volume of industrial production is 73% in Azerbaijan (16% in 1991), 42% in Kazakhstan (13% in 1991), And 37% in Turkmenistan (34% in 1991 –)]; a hope for quick revival and economic wellbeing; and a method of attracting wide-scale investments.

The significance of the Caspian oil, if oil resources permit, is that it could be considered a potential world reserve that might be utilized in the case of depletion of oil resources in the North Sea, the Persian Gulf and other world regions or in the case of changes in price level on the world markets.

The other, and historically most essential, natural resource of the Caspian is its biological resource. The most important of the Caspian's biological resources is its fish stocks:about 123 fish species and subspecies. Their composition has been shaped by the course of the historical evolution of the sea: isolated from other seas and the Atlantic Ocean, it contains species originating both in the north and south (the Mediterranean).

The Caspian is biologically unique, because together with the rivers that flow into it—first and foremost the Volga—it contains the world genofund of the sturgeon and is the world's only repository of a diversity of species of sturgeon. Until recently, catches of sturgeon in the Caspian Sea accounted for up to 90 percent of the total world catches.

The fresh, shallow water in the Northern Caspian is especially significant. The inflow of river water rich in food, uniquely favorable conditions for spawning and the growth of fry, and the limited role of carnivorous predators make the region a kind of nursery for this most valuable fish species. It was no accident that in the 1970s an area of the Caspian Sea lying to the north of 44° 12′ N.L. was declared a nature reserve.

Sturgeon are valuable for, among other things, their caviar, an expensive delicacy in high demand on the world market. However, diminishing catches in the Caspian have led to a drop in caviar production. In 1989 the Soviet Union produced 1365.6 t of black caviar and Iran 282 t tons. By the late 1990s Russia produced only 40 t per year, other new sovereign Caspian states (excluding Turkmenistan) 34.8 t, and Iran about 150 t. Already, even before full-scale production of hydrocarbons in the Caspian Basin has begun, the situation of the sturgeon in the Caspian Sea is catastrophic, so much so that some experts speak in terms of the Caspian losing its fishery significance.

Today the Caspian region, in terms of its socioeconomic development, has become a zone of national interest to the Caspian states; however, some interstate differences and the ongoing uncertainty of the legal status of the Caspian, issues that have arisen since the breakup of the Soviet Union, complicate the settlement of issues about rational nature management and environmental security. The situation requires the adoption of a coordinated, mutually beneficial legal status taking into account the common interests of the Caspian states.

3 International Legal Status

For more than 250 years the Caspian Sea was shared by two states: Russia (the Soviet Union) and Persia (Iran). After the disintegration of the USSR, the newly independent states of Azerbaijan, Turkmenistan and Kazakhstan laid their claims to the wealth of the Caspian. As a result, the issue of the equitable sharing of that wealth has emerged.

It's hardly necessary to prove that a body of water washing the shores of five sovereign states and being an object of competing international interests requires a legal status that will determine the legal regimes required for its shared utilization for shipping, fishing, monitoring and exploration of mineral deposits. An agreement is also required on environment protection.

Until the demise of the Soviet Union the legal status of the Caspian Sea was established by the Treaty of Friendship between the Socialist Federate Republic of Russia and Persia on February 26, 1921, and the Treaty of Commerce and Navigation between the Union of Soviet Socialist Republics and Iran on March 25, 1940. Prior to the conclusion of the 1921 Treaty, the status of the Caspian was subject to norms of international law. We are referring to the Gulistan Treaty of 1813 which was concluded after the Russian–Persian War of 1804–1813. By this Treaty, Russia was granted an exclusive right to have its Navy on the Caspian, which was confirmed in the Turkmanchai Treaty (after the war between Russia and Persia of 1826–1828), while Persia was without such a right (Article 8 of the Treaty). This Treaty remained in force till 1917. According to Barsegov [5], "the very idea of absence of the international legal status of the Caspian is quite absurd: at the closing of the twentieth century there are no and cannot be sea spaces that have no such status". However, the newly independent states were not participants in these Treaties, and naturally enough, the provisions of the Treaties did not meet their political and economic interests in the new geopolitical situation. Moreover, until a new legal status is adopted, all of the above-mentioned international documents remain valid, because all former Soviet republics and the new sovereign states recognized the treaties signed by the USSR.

This means that, as before, the Caspian Sea remains in common use, but this time not of two powers as in the past, but of five Caspian states. It also means that these five states have equal rights to its utilization; that they can enjoy the rights of free shipping and fishing; that, except for a 10-mile offshore zone, the division of the Caspian waters is inadmissible; and that navigation over the sea by ships under flags of non-Caspian countries is prohibited.

Naturally, the two Treaties say nothing about mineral resources of the Caspian, because in neither 1921 nor 1940 did mankind have technical facilities for oil and gas extraction from the sea bed at its disposal. Since the signing of the last Soviet–Iranian Treaty, the Caspian region has changed radically. As a result, the existing legal status fails to comply with the current political situation and, therefore, cannot regulate the relationships among the Caspian states in full measure.

Both "old" and "new" littoral states were quick to detect the inconsistencies of the acting legal status of the Caspian and, consequently, of the Soviet–Iranian Treaties. They have come to terms in relation to the starting point from which they would move towards the elaboration of a new legal status. They now agree that the exclusive rights to the Caspian Sea belong not to two, but to five countries; that these rights are applicable also to the Caspians natural resources , including hydrocarbon deposits, and that there is a need to adopt a convention on the legal status of the Caspian Sea and that such a convention must be passed only on the basis of general consensus.

Russian diplomats proceeded from the principle of a joint use of the Caspian Sea resources (a condominium) and that all issues related to activities (including development of the Caspian resources) should be settled with the participation of all Caspian states.

The Caspian is a closed (inland or terminal) water body having no natural links with the world ocean. In legal terms it cannot be considered either as a sea or as a lake. In particular, it is not subject to the action of the 1982 UN Convention on the Law of the Sea with, for example, its provisions relating to exclusive economic zones and continental shelf. Here, a precedence right shall operate. Consequently, there are no grounds for unilateral claims in relation to the establishment of similar territorial claims in the Caspian, or for the introduction of the elements of their various legal regimes. At present there is no sectoral, or for that matter any other, accepted delineation of the Caspian seabed.

However, because such a state of affairs was unsatisfactory for the Caspian states, they chose to pursue unilateral actions. This was caused, in part, by the deteriorating economic situation in the Caspian states—CIS members—which made them greatly dependent on the Caspian and its resources. The so-called "oil syndrome" appeared. Elaborate national programs for the development of the Caspian Sea were often in conflict with each other.

Not waiting for the final decision on the legal status of the Caspian, the major oil companies, including those from Russia, became active in this region. They were in a hurry to secure big business deals in Caspian oil and were seeking to widen their participation in projects on the Caspian by forming consortia for the development of rich oil and gas deposits.

It is interesting to note that in the early 1990s, the big Russian oil companies in the Caspian were seeking to take on the functions of a state in such issues as foreign relations and security. "The origin of this should be sought in the fact that the traditional diplomacy had failed to adjust to the new changes that were occurring in the region and was slow to respond to the new tendencies with the growing number of multifarious parameters" [6].

At first, Russia advocated that the new Convention maintained the principle of "common water" (the regime of common use of Caspian waters) stipulated in the Soviet–Iranian Treaties, having extended it to mineral resource use. This approach found support only from Iran, while Azerbaijan, Turkmenistan and Kazakhstan were against it; Azerbaijan and then Turkmenistan formulated a joint demand on division of the sea into national sectors, each sector being in the complete sovereignty of each Caspian state, i.e., a division of the sea with state borders along the whole sea profile: the seabed, water body, water surface, air space. Azerbaijan and Turkmenistan adopted relevant legal acts unilaterally.

For Russia the division of the sea into national sectors was and remains an unacceptable approach. It would undermine the rational system of economic activities on the sea that had been established over decades, and enable each littoral state, on different pretexts, to impose in its sector any restrictions it chooses for other Caspian countries. In the course of sea division a host of new problems will arise and territorial disputes and conflicts would most likely develop.

Adhering strictly to the consensus principle, Russia started seeking a compromise. It was necessary to draw closer two quite polar positions, one in favor of condominium and the other in favor of division of the Caspian into national sectors. "Finally, it was proposed to put the principle of 'seabed division–common water' in the basis of a new legal status of the Caspian. In particular, this would mean a division of the Caspian seabed among neighboring and opposite-lying states, as a result of which each of them would have at its disposal a section of the seabed for exercising its sovereign rights to mineral resources. In such a situation it is understood that a dividing line would not be a territorial border, but would only delimitate the resource jurisdiction of the littoral states" [7].

In search of a compromise with regard to the absence of appropriate norms in the Soviet–Iranian treaties and bearing in mind that the Caspian states had already started development of hydrocarbons in the seabed, Russia put forward a proposal to settle the status issue. It proposed the following formula: "to divide the seabed for mineral resource utilization– common water" and a stage-by-stage principle by which the Caspian legal status would be formed : easily resolvable issues to be addressed first, while the settlement of more difficult ones were to be left for future consideration. Among priority issues waiting for their settlement are (1) subsoil resource use, (2) fishing and (3) ecology.

A new Russian position was made public in January 1998 in a Joint Declaration of the Presidents of Russia and Kazakhstan: "A consensus should be found in relation to a just division of the Caspian seabed leaving in common use the water surface, including free shipping, agreed fishing norms and environment protection". Thus, Russia agreed to division of the Caspian seabed among the Caspian littoral states.

On July 6, 1998, the Russian and Kazakh governments signed the *Agreement between the Russian Federation and the Republic of Kazakhstan on Seabed Division of the Northern Caspian for Exercise of Sovereign Rights to Mineral Resource Utilization*. Article 1 of this Agreement states that "the seabed of the Northern Caspian and its subsoil resources, leaving the water surface in common use, including free navigation, agreed fishing quotas and environment protection, is divided between the Parties along a "median line", modified on the principle of justice and upon agreement of the Parties. The Agreement also stated (Article 2) that "the Parties exercise their sovereign rights in relation to exploration, development and management of the seabed and interior resources of the Northern Caspian within their seabed sections to a delineation line".

It is necessary to understand the difference between a median line and its modification. A median line, used for delimitation of territorial waters among neighboring (adjacent and opposite-lying) states, is a line each point of which is equally distant from the respective nearest points on coasts of these states. A modified median line includes also all parts that are not equally distant from the coasts of the Parties and is delineated with regard to islands, geological structures and other specific circumstances and costs incurred for geological studies [8].

In addition, a very important point was stressed. "The seabed is not considered a territory, but is seen as promising structures and fields, i.e., a state border is not drawn under water. Being unlike territorial jurisdictions, such jurisdiction in respect of the sea is called resource jurisdiction.

A modified line is a political line drawn on the basis of the classic line, but, in its absence, in favor of this or that party so that, taking into consideration the whole complex of circumstances, including already sustained expenditures, to justly divide the resources under the seabed" [9].

The Russian position with respect to the legal status of the Caspian was to avoid dead-end negotiations by suggesting compromise options. At the same time the Russians stood firm on the issue of leaving the water body in the common use of all Caspian states.

New approaches by Russia and Kazakhstan reflected considerable geopolitical shifts for the region. It had become obvious that a strict adherence to previous positions, i.e., with orientation only to the Soviet–Iranian Treaties without seeking new forms of cooperation, could lead to the isolation of Russia and restriction of its participation in many regional processes.

By early 2000 the Caspian region had secured a more prominent place in the foreign policy of Russia. At one of the meetings of the Security Council (SC) of the Russian Federation (RF) it was stated that the Caspian is "a traditional zone of the Russian national interests" [10].

The RF Foreign Policy Concept adopted in summer 2000 stressed the need to enhance the diplomatic influence of Russia on the Caspian in addition to its economic interests: "Russia will seek elaboration of such status of the Caspian Sea that will enable the littoral states develop mutually beneficial cooperation in management of the regional resources on an equitable basis taking into consideration the lawful interests of each other" [11].

The same provisions were also included in the Marine Doctrine of Russia. It states, in particular, that the Caspian region possesses unique, in terms of volumes and quality, mineral and biological resources, and the following long-term tasks should be addressed: "the identification of the optimal benefits for the Russian Federation of the international legal regime of the Caspian Sea and a procedure of fish, oil and gas resources management; in cooperation with other littoral states, conservation of the marine environment; creation of conditions, including utilization of the capacities of the subjects (Astrakhan Region, Kalmykia and Daghestan) of the Russian Federation, for establishing the base of operations and utilization of all elements of the marine environment; updating merchant sea and mixed-type (river–sea) ships and the fishing fleet; not admitting the forcing out of the Russian fleet from the marine transportation services market; organization of ferry lines as a part of an intermodal conveyance system linking with the Mediterranean and Baltic Seas; development, rehabilitation and specialization of existing ports" [12].

Russia suggested a new idea, a stage-by-stage approach to the settlement of Caspian problems not linking them with an agreement on the legal status of the Caspian. As concerns development of disputed oil fields, application of the 50 × 50 principle was proposed. This demonstrated that Russian policy towards the adoption of the Caspian legal status was becoming more pragmatic in economic terms.

Such proposals from Moscow inspired, in the second half of 2000, the negotiating processes among the Caspian states. Thus, on 9 October 2000 in Astana, Kazakhstan, the *Declaration of the Russian Federation and the Republic of Kazakhstan on Cooperation on the Caspian Sea* was signed. This Declaration confirmed "readiness of these countries to facilitate effective improvement of the five-lateral negotiation process on the Convention on the Legal Status of the Caspian Sea and to organize systematic activity of the Ad-Hoc Working Group of deputy ministers of the Caspian states". At the same time Russia and Kazakhstan stressed "their assurance that on the basis of consensus on the new legal status of the Caspian Sea it is advisable to use a compromise proposal on delineation of the seabed among neighboring and opposite-lying states along the median line modified on the basis of arrangements attained among the parties with a view of exercising their sovereign rights to utilization of subsoil resources leaving the water body in common use to ensure free shipping, coordination of fishing quotas and environment protection" [13].

The orientation of Russia towards bilateral cooperation with Kazakhstan on the issues of the legal status of the Caspian made it possible to formulate and develop new principles for the Caspian delineation. The Russian–Kazakh alternative solution to the Caspian legal status seemed attractive to other Caspian states as well.

In January 2001 the Russian Federation and the Azerbaijan Republic signed the *Joint Declaration on the Principles of Cooperation on the Caspian Sea*. This document stated "on the first stage, to delimit the Caspian seabed among neighboring and opposite-lying states into sectors/zones on the principle of a median line drawn with regard to equidistant points and modified as arranged between the parties and also with regard to the general recognized principles of the international law and the established practices on the Caspian". In addition, "the Parties agreed that each littoral state in the sector/zone, formed as a result of such delimitation, would exercise exclusive rights in relation to mineral resources and other lawful economic activities on the seabed" [14]. In contrast to the Russia–Kazakh Declaration, the new term "sector/zone" is used and the phrase "other lawful economic activities on the seabed" is added. This Agreement has shown that Russia decided, primarily, to settle the status with its neighbors, Kazakhstan and Azerbaijan.

The principle of a just seabed delineation was "the most for which Russia was ready". A modified median-line principle is compromise in its essence, because it admits deviations from the ordinary median line in favor of this or that party upon mutual arrangement.

Russia's approach was based on two principal issues:

1. The Caspian Sea is a unique landlocked body of water, and in international terms it cannot be treated either as a sea or a lake.
2. Not to link the adoption of the general convention with the documents on conservation and utilization of bioresources, believing that the priority issue here should be the biological resources that are the greatest value of the Caspian Sea.

Therefore, Russia tried to find some consensus that would combine the equitable division of hydrocarbon resources of the Caspian and, at the same time, leave in common use the water area, including guarantees of free shipping, agreed fishing quotas and environment protection.

The evolution in the Russian policy towards bilateral agreements with different Caspian countries did not pass unnoticed in Iran, which still adhered to the earlier approaches. Russia was interested in pushing for settlement of the Caspian legal status issue on a multilateral basis.

The Iranian Presidents visit to Moscow in March 2001 brought Russian–Iranian relations to a higher level. In the *Joint Declaration of the Russian Federation and the Islamic Republic of Iran on the Legal Status of the Caspian Sea* it was said that "as concerns the widening of bilateral cooperation in all spheres related to the Caspian Sea, the Parties recognize the Treaty of Friendship between the Socialist Federate Republic of Russia and Persia on February 26, 1921 and the Treaty of Commerce and Navigation between the Union of Soviet Socialist Republics and Iran on March 25, 1940 with supplements as the legal base regulating at present the activities on the Caspian. The Parties, until improvement of the legal regime of the Caspian Sea, officially do not recognize any borders in this sea and with regard to the mentioned base are developing cooperation on the Caspian Sea in various fields by elaboration of the required legal mechanisms". Further on it was noted that "all solutions and arrangements concerning the legal status and regime of utilization of the Caspian Sea will come into force should they be adopted on the mutual consent of the five Caspian states" [15].

However, this Declaration failed to change approaches of the two countries in relation to a final settlement of the legal status of the Caspian Sea. The RF Foreign Ministry noted that "Iran must not feel left out, at least due to the fact that its position on the Caspian, in comparison to its position in the time of the USSR, has essentially strengthened. In the 1930s unilaterally and without any legal validation the so-called Astara–Gassan–Kuli line was drawn which the Iranians did not cross. That part of the Caspian which was under the control of Iran, when its northern neighbor was only the USSR, was smaller than the share that Iran can receive now. And it will be assigned to Iran by the norms of the international law". And further on: "If the Caspian is to be divided, say, from a blank sheet, quite likely there will be no problem with these 20% at all. Each will get its equal share. But each Caspian state has its own multicentury history, has its own established economic practice" [16].

Simultaneously with these multilateral negotiations, Russia is energetically discussing the issue of the Caspian legal status on a bilateral basis with Kazakhstan and Azerbaijan within the framework of the earlier attained arrangements. One proof that the Caspian countries are seeking to settle the issue of the Caspian legal status is that there has been a greater number and greater frequency of meetings of the Ad-Hoc Working Group at the level of deputy foreign ministers of the Caspian states for the elaboration of the Convention on the Legal Status of the Caspian Sea (for the period from 2001 to 2005 12 meetings were held).

In November 2001 in Moscow the *Agreement between the Republic of Kazakhstan and the Azerbaijan Republic on Division of the Caspian Sea between the Republic of Kazakhstan and the Azerbaijan Republic* was signed. It stated that "the Caspian seabed and its subsoil are divided between the Parties along the median line drawn on the basis of equidistance from the initial base points on the shoreline and islands".

The development of joint positions was facilitated by resorting to both the multilateral and bilateral formats. In Moscow on 13 May, 2002 the Presidents of Kazakhstan and Russia signed the *Protocol on the Bilateral Agreement on the Delimitation of the Seabed of the Northern Caspian with a View to Exercise Their Exclusive Rights to Subsoil Resource Utilization of 6 July 1998.* According to this Protocol, Russia has received in its resource jurisdiction the Khvalynsky field and the Tsentralnaya structure and Kazakhstan received the Kurmangazy structure (Kulalinskaya). The hydrocarbon resources of these fields will be developed jointly. It was also noted that "The Protocol fixed the route of a modified median line delineating the subsoil resource use of two states, outlined the resource jurisdiction of the parties over key fields and structures located at the juncture of these zones, and stated that their hydrocarbon resources would be developed jointly on the 50 × 50 principle" [17].

In this way the "principle of resource division" was given life. After the aforementioned Agreement, the *Agreement between the Russian Federation and the Azerbaijan Republic on Delimitation of Neighboring Parts of the*

Caspian Seabed of September 2002 was signed that, in fact, finalized the division of the Northern Caspian.

An agreement was also signed on the delimitation of neighboring parts of the Caspian seabed. Article 1 of this document stresses that "the Caspian seabed and its subsoil are divided among the Parties on the median line principle and also in compliance with the generally recognized principles of the international law and the practice established on the Caspian". Article 2 notes that "the mineral resources of structures crossed by a delimitation line will be developed on the basis of the best international practice applicable in case of the development of transborder fields by authorized organizations appointed by the Governments of the Parties". The Agreement also specifies the accurate coordinates of a dividing line and notes that "the Parties exercise their sovereign rights in relation to mineral resources and other lawful economic activities related to subsoil resource utilization on the seabed within their sectors to the delimitation line determined in Article 1 of this Agreement". Pursuant to this Agreement, "the Parties will facilitate the attainment of the general consensus of the Caspian states on the division of the Caspian seabed abiding by the principles of this Agreement". Proceeding from the Agreement on the Seabed Delimitation signed in late 2001 between Azerbaijan and Kazakhstan, the following conclusion can be drawn: "an interaction of three countries in relation to the development of mineral resources of the seabed in the Northern and Central Caspian acquires a firm international legal base".

In retrospect, "2002 was a year of breakthrough for the settlement of international legal activities on the development of mineral resources of the Caspian" [18]. Russia, Kazakhstan and Azerbaijan came to terms with respect to the principle of the Caspian Sea division, i.e., to divide only the seabed, leaving the water column in common use.

On 14 May, 2003 in Almata the *Agreement between the Russian Federation, the Azerbaijan Republic and the Republic of Kazakhstan on a Junction Point of Lines Dividing the Neighboring Parts of the Caspian Seabed* was signed. Thus, the development of mineral resources in the seabed of the Northern Caspian was given an international legal dimension.

At present the Caspian littoral states agree that determination of a new legal status is a multilevel and versatile process in which the elaboration of the Convention on the Status as a basic document goes parallel with preparation and conclusion of agreements on individual kinds of activities on the Caspian. An international legal basis of cooperation is formed so that both "on the top" and "in the bottom" each new arrangement becomes a kind of a "brick" placed into the final constructed building of a status.

In order to become comprehensive and effective, the developed legal status should combine provisions reflecting new geopolitical, operational–economic, environmental and other realities with the norms and principles of the Soviet–Iranian Treaties of 1921 and 1940, the adequacy of which was proven over decades of their application on the Caspian. There is one more source of "Caspian law": the economic activities on the Caspian established for many years that cannot be neglected and that should be reflected in legal norms included both into the Convention on the Status and in departmental agreements.

References

1. Institut de Relations Internationales et Stratégiques (2002) L'Anneé Stratégique. l'Etudiant, France
2. Nikolaev AI (2003) Rossiyskaya Gazeta 28.01.2003:1
3. Kalyuzhny VI (2003) Caspian Sea Bull 3:7
4. (2004) Baku Today/Turan 17.11.2004
5. Barsegov Yu (1998) Caspian in the international law and international policy. Institute of World Economics and International Relationships, Moscow
6. Tsyganov PA (2002) Theory of international relations. Gardariki, Moscow
7. Kalyuzhny VI (2002) Caspian Sea Bull 2:5
8. Urnov AJu (2000) Caspian Sea Bull 3:2
9. Urnov AJu (2000) Caspian Sea Bull N 3:11
10. (2000) Nezavisimaya Gazeta 11.07.2000
11. Foreign Policy Concept of the Russian Federation (2000) Caspian Sea Bull 4:5
12. Russian Federation Government (2001) Resolution of the RF Government of July 31, 2001. No. 566
13. Declaration of the Russian Federation and the Republic of Kazakhstan on Cooperation on the Caspian Sea (2000) Caspian Sea Bull 6:3
14. Russian Federation and Azerbaijan Republic (1998–2003) Joint declaration of the Russian Federation and Azerbaijan Republic on principles of cooperation on the Caspian Sea. Diplomatic documents on the international legal status of the Caspian Sea (1998–2003)
15. Russian Federation and the Islamic Republic of Iran (2001) Joint declaration of the Russian Federation and the Islamic Republic of Iran on the legal status of the Caspian Sea
16. Kalyuzhny VI (2001) Nezavisimaya gazeta. October 02:3
17. Kalyuzhny VI (2002) Caspian Sea Bull 6:1
18. Kalyuzhny V (2002) Speech in Iran on December 3, 2002

Hdb Env Chem Vol. 5, Part P (2005): 257–268
DOI 10.1007/698_5_014

Published online: 13 October 2005

Conclusions

Aleksey N. Kosarev[1] · Andrey G. Kostianoy[2] (✉)

[1]Geographic Department, Lomonosov Moscow State University, Vorobievy Gory, 119992 Moscow, Russia

[2]P.P. Shirshov Institute of Oceanology, Russian Academy of Sciences, 36 Nakhimovsky Pr., 117997 Moscow, Russia
kostianoy@online.ru

Abstract This conclusion completes one more review concerning the natural condition of the Caspian Sea and the principal trends of change in the second half of the 20th century. The change in the mean sea level variations that occurred in the mid-1970s, when the long-term fall was replaced by a rapid and significant rise, represents an important indicator of the changes in the natural regime of the Caspian Sea, including thermohaline structure, and nutrients and oxygen distribution. The most important consideration is chemical pollution (oil and oil products, phenols, detergents in the North Caspian), which grows with the intensification of human activity on the sea coasts and in the sea proper and represents one of the most hazardous kinds of anthropogenic impact on the Caspian ecosystem. The biological section of the monograph discusses the main problems concerning the biological diversity, the introduced species, and the biological resources. One of the most catastrophic aftereffects of anthropogenic intervention into the Caspian ecosystem is related to the population of sturgeons. Special attention is paid to socioeconomic, legal, and political problems in the Caspian Sea region. The unsettled delineation of the Caspian Sea and the uncertainty in its legal status are the main obstacles to successfully coping with many issues, including environmental protection and preservation of the biological resources.

This conclusion completes one more generalization concerning the natural condition of the Caspian Sea and the principal trends of the changes in the second half of the 20th century. The monograph represents the results of the studies of the Caspian, both new and those supplementing the previous data. In this respect, we should note that, if the rate of the studies is retained at the present-day level (which is quite probable), it might take a long time before the next monograph will be issued.

In popular science, the regime of the Caspian Sea is often referred to as "mysterious". Actually, an analysis shows there are no processes in the Caspian Sea waters that cannot be learnt about. Meanwhile, this sea is indeed distinguished by its particular physicogeographical conditions such as the isolation from the World Ocean and inland position. Precisely these features define the formation and variability of the structure and dynamics of the Caspian waters and its biological features. This monograph was preceded by a series of review publications devoted to the Caspian Sea issued in

1992–1996 ([1–3]). As compared to the monographs listed, this new one is distinguished by the following important particularities. In it, new materials from field observations up to the 2000s are used. The scope of the problems is significantly extended due to assessment of the ecological and socioeconomic aspects; methods of remote satellite observations are applied. A comparison between the published books and this new one shows that many principal features of the natural regime of the Caspian Sea have been retained. Along with this, significant changes have also been recognized. On the whole, they may be determined as the growth in the anthropogenic impact on the coastal areas and the sea proper. It is manifested first in the morphological and other changes occurring in the near-mouth and other shallow-water regions of the Caspian Sea and in the increase in the basin pollution. The technogenous impact affects many abiotic and biotic components of the Caspian ecosystems.

The change in the tendency of the mean sea level variations that occurred in the mid-1970s, when the long-term level fall was replaced by its rapid and significant rise, represents an important indicator of the changes in the natural regime of the Caspian Sea. This situation also affected the character of the natural conditions in the sea. It should be noted that, despite the regional differences, the Caspian Sea as a whole responds to the external forcing as an integer basin with a common hydrological regime. Below, in a concise form, we present the most interesting results and inferences obtained in this monograph.

While considering the general physicogeographical conditions of the Caspian Sea, we will focus our attention on the issues concerning the sea level oscillations. At the present stage of their studies, one can highlight the following aspects of this problem. First of all, we note the growing recognition of the priority of water–climatic control over the long-term level oscillations that are observed in the publications. At the same time, the number of papers in which the geological and geophysical factors are regarded as the principal reason for these oscillations decreased. The existing interannual variations of the Caspian Sea level are rather justly correlated with the changes in the total water balance over the given time interval. Therefore, there is no need in assessing any other (including "exotic") factors.

One more aspect of the analysis of the interannual sea level variations lies in the extension of the number of factors involved in the problem. For example, in the studies referring to the middle of the 20th century, the regularities of the Caspian level changes were examined based on an analysis of the climatic and circulation factors, mostly in the basin of the Volga River. During the past decades, with the development of remote sensing for these purposes, the scientists began to recruit data on the large-scale climatic processes, which cover significantly vaster areas both in the Northern and in the Southern hemispheres of the earth such as the North Atlantic Oscillation (NAO), El Niño–Southern Oscillation, and other major climate-forming systems. As a result, the principal conclusion was justified that the regime of the

Caspian Sea level is mostly controlled by the proportions of the cyclonic and anticyclonic activity in the basin of the sea (mainly, in the Volga River region) and the related precipitation regime.

On the whole, the oscillations of the Caspian Sea level represent a result of mutually related hydrometeorological processes, which proceed not only in the sea catchment area but also far beyond it. Therefore, long-term forecast of the sea level changes represents an extremely difficult task. At present, only probability estimates for the forthcoming decades are feasible. According to these data, at the present-day climatic regime, the basic mark of the mean level of the Caspian Sea lies within the limits $-28 \pm 1, 5\text{–}2$ m. The possibility of such level oscillations over the future decades should be taken into account when planning the industrial activity in the coastal zone and in the sea proper.

The significant sea level rise observed from the end of the 1970s leads to changes in the structure of the coastal zone including the river deltas. Obviously, as sea level rises, the coastline moves landward and abrasion and shore flooding occur. Meanwhile, in different areas of the coasts, these general tendencies are manifested in different ways, depending on the type of the coast and the influence of other factors. In this book, we briefly discuss the influence of the sea level rise on the geomorphological changes in three regions of the Russian coasts (the Volga River delta, Kalmykia, and Dagestan) and in the coastal area of Azerbaijan.

In the Volga River delta, the influence of the present-day level rise was weaker than might be expected, since the presence of a vast shallow-water seaside off the river mouth suppresses this influence. During the decades of the sea level rise, the depths on the Volga seaside increased only by about 1 m. However, in the case of a further level rise, its influence in this region may significantly grow; in particular, storm surges may more freely penetrate into the delta and the flooding areas may increase. The coasts of Kalmykia, especially in its southern part, experience a landward receding of the coastline and the flooding of the shore area at a rate of approximately 200 m per year; in so doing, coastal bars are washed away. On the coasts of Dagestan, one observes an enhancement of the coastal erosion including marine terraces; the erosion of the marine edge of the Sulak River delta also became stronger. The southern near-shore areas of the coasts of Azerbaijan most suffered from the level rise; the overall flooded area of the Azerbaijan coastal zone comprised about 500 km^2. The flooding hazard is urgent for many industrial and municipal objects.

When analyzing the physicogeographical conditions of the Caspian Sea, we highlighted the important role of the delta regions of the entering rivers. In the delta areas, the changes in the natural conditions proceed especially intensively; at the same time, these regions are the most interesting from the point of view of biological resources. In the river deltas and in the seaside areas off them, due to the combination of a series of factors, one observes crucial qualitative changes in the chemical characteristics of the riverine waters supplied to the sea and in the conditions of the transport and sedimentation of organic substances. Due to these reasons, the near-mouth areas of the rivers represent a kind of geochemical barrier.

The existing concepts concerning the character of the hydrological structure of the Caspian waters are essentially refined. The principal changes in the thermohaline structure refer to the interannual variability. The most radical transition in the sea regime occurred at the end of the 1970s, when the long-term level fall was replaced by its rapid rise by more than 2 m. Previously, it has already been established that, in the course of this process, the horizontal thermohaline structure (which reflects the changes in the rate of the Volga water supply), the winter severity, and the character of the atmosphere circulation over the sea also changed. In this monograph, in addition, we assessed the large-scale variations in the vertical structure of the entire water column.

The analysis was performed assuming that both the circulation and thermohaline regimes of the sea are mostly formed under the action of the atmospheric processes over the sea area and its vast watershed. The waters of the entire water column of the Caspian Sea are formed in the surface layer; only the winter mode of the surface waters can penetrate down to great depths driven by the deep convection and the sliding of the waters over the continental slope.

In the Caspian Sea, synchronously with the sharp turn in the level history at the end of the 1970s (from a fall to a rise), the thermohaline structure of the waters of the surface layer also changed under the influence of significant anomalies of the riverine water inflow and other thermohydrodynamical factors. The most significant changes in the external impacts happened between the decades 1968–1977 and 1978–1987. As a result, in 1978–1987, the sea surface salinity values in the summertime decreased by 0.2–0.3 psu in the South Caspian and by 1.0–1.5 psu in the North Caspian. At the same time, the water temperatures in February increased by 0.5–2.0 °C over the entire sea area.

The long-term variability in the annual volumes of the riverine runoff, winter severity index, and water temperature and salinity observed in 1956–2000 suggests the following tendencies in the character of the vertical thermohaline structure of the Caspian Sea. Starting from the second half of the 1970s, water salinity dropped over the entire water column of the Caspian Sea. In so doing, a stable vertical salinity stratification was formed. By the middle 1990s, the thickness of the summer thermocline decreased three- to fourfold, while the vertical gradients in it increased to the same extent. The temperature of the deep waters in the Middle Caspian also noticeably increased (by 0.5 °C).

The radical transformation of the vertical salinity structure of the Caspian Sea waters is the most prominent event in the multiannual variability of its hydrological condition. In the middle 1970s, the vertical salinity structure featured a subtropical type. Every autumn, due to the erosion of the seasonal thermocline, conditions favorable for the development of a deeply penetrating convection were formed. The salinated surface waters were forced to downwell to the deep-water depressions of the sea. By the middle 1990s, the vertical salinity structure of the Caspian Sea acquired a subarctic type characterized by a salinity growth with depth. Under these conditions, at the autumn destruction of the thermocline, the salinated surface waters could not penetrate deeper than 200 m. From the middle 1980s, the enhanced accumulation of the winter waters in the intermediate layers resulted in an upward forcing of the seasonal thermocline and reduction of its thickness.

Thus, one can distinguish two types of regimes in the multiannual trends of the variability in the thermohaline structure of the Caspian Sea waters. One of them, observed in the 1960s–1970s, is characterized by a combination of a reduced riverine runoff and severe winters. Under these conditions, the entire water column is intensively ventilated. This leads to a leveling of water salinities over the vertical, their growth over the entire water column, an increase in the thermocline thickness, and a temperature drop in the deep-water layers. The other type of regime, observed in the 1980s–1990s, is characterized by an inverse combination of the external factors with an enhanced riverine runoff and domination of mild winters. In this case, a stable vertical salinity stratification is formed, which leads to the suppression of the vertical diffusive exchange in the water column. The systematic ventilation of the near-bottom waters is distorted and the mixed zone is restricted by the layer from 200 to 300 m. As a result, the temperature at great depths becomes somewhat increased.

The multiannual hydrological changes in the Caspian Sea water column affect its hydrochemical structure and the formation of the biotic component of the ecosystem. In this respect, the second type of the hydrological processes distinguished is far less favorable, since the enhancement of the vertical density stratification is accompanied by a reduction of the oxygen content in the deep-water layers and leads to a poorer supply of the upper euphotic layer with nutrients. The weakening of the winter convection and the sharpening of the summer thermocline result in a growth in the concentrations of pollutants in the surface water layers. These phenomena may cause very unfavorable aftereffects for the Caspian biota.

The assessment of the hydrological structure of the Caspian Sea waters is crowned with an analysis of the particular features of the summertime upwelling characteristic of the shelf zone off the eastern coast of the sea. In so doing, we traced remote links of the 4–5-year variability in the upwelling intensity with the manifestations of the El Niño–Southern Oscillation cycling.

For the analysis of the hydrological characteristics of the Caspian Sea, we also used the data of satellite observations. Their assessment allowed us to implement monitoring of the sea surface temperature (SST) in the Middle and Southern Caspian over 1982–2000. This provided the recognition of a positive trend in the sea surface temperature of 0.05–0.10 °C/year. It was shown that the majority of the winter and summer SST anomalies in the Caspian Sea were related to the El Niño events. A long-term monitoring of the thermal regime of the Caspian Sea is required, based on satellite observations and studies of the response of the sea to the manifestations of the global atmospheric interactions in the region.

Remote sensing also helped to obtain the chlorophyll concentration distribution in the surface layer of the Caspian Sea in 1997–2004. This is a very characteristic parameter because it allows one to estimate water productivity in various regions of the sea. The chlorophyll content was especially high in the shallow-water North Caspian, distinguished by its high biological productivity. The interannual variations in the chlorophyll contents were related to the changes in the Volga River runoff; among the climatic factors, the influence of the index of the North Atlantic Oscillation was the strongest.

In the section considering water chemistry, we showed the particular features of the salt composition in the waters of the Caspian Sea, which differs from that of the World Ocean waters because it was formed under conditions of isolation from the Ocean. The spatial changes in the salinity/chlorine ratio make difficult the application of a common equation of state. In order to compile a practical salinity scale, similar to that compiled for the World Ocean, and a refined equation of state, it is necessary to carry out additional studies.

The chemical structure of the Caspian Sea waters is subjected to a multiannual variability, which is mainly defined by the hydrological conditions of the sea. The observations show that, over the past decades, the hydrochemical regime of the Caspian Sea again approached the condition that had been observed at the beginning of the preceding century, before the long-term sea level fall started. In particular, high concentrations of mineral forms of nutrients and low dissolved oxygen contents in the deep-water layers of the Middle and South Caspian are again observed. These phenomena are noted even in the near-bottom layers of the North Caspian shelf. In the spring and summer, a significant oversaturation of the euphotic layer with oxygen (up to 120–130%) takes place due to the photosynthesis of phytoplankton.

The oxygen regime in the sea mostly depends on the processes of mixing and is very important for the formation of biological productivity. During the past decades, the increase in the Volga River runoff caused an enhancement of the stability of the vertical water stratification in the deeper zones of the North Caspian shelf. The supply of nutrients, especially of their mineral forms, to the North Caspian also increased. The subsequent eutrophication of the waters of the North Caspian led to a significant drop in the near-bottom concentrations of dissolved oxygen in the deeper zones of the North Caspian, sometimes down to the hypoxy threshold. In 1996–2003, the oxygen content in the summertime decreased in the eastern part of the North Caspian as well. On the whole, the area of the zones where the oxygen concentration in the near-bottom layer in the warm season layer was less than 80% increased threefold. The deterioration of the oxygen regime in the North Caspian negatively affected the dwelling conditions of benthos.

In the depressions of the Middle and South Caspian, with the decrease in the winter severity, conditions for the mixing in the deep layers became less favorable. In 1998–2003, the oxygen contents in the near-bottom layers of the depressions equaled 0.5–1.5 mL/L (only 10–20% of the saturation value), which is significantly lower than in 1964–1980. On the whole, the present-day oxygen conditions in the deep-water depressions of the Caspian Sea are unfavorable for marine biota. This is related to the weakening of the deep-water ventilation and to the growth in the oxygen uptake for the organic matter oxidation. The present-day oxygen regime of the Caspian Sea is close to that observed at the beginning of the 20th century, when the volumes of the riverine runoff and the thermohaline water structure in the Caspian Sea were comparable with those observed at present.

The principal source of the nutrient supply to the sea is represented by the riverine runoff, which provides about 95% of the total amount of nitrogen and phosphorus in the North Caspian. The proportion of the Volga River in the nutrient supply to the sea makes up 80–90%. Attention should be focused mainly on the mineral forms of nutrients, which provide the abiotic basis of the biological productivity. The contents of mineral forms of nitrogen and phosphorus in the sea are subjected to a strong interannual variability determined by the complicated character of biogeochemical links in the Volga River runoff and in the near-mouth area. In 1965–1985, the share of mineral phosphorus in its total content comprised about 10–20% and that of nitrogen was about 30–45%. In 1986–1999, the proportions of mineral phosphorus and nitrogen in their total contents increased up to 30 and 50%, respectively. Almost all the mineral phosphorus and nitrogen delivered to the sea precipitates on the bottom sediments of the North Caspian shallow-water areas, while the remainder is utilized by phytoplankton. In the deeper zone of the North Caspian, the amount of nutrients may be two to three times as small as that in the shallow-water regions of the Volga River seaside; these differences are especially significant during flooding periods. In the Middle and South Caspian, the nutrient distribution is determined not only by the biochemical (production–destruction) processes but also by the advective–diffusive water transport (currents and turbulence).

The multifold growth in the delivery of mineral forms of nutrients to the Caspian Sea in the 1980s–1990s led to the eutrophication of its euphotic layer and to changes in the vertical biogenic structure in the deep-water basin of the sea. In 1985–1991, the amount of mineral phosphorus in the upper 100-m layer was twice as great as that in 1964–1981; in the deep-water layers, the corresponding growth was 1.5-fold. In 1985–1991, the parameters of nitrates and phosphates sharply increased as compared to the end of the 1970s. The greatest growth was confined to the layer 200–600 m in the Middle Caspian and to the layer 200–800 m in the South Caspian, that is, to the zone deeper than 200 m.

The recently observed low nutrient concentrations in the upper euphotic layer of the sea are related to the active consumption of nutrients by phytoplankton. It was favored by the extremely intensive development of the ctenophore Mnemiopsis, which started in the Caspian Sea in 2000 and which eats about 80% of zooplankton. This led to the shortening of the trophic web and to the acceleration of the biochemical cycling of nitrogen and phosphorus in the euphotic layer. Under the present-day conditions, a kind of dynamic isolation of the euphotic layer from the deep waters is observed caused by the sharpening of the summertime thermocline and weakening of the vertical diffusive nutrient flux from the sea depths to the upper layer.

On the whole, the present-day condition of the nutrient (and oxygen) regimes in the Caspian Sea (background levels and vertical distributions) is close to that observed before the 1930s and strongly differs from their condition characteristic of the 1960s–1970s. Thus, as far as nutrients are concerned, we can also suppose the existence of two types of multiannual regimes of the sea similar to the cases of other hydrochemical characteristics.

Finally, to conclude the section devoted to natural chemistry, we can underline the following. The salt (ionic) composition of the Caspian Sea significantly differs from that of the World Ocean; it is significantly more irregular in space and time. The instability of the hydrochemical regime of the Caspian Sea waters also represents its characteristic feature, which is especially clearly manifested in the near-mouth areas of the rivers. In the hydrochemical regime of deep-water parts of the sea, a greater role belongs to the biochemical (production–destruction) and dynamic (advective–diffusive) processes. Meanwhile, the principal reason for the long-term changes in the natural chemical condition of the waters is related to the character of the delivery of chemical compounds with the riverine runoff.

The Caspian Sea represents the most vast inland catchment in Eurasia, which covers the major industrial regions of Russia and Caucasus and suffers an increasing anthropogenic impact. In the Caspian Sea, chemical pollution is the most important. All of its principal constituents may be observed in the sea; the most important pollutants are oil and oil products, phenols, and, in the North Caspian, detergents (surface-active substances). The chemical pollution grows with the intensification of the human activity on the sea coasts and in the sea proper and represents one of the most hazardous kinds of anthropogenic impact on the Caspian ecosystem.

The main sources of pollution of the Caspian natural environment are transborder atmospheric and water transfer of pollutants from other regions, washing off with river flows, discharge of untreated industrial and agricultural wastewaters, municipal–domestic wastewaters from cities and settlements in the coastal zone due to the insufficient number of treatment facilities, oil and gas operations on land and offshore, oil transportation via sea, river and sea navigation, secondary pollution during bottom dredging operations, and sea level rise. The increased concentrations of pollutants are characteristic of the near-mouth areas of the rivers. They are observed not only off the Volga River but also off the rivers of the western coast of the sea such as the Sulak, Terek, and Samur rivers. Another particular feature is related to the fact that the degree of pollution of the eastern shelf of the Caspian Sea is lower than that of the western shelf because, in the latter case, the amount of pollutants is reduced because of the small number of sources – rivers and industrial enterprises.

With the increase in the economic potential of the Caspian countries due to hydrocarbon extraction, construction of new sea ports, rehabilitation of existing ports, revival of the merchant and tanker fleet, enhancement of the navy component, and construction of oil and gas pipelines this environmental stress may grow. The risk of the negative effects of the hydrocarbon field development in the bottom and coastal regions of the Caspian Sea is especially great in the shallow-water Northern Caspian, which is exclusively important for development of the unique commercial biological resources of the entire Caspian Sea, and which is at the same time a nature-reserve zone. Oil and chemical pollution of soils is observed over all the territories of the oil and gas fields. The main sources of this sort of pollution are breaks in oil pipelines, emergency flowing of exploratory wells, violations of the technologies of storage, accumulation, separation, and transportation via pipelines, and inadequacy of the constructions and equipment used in oil production and transport. The highest values of oil product pollution are noted in the areas near major cities, ports, and industrial regions such as Makhachkala, Neftyanye Kamni, Baku, and Turkmenbashi. Over the entire area of the Caspian Sea, there is a tendency to a decrease in the concentrations of pollutants down the water column. For example, the contents of oil hydrocarbons, phenols, and surface-active substances are noted to decrease from their maximum values at the surface down to almost zero values in the 500–1000-m layer.

Over the recent decades, the interannual variations in the contamination of the waters and sediments of the Caspian Sea were mainly caused by the influence of the sea level rise, the accidental oil losses, and the general fall in the industrial activity in the Caspian basin. In 1978–1995, a greater contamination was observed in the waters of the North Caspian subjected to the runoff of the Volga River. The mean annual values of the concentrations of oil products, phenols, and surface-active substances were 0.19, 0.006, and 0.08 mg/L, respectively. Smaller respective values were characteristic of the waters of the South Caspian, where the principal oil and gas fields are located (0.13, 0.005, and 0.045 mg/L) as well as of the waters of the Middle Caspian (0.08, 0.005, and 0.04 mg/L).

At the same time, the contents of contaminants in the sediments of the Caspian Sea varied within wider limits (oil products from 0 to 226 mg/L, phenols from 0 to 40 μg/L, and mercury from 0 to 4, 7 μg/L). The maximum values were detected in Baku Bay and off the cities of Sumgait (north of the

Apsheron Peninsula) and Turkmenbashi; the minimum values were concentrated in the deep-water regions of the sea. In shallow-water areas, bottom sediments represent sources for secondary contamination – at sea level rises under the influence of dynamic processes additional pollutants are supplied to the near-bottom layer of the sea.

Although in recent years the water level in the Caspian has remained relatively stable, a further rise of the sea level may still lead to emergency situations in oil production areas, flooding of drilling sites located in lowlands, breaking of protection embankments and levees around drilling sites, breach of on-field pipelines, and pollution of underground waters, which will subsequently contribute to the additional sea pollution.

The biological part of the monograph includes three sections concerning the biological diversity, the introduced species, and the biological resources. Since the Caspian Sea is a brackish-water basin, the distribution of the species composition in it is controlled by the response of the organisms to the water salinity. The quantitative distributions of phyto- and zooplankton, benthos, and fishes are rather stable. The seasonal changes are greater than the interannual variations; the latter may vary only in magnitude rather than in the general character of distribution. The greatest changes are caused by the invasion of alien species (*Mytilaster lineatus, Nereis diversicolor, Abra ovata, Mnemiopsis leidyi, Pseudosolenia calcar-avis*, and others). Besides, the changes in the Caspian biocommunity are caused by the sea level interannual variability and are mostly manifested in the North Caspian, in the zone of the interface between the riverine and sea waters.

The changes in the biological diversity of the Caspian Sea and in the general ecology of the basin over the past century are subjected to the strong influence of species invaders. In the past century, the Caspian Sea accepted a great number of new alien species of algae, invertebrates, and fishes; they occasionally penetrated into the basin due to the human activity (with coacclimation, biofouling, ballast waters of ships, etc.) or were purposively introduced. Of these species, 46 species successfully acclimated, 18 species disappeared after certain delays, while the fate of the rest of the species is still unclear. In the first half of the 20th century, nine new species have appeared, mostly in the course of acclimation. The most successful was the introduction of two invertebrate species (the worm nereis and the bivalve abra), which significantly improved the food base of sturgeons, and two mullet species (the gray mullets *Liza saliens* and *L. aurata*). With the opening of the Volga–Don canal in 1952, a new stage of the colonization of the Caspian Sea by invaders began. First, they were fouling and related benthic organisms (20 species) and then, starting from the 1980s, pelagic inhabitants began to prevail. This process goes on at present as well; the most illustrative example is the invasion of the ctenophore *Mnemiopsis leidyi*.

In 1999, in the region at the boundary between the Middle and South Caspian, previously unknown gelatinous animals – ctenophore mnemiopsis and medusa aurelia – were encountered. Most probably, the ctenophore was carried with ship ballast waters from the Black Sea, where it caused significant changes in the ecosystem of the sea. In 2002, the abundance of mnemiopsis in the Caspian Sea essentially increased and was several times as high as that observed in the Black Sea in the period of its maximum development in 1989. The aftereffects of the invasion of mnemiopsis into the Caspian Sea are extremely strong. In 2000, it populated the entire sea area. This resulted in a decrease of the zooplankton biomass (since mnemiopsis is a zooplanktophage) as compared to the preceding years; in all the groups, this decrease was 5–6-fold, while selected groups featured a 6–12-fold drop. Subsequently, the decrease continued. Zooplankton serves as food for three sprat species, which are the most abundant fishes and have a high commercial value; their catches significantly reduced. As a result, the decrease in the zooplankton abundance affected all the trophic levels. In order to suppress the growth in the abundance of the mnemiopsis population, the possibility of acclimation in the Caspian Sea of the predator ctenophore Beroe feeding mostly on other ctenophores is considered.

The introduction of abra and nereis significantly enhanced the food resources for sturgeons; planctivorous fishes feed on the invaded copepods, which replaced the Caspian species; by the way, the latter were also consumed by them. Nevertheless, from the point of view of biological diversity, the invasion of numerous species is a negative factor because they replace aborigine organisms. In the future, a kind of "Mediterranization" of the Caspian Sea is possible, i.e., the replacement of autochtonous Caspian species by Mediterranean invaders. Thus, in order to conserve the unique Caspian brackish-water fauna and its original community, it is necessary to carry out quarantine procedures, which can provide protection of the Caspian Sea with its aboriginal organisms from the invasion of more competitive marine species.

One of the most catastrophic aftereffects of the anthropogenic intervention into the Caspian ecosystem is related to the population of sturgeons, which are the most valuable commercial fishes of the Caspian Sea. In the past century, their catches were maximum in the middle 1970s ($26–27 \times 10^3$ tons); this occurred after the legal prohibition of their marine fishery issued in 1960. However, later, sturgeon abundance catastrophically dropped. First, from the end of the 1980s, a disease (miopathy) developed, which involved the greater part of the sturgeon population. The origin of this disease is still unclear. In addition, after the disintegration of the USSR and the formation of new independent states, a non-controlled fishery has started in the sea. As a result, the commercial resources have multifold decreased as compared to the 1970s. In addition to sturgeons, two sprat species were of significant commercial importance; however, after the mnemiopsis invasion, their catches have sharply dropped. Also important from the commercial point of view are roaches and bream as well as herrings, carps, pike-perches, and other species. One of the principal problems of the Caspian ecology consists of the necessity of elaboration of efficient measures for protection of the marine environment from negative anthropogenic impacts, first of all, the rehabilitation of the unique sturgeon community.

The threats of the deterioration of the environmental situation in the Caspian region and of the depletion of its natural resources directly depend on the condition of the economy and awareness of the society about the global character and importance of these issues. This threat is especially great because of the excessive development of the fuel-power industries, drawbacks of legal foundations of the nature conservation activities, restricted application of the nature-saving technologies, and low ecological culture, which increases the risk of technogenous catastrophes.

The unsettled delineation of the Caspian Sea and the uncertainty in its legal status are the main obstacles for successfully coping with many issues, including environmental protection and preservation of the biological resources. Here, the key issue should be the provision of national and international environmental safety. This means elaboration of a system of coordinated state and interstate mechanisms, actions, and guarantees based on the compliance, by one and all states, with the common humanitarian principles and norms of the international legislations that are called to guarantee effective solutions or to prevent emergence of environmental problems of interstate and world community dimensions. Rapid settlement of the legal status of the Caspian is necessary for a transition to sustainable development capable of ensuring a balanced solution of the socioeconomic and nature-conservation issues in the interests of the Caspian countries and the whole world community.

References

1. Terziev FS, Kosarev AN, Aliev AA (1992) (eds) Hydrometeorology and hydrochemistry of the seas, vol 6. The Caspian Sea, Issue 1. Hydrometeorological conditions. Gidrometeoizdat, St. Petersburg [in Russian]
2. Kosarev AN, Yablonskaya EA (1994) The Caspian Sea. SPB Academic, The Hague
3. Terziev FS, Maksimova MP, Yablonskaya EA (1996) (eds) Hydrometeorology and hydrochemistry of the seas, vol 6. The Caspian Sea, Issue 2. Hydrodynamical conditions and oceanological background of the formation of the biological productivity. Gidrometeoizdat, St. Petersburg [in Russian]

Subject Index

MIX
Papier aus verantwortungsvollen Quellen
Paper from responsible sources
FSC® C105338

If you have any concerns about our products,
you can contact us on
ProductSafety@springernature.com

In case Publisher is established outside the EU,
the EU authorized representative is:
Springer Nature Customer Service Center GmbH
Europaplatz 3, 69115 Heidelberg, Germany

Printed by Libri Plureos GmbH
in Hamburg, Germany